SECOND EDITION

Biomedical
Engineering
Principles

SECOND EDITION

Biomedical Engineering Principles

Arthur B. Ritter
Vikki Hazelwood
Antonio Valdevit
Alfred N. Ascione

CRC Press
Taylor & Francis Group
Boca Raton London New York

CRC Press is an imprint of the
Taylor & Francis Group, an **informa** business

CRC Press
Taylor & Francis Group
6000 Broken Sound Parkway NW, Suite 300
Boca Raton, FL 33487-2742

First issued in paperback 2018

© 2011 by Taylor and Francis Group, LLC
CRC Press is an imprint of Taylor & Francis Group, an Informa business

No claim to original U.S. Government works

ISBN 13: 978-1-138-07324-1 (pbk)
ISBN 13: 978-1-4398-1232-7 (hbk)

Library of Congress Cataloging-in-Publication Data

Ritter, Arthur B., author.
  Biomedical engineering principles / Arthur B. Ritter, Vikki Hazelwood, Antonio Valdevit, and Alfred Ascione. -- Second Edition.
       p. ; cm.
    "A CRC title."
    Includes bibliographical references and index.
    Summary: "Linking methodologies in engineering, medicine, biology, and physics, this reference clearly defines basic principles in image processing and biomechanics, the modeling of physiological processes, and bioelectric signal analysis for a solid understanding of devices and designs for improved functioning of the human body-including cutting-edge discussions of tissue engineering applications such as skin equivalents, cardiovascular components, bone regrowth, muscle tissue, and the regeneration of nerves. The author also discusses technologies at the forefront of cardiac care including total artificial hearts, left ventricular assist devices, defibrillators, pacemakers, and stents, as well as issues related to minimally invasive and robotic surgery, next-generation imaging devices, nanomotors, and nanodevices"--Provided by publisher.
    ISBN 978-1-4398-1232-7 (hardcover : alk. paper)
    1. Biomedical engineering. I. Hazelwood, Vikki, author. II. Valdevit, Antonio, author. III. Ascione, Alfred, author. IV. Title.
    [DNLM: 1. Biomedical Engineering--methods. 2. Biomechanics. 3. Physiological Phenomena. 4. Signal Transduction. QT 36]

R856.R58 2011
610.28--dc22
                                                                                              2010051442

Visit the Taylor & Francis Web site at
http://www.taylorandfrancis.com

and the CRC Press Web site at
http://www.crcpress.com

Contents

Part II Signal Processing

Part III Biomechanics

Part IV   Capstone Design

Preface

This book is intended to be an introduction to the field of biomedical engineering for students with undergraduate training in engineering, physics, and mathematics. We assume that the student has no more than a passing acquaintance with molecular biology, physiology, biochemistry, and signal processing. Each chapter therefore contains a discussion of the biological, physiological, and biochemical principles that are pertinent to the subsequent engineering analysis. Moreover, each chapter has been either extensively updated or rewritten. We have also eliminated the part on tissue engineering as this field of biomedical engineering has grown so large and fast that it deserves a textbook of its own. Similarly, the whole field has opened up so that any speculation on future developments is likely to be obsolete even before the book is published.

Since the field of biomedical engineering attracts students from a variety of backgrounds, it seems prudent for us to tell the reader how we have organized the material in this book. This will allow one to organize a course on biomedical engineering based to some extent on the background and interests of the students.

Systems Physiology, Transport Processes, Cell Physiology, and the Cardiovascular System
Part I consists of Chapters 1 through 5. Chapter 1 contains an introduction to systems physiology. Just from a simple immersion in the field of physiology, one can easily discern that homeostasis, the maintenance of a stable internal environment in the body, requires the collaborative efforts of millions of cells and dozens of organs. This chapter takes a first look at treating homeostasis as the interaction of systems.

Chapter 2 contains an overview of and an introduction to engineering analysis of physiological systems, the nature of biological data, and the role of models and simulation in experimental design. This chapter introduces the concepts of electrical and mechanical analogs of physiological processes, conservation of mass, compartments, convection, and diffusion, and development of pharmacokinetic models for drug distribution.

Chapter 3 covers cell physiology. It introduces the primary mechanisms by which water and solutes get into and out of cells. It discusses diffusion, facilitated transport, and active transport, as well as the generation of the resting membrane potential and the propagation of the action potential. It introduces channels, carriers, pumps, and cellular signaling. Finally, the processes by which muscles develop force and use metabolic energy are discussed in terms of the sliding filament model.

Chapter 4 covers fundamentals of hemodynamics and the nature of blood and blood vessels as engineering materials. Examples of flow in large vessels and in the microcirculation are presented. The simulation of pulsatile flow in patient-specific distensible vessels using computational fluid dynamics (CFD) software with fluid–structure interaction (FSI) is introduced with some current examples.

Chapter 5 is an introduction to the cardiovascular system. It covers the cardiac conduction pathway, control of heart rate, EKG measurement and interpretation, cardiac output, cardiac work, and autonomic and local regulation of blood flow. It also covers measurements of pressure, flow, and impedance in cardiac catheterization and vascular function labs. A whole-body compartmental model of flow and pressure drop in the cardiovascular system is developed. Finally, the failing heart is discussed in terms of the changes in hemodynamic parameters and mechanisms that compensate for left heart failure.

Signal Processing
Part II reviews the concepts of biomedical signal processing. Ultimately, any engineering analysis must be compared with experimental data both for validating the model and for estimating the sensitivity of the model to small changes in parameter values. If the model relies on specific mechanisms (rather than being entirely empirical), then a necessary (though not sufficient) condition for validating the theoretical analysis is that the model be able to predict established data and trends using physically realizable values of the model parameters. Modern biomedical instrumentation predominantly produces electrical signals as outputs, no matter which physical variable is being measured. Also, data acquisition is invariably linked to a digital computer. This part, therefore, presents the elements of signal processing necessary for a biomedical engineer, both from a theoretical as well as from a practical point of view.

Chapter 6 introduces a major digital signal processing concept, the time to frequency domain transformation. Euler's Identities are presented, enabling the reader to grasp the concept in the analog domain. Then biomedical signals are introduced along with techniques for representing these signals in the digital domain. Digital concepts such as quantization error and sampling theory are reviewed. Topics include sampling theorem, sampling rate, and aliasing.

Chapter 7 introduces the digital engines capable of transforming biomedical signals into the frequency domain. These include the discrete Fourier transform and the fast Fourier transform. Additional topics include frequency resolution, zero padding, the power spectrum, and coherence.

Chapter 8 introduces major system level concepts, such as superposition. Then digital filtering techniques are demonstrated. The requirement to window data and the associated advantages are also discussed.

Chapter 9 discusses techniques and examples for physiological signal processing. Topics include autoregressive modeling, time–frequency analysis, short-time Fourier transforms, and quadratic distributions. The chapter continues on to give examples of physiological signal processing. These include physiology of the autonomic nervous system and heart rate variability, measurement of physiological stress, circadian rhythms, and spectral analysis of biomedical signals.

Biomechanics

Part III provides an introduction to and practical applications of biomechanics. Chapter 10 is an introduction to the principles of biomechanics and mechanical properties of bone. It starts with the structure of materials, internal and external loading, and musculoskeletal motion. Since this is an introductory text, the chapter focuses on statics and mechanical characterization of materials, the structural and functional characteristics of bone and its role in the musculoskeletal system, and a discussion of cortical and trabecular bone and their mechanical properties. It includes a quantitative discussion of bone fracture and crack propagation and contains topics such as arm rotation, muscle mechanics, muscular control of posture, equilibrium, measurements of muscle force, electrical stimulation of skeletal muscle, mechanical characteristics of biological materials (bone, muscle, skin, and ligaments), bone remodeling, body cycles, thermal regulation, and hypothermia.

Chapter 11 covers the structure and mechanical properties of the musculoskeletal tissue, connective tissue, and spine. Connective tissue consists of ligaments, tendons, muscle, and cartilage. This chapter includes a quantitative treatment of muscle contraction using a mechanical model. It also looks at the extremities, including the knee, tibia, femur, and patella in terms of their range of motion and mechanical properties. The overall motion of the knee joint (including anterior and posterior cruciate ligaments) is presented from a statics point of view. The hip is then added and an analysis of the system in terms of human gait is presented. The chapter then goes on to consider the shoulder with the humerus, scapula, and glenohumeral joint as contributing entities in the great range of motion of the shoulder. Finally, the spine is discussed in terms of its functional anatomy, mechanical properties, load bearing, range of motion, and center of rotation.

Capstone Design

Part IV consists of Chapters 12 and 13. Chapter 12 discusses the differences between design for living and nonliving systems and the role of the FDA and presents an approach to the capstone design project for biomedical engineers. It outlines a paradigm for vetting projects, setting goals, and evaluating team progress, including identifying unmet needs, the stage-gate process and the design cycle as an iterative process, teamwork, and entrepreneurship. It gives practical advice to students for using time management to complete the design project in the limited time available.

Chapter 13 discusses taking an idea for a medical device from inception to proof of concept in 32 weeks (two semesters). It presents a summary of major milestones and deliverables in a two-semester capstone design sequence, including tips for maintaining a lab book, weekly group plans, and project review meetings. It also discusses the idea of a "proof of concept," and presents the elements necessary for a proper group presentation and evaluation criteria. The chapter then goes on to take the student through the timing and deliverables necessary to ensure a successful outcome for a group capstone design project.

Because of the current demand in the biomedical sciences to put more emphasis on understanding basic mechanisms and problem solving rather than on empiricism and factual recall, we feel that there is a very definite need for a book of this kind. Moreover, the subject matter is sufficiently basic so that it cuts across several traditional engineering and biomedical science disciplines. Our thought is that knowledge of the basic laws of mass and momentum transport as well as model development and validation, biomedical signal processing, biomechanics, and capstone design have important, if not indispensable, roles in the engineering analysis of physiological processes. For example, patient monitoring and anesthetic delivery during surgery might involve monitoring of blood flow and organ perfusion, a model for the distribution of the anesthetic agents so that the optimal time course between maintenance doses can be determined, and monitoring of patient response (EKG, blood pressure, cardiac output, etc.) as well as signal processing of the measured variables. This is an important application that illustrates the integration of material from several of the sections of the book.

Obviously, there is more material in this book than can be conveniently used in an introductory course. As a guide to prospective instructors of biomedical engineering, Chapters 1, 3, 5, 7, 10, and 12 would be suitable for a well-balanced three semester hours undergraduate introductory biomedical engineering course. Having some additional material in the book should prove useful to instructors and advanced students. In addition, it would allow some flexibility in tailoring a particular course to the backgrounds of the students.

We have tried to use consistent notation throughout the book. Unfortunately, the material cuts across many disciplines, each of which has independently developed a particular set of nomenclature. As an aid to reading the literature, we have tried to use the notation that is most common to the particular field being presented. Generally, our notation represents a compromise between that used in the biological literature and that used by engineers and physicists.

The illustrative examples in each chapter were chosen specifically for this book. The problems and discussion questions were written from the perspective of a biomedical engineer. In general, the major objective was to provide an informative textbook for beginning students.

For MATLAB® and Simulink® product information, please contact:

The MathWorks, Inc.
3 Apple Hill Drive
Natick, MA, 01760-2098 USA
Tel: 508-647-7000
Fax: 508-647-7001
E-mail: info@mathworks.com
Web: www.mathworks.com

Acknowledgment

We would like to close by repeating what a very smart person advised us to do to maintain our personal homeostasis, that is, eat right—plenty of fresh fruit and vegetables, get some exercise every day, and, above all, don't be lazy.

Authors

Dr. Arthur B. Ritter is a distinguished service professor and director of biomedical engineering, Stevens Institute of Technology, Hoboken, New Jersey. He received his BChE from the City College of New York, and his MS and PhD in chemical engineering from the University of Rochester, Rochester, New York. Before returning for his PhD, he had over 10 years of industrial experience in the aerospace industry for the U.S. Navy and United Aircraft in solid rocket propellant development and as a development engineer for the Mixing Equipment Company and the DuPont Co. Before founding the biomedical engineering program at Stevens, he was a tenured faculty member in the Department of Physiology and Pharmacology at the University of Medicine and Dentistry of New Jersey–New Jersey Medical School.

Dr. Vikki Hazelwood has a PhD in biomedical engineering from Stevens Institute, where she specializes in translational research in medicine. She also holds a BS in chemical engineering from Rutgers and an MS in biomedical engineering from the New Jersey Institute of Technology.

Dr. Hazelwood arrived at Stevens after having served 25 years in industry. Most recently, she has held executive positions in sales and business development for medical device companies focused on drug delivery technology and specialty biomaterials (metals and polymers) used primarily in the fields of orthopedics and interventional cardiology. Her experience includes many years of clinical interface with surgeons and New York metropolitan area hospitals, as well as successful business collaborations with senior decision makers of global medical device companies.

In her earlier years, Dr. Hazelwood was a project manager for process systems, i.e., capital equipment used to manufacture pharmaceuticals, medical products, and other specialty chemicals. Throughout her career, she has proven that her tactical skills and expertise match her unique and creative strategic insights.

Since she has been at Stevens, Dr. Hazelwood has successfully brought entrepreneurship into the educational process, and, in just three years, has worked with her students to develop a venture capital–backed start-up company, SPOC Inc. (a point-of-care product for pain management), for which she served as president and CEO, and her student founders as coinventors and full-time employees. The company has received $1 million in extramural financial backing and has recently received a 510 clearance from the FDA.

Dr. Antonio Valdevit is an assistant professor of biomedical engineering, Stevens Institute of Technology, Hoboken, New Jersey. He was originally trained as a physicist having received his BSc from McMaster University (Hamilton, Ontario, Canada) and his MSc from Queen's University (Kingston, Ontario, Canada). He has worked in academic institutions such as Johns Hopkins and the Cleveland Clinic prior to obtaining his PhD in biomedical engineering at the Stevens Institute of Technology. He has obtained 14 patents in the biomedical field and his current research interests include bone regeneration and spinal biomechanics.

Professor Alfred N. Ascione is a distinguished member of technical staff at Alcatel-Lucent in Murray Hill, New Jersey, and an adjunct professor in the Department of Biomedical Engineering at Stevens Institute of Technology, Hoboken, New Jersey. He has over 15 years of experience in the electronics countermeasures industry and over 12 years of experience in the wireless communications industry. Between instructing at both Stevens and the New Jersey Institute of Technology (NJIT), he has accumulated over 20 years of teaching experience. Professor Ascione received both his BSEE and MSEE from NJIT.

Part I

Systems Physiology, Transport Processes, Cell Physiology, and the Cardiovascular System

1

A Systems Approach to Physiology

1.1 Introduction

Biomedical engineering is not one discipline, but several interacting disciplines that coexist under the same umbrella. A classic description of biomedical engineering is the application of the principles of engineering design and analysis to problems in medicine and the life sciences. With this general description, "biomedical engineers" may have widely different backgrounds depending on the particular subspecialty that they are involved with. In keeping with this fact, biomedical engineering programs usually offer several different tracks or concentrations that build on one or more engineering disciplines.

We can also make the statement that the human body does not work like a mechanical or electrical system. Biological systems are constantly evolving, usually redundant, and can change set points to adapt to changes in their environment. They can differentiate between self and nonself and have a variety of defense mechanisms to attack foreign substances. They have especially acute sensors, including vision, hearing, touch, pain, and smell, that allow their main computer (*the brain*) to anticipate events and take action to avoid injury (*anticipatory control*) rather than always reacting to stimuli. This implies that organ systems such as the heart and lungs interact with each other (and many others) in a dynamic way and depend on the functioning of each other to maintain the organism in a condition called *homeostasis*.

Homeostasis refers to the fact that just about all of the organs and tissues of the body function in such a way as to maintain the body's internal environment within narrow ranges (temperature, pressure, pH, ion concentrations, etc.). For engineers, the second important point, referred to in the previous paragraph, is that biology is capable of fighting back. So, any engineering device that is to spend time in the body (e.g., implants) must be made of material(s) that is *biocompatible*. That is, materials that do not cause an immune or inflammatory reaction by the tissues they are in contact with.

Figure 1.1 is an example of the interaction of several organ systems that are involved in digestion of food, elimination of waste, and synthesis of energy stores (adenosine triphosphate [ATP]). What this simple diagram does not show are the myriad of other processes that are taking place simultaneously.

3

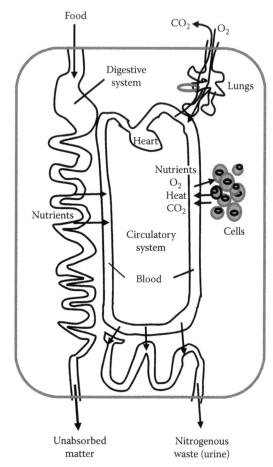

FIGURE 1.1
Nutrients are provided by the digestive system and oxygen is provided by the respiratory system. Nutrients and oxygen are distributed by the heart and circulatory systems. Metabolic wastes are eliminated by the respiratory (CO_2) and urinary (urea and uric acid) systems.

For example, the cells in the right-hand side panel are utilizing oxygen and glucose transported from the blood to aerobically synthesize ATP that provides the metabolic energy for almost all work done by the cells, tissues, and organs. In the process, the cells produce CO_2, which is taken up by the red blood cells in the blood and converted to bicarbonate (HCO_3^-). The heat generated is carried away by the flowing blood. The HCO_3^- in the red blood cell is exchanged for a Cl^-, which remains in the plasma until it reaches the lung, where it is reconverted to CO_2 and transported to the atmosphere by the pulmonary system. While in the plasma phase, HCO_3^- acts as one of the major buffering systems to regulate blood pH. The heat is transported from the blood to the atmosphere through the skin (large surface area) and through the expiration of water vapor and gases that are at body temperature by the lung. In the meantime, glucose levels are monitored by the pancreas and insulin is secreted by the β-cells to bring blood glucose levels back to control values. Body temperature is regulated by heat-sensitive neurons in the hypothalamus and cold-sensitive neurons in the midbrain. These areas institute appropriate heat-increasing or -decreasing actions.

1.2 An Analogy

The human body is constantly working to maintain homeostasis, awake or asleep. The best analogy is to consider an automobile or any other complex piece of machinery. One can independently discuss the design and operation of the motor, the transmission, the differential gear, etc. as systems within a car. However, when all is said and done, for the automobile to function properly, all of the systems must perform their separate functions as an integrated whole. Anyone who has had to deal with a blown motor or non-functional transmission or even a flat tire can testify to this. There are two important differences between the human body and a piece of machinery. First, in an automobile, the driver acts as the controller—braking, accelerating, and negotiating through traffic. In the human body, this function is distributed among reflexes and areas of the brain. Some of these controllers operate autonomously (i.e., without conscious thought—think of beating of heart and breathing) and some operate consciously (playing a musical instrument). Second, one can purchase parts for just about any failed component of an automobile that will allow the system to again function properly. The same cannot be said of replacement of human parts (see biocompatibility).

1.3 Physiological Control Systems

There are four major control systems in the body that act systemically as well as numerous autoregulatory processes that act locally to maintain homeostasis:

- The *nervous system* regulates the musculoskeletal system through somatic neurons to control locomotion, secretory activities of glands and cells, as well as balance. Through the autonomic division of the peripheral nervous system, it also regulates the activities of many internal organs. In addition, it allows us to obtain feedback from sensory and motor neurons so that we can adapt to changing conditions. These controls also include reflex responses and are usually fast, direct, and short acting.

Figure 1.2 shows the overall organization of the nervous system. Sensory input is integrated with information from memory and higher centers of the brain. Stimuli are transferred through the neural processes to the motor neurons and appropriate action is taken.

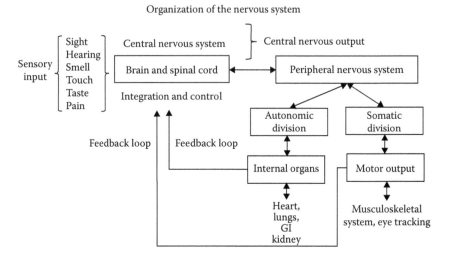

FIGURE 1.2
A schematic diagram of neural integration and control.

Sensory input. There are five basic types of sensory receptors:

1. *Mechanoreceptors.* These detect the mechanical deformation of the receptor membrane or the tissue surrounding the receptor and include both tactile (touch) and body position and movement (proprioceptors) sensing. Examples are muscle spindles (primary and secondary ending), muscle tendon (Golgi tendon organ), hair receptors, vibration (Pacinian corpuscle), high discrimination touch (Meissner's), and crude and deep pressure and touch.

2. *Nociceptors.* These are the pain receptors and detect damage to the tissue surrounding the receptor or the release of specific pain-mediating molecules (e.g., prostaglandins and endorphins). Examples are pricking and aching pain receptors.

3. *Chemoreceptors.* These are responsible for detecting changes in the chemical composition of blood and for taste and smell. Examples are osmoreceptors (osmotic pressure) and aortic and carotid bodies (P_{O_2}, P_{CO_2}, and pH).

4. *Temperature receptors.* These detect changes in the temperature of the receptor (hot or cold).

5. *Photoreceptors.* These detect the concentration of photons that strike the retina of the eye.

Integration and control. The brain can store information, determine reactions that the body performs in response to sensory input, and plan for execution of learned behavior (gymnastics, diving, piano playing).

Two-way transfer of information. Appropriate signals are transmitted from segments of the brain and spinal cord through motor output portions of the peripheral nervous system to carry out desired activity. Peripheral sensors then feed back the current state of the motor activity and body balance to the association areas of the brain. Here, integration of the activity occurs and correction is taken if necessary. The spinal cord is also the execution center for reflex activity. Reflexes bypass the brain during execution, though sensors transmit the current position and balance of the body back to the brain.

- The *endocrine system* consists of major glands that secrete chemical substances called hormones that regulate many of the metabolic activities of cells. These include rate of metabolism, growth, and sexual reproduction. Hormones are secreted into the bloodstream where they are carried to specific target cells. Hormones are a system of regulation that complements the nervous system.
  - For example, the pancreas has an anatomical structure called the *islets of Langerhans.* The islets contain four types of cells that secrete hormones. The β-cells (60% of the cells) secrete *insulin* in response to increased blood glucose levels. Insulin regulates glucose uptake and metabolism in all cells. The islet α-cells (25% of the cells) secrete *glucagon,* which acts mainly in the liver and opposes the effects of insulin.
  - *Adrenocorticol* hormones, secreted by the *adrenal gland,* control sodium and potassium ions and protein metabolism.
  - The *parathyroid gland* secretes *parathyroid* hormone (PTH) that controls calcium absorption in the gastrointestinal (GI) tract.

These controls are generally slow, indirect, and long-lasting.

- The *immune system* provides defense mechanisms that protect the body against bacteria, fungi, and algae as well as dust, pollen, etc.
- The *integumentary system* (mainly skin) provides protection against minor bruises, cuts, and abrasions and directs the defense against foreign invaders. It also protects the underlying tissue against dehydration and is important in temperature regulation by providing a large surface area for heat transfer and the inclusion of sweat glands.

Physiological control systems act in two fundamental ways: negative feedback and positive feedback. A negative feedback control system acts in such a way as to oppose a homeostatic imbalance. For example, a fall in blood pressure (BP) triggers responses that raise BP back to control (normal) values.

A positive feedback control system acts in such a way as to reinforce a homeostatic imbalance. In general, positive feedback systems can be destructive unless they are part of an overall negative feedback process. For example, one of the steps in blood clotting involves formation of a platelet plug to stop blood flow prior to clot formation. Injured tissue releases chemical signals that attract and activate platelets at the injury site. Activated platelets in turn activate additional platelets in a positive feedback mechanism that continues until the platelet plug grows large enough to stop the flow.

1.4 A Generic Negative Feedback Control System

A single input–single output, negative feedback control system that regulates some variables (e.g., pH, BP, and osmotic pressure) is illustrated in Figure 1.3. The arrows indicate the direction of information flow.

Here, y is the *controlled variable* (e.g., pressure). Going clockwise around the loop gives us the following:

- The *sensor* (e.g., baroreceptors) senses the current value of the controlled variable and sends a stream of action potentials (APs) to the cardiovascular (CV) control center in the medulla, whose frequency varies with the deviation of the measured variable from the set point (the signal is frequency modulated).

- At the *controller,* the measured value is compared with a set point (see note below). When the measured value deviates from the *set point,* an error is generated. In the case of a negative feedback control, the error is *positive* when the set point is higher than the measured value and *negative* when the measured value is lower than the set point. A positive error drives the controller to increase control activity to raise the value of the measured variable to obtain homeostasis, while a negative error drives the controller to decrease the value of the measured variable to obtain homeostasis. A disturbance entering the system will drive the control variable away from its homeostatic value and initiate an error signal through the feedback loop.

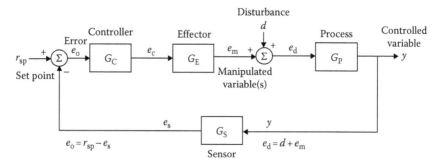

FIGURE 1.3
A block diagram for a basic, generic, negative feedback control system.

NOTE: Unlike analog or digital electronic or pneumatic controllers, the set point for many physiological variables, including BP, is not necessarily constant. Systemic BP increases *appropriately* during exercise, fight or flight, physical or emotional stress, and hypertension. The set point is modified by increased metabolic activity, changes in hormonal activity, and input from the higher centers of the brain, and the *revised set point is appropriately defended* by the control processes.

- Based on the *direction of the error signal*, the controller modifies its neural output to the *effector organs*. In the case of beat-to-beat or short-term BP regulation, sympathetic neural outputs to the *heart* and *arterioles* (effector organs) and *vagus nerve* output to the heart are changed. Changes in heart rate (HR), force of contraction (stroke volume [SV]), and peripheral resistance are altered in such a way as to bring about homeostatic control. Control action continues until the error signal is zero.

Equation 1.1 shows the relationship between mean arterial pressure, $\bar{P}_{art}$, SV, and total peripheral resistance (R_{TPR}):

$$\bar{P}_{art} = HR \times SV \times R_{TPR}$$

$$= (beats/s) \times (mL/beat) \times (mm\,Hg/mL/s) = mm\,Hg \tag{1.1}$$

where
$\bar{P}_{art}$ is the mean arterial pressure, that is, the integrated average of the pulsatile pressure change with time over one heart beat in the large arteries (mm Hg)
HR is the number of heart beats averaged over 60 s (beats/s)
SV is the volume of blood ejected from the left (or right) ventricle in one heart beat (mL/beat)
R_{TPR} is the total resistance to flow of blood in all the blood vessels between the aorta and the right ventricle (mostly arterioles) in mm Hg/mL/s

Changes in any of the variables on the right-hand side of Equation 1.1 will change $\bar{P}_{art}$.

- Long-term regulation of BP involves the kidney and the renin–angiotensin–aldosterone system. The kidney regulates blood volume through urine output and the osmotic pressure of the blood through water and sodium reabsorption.

In a compliant system such as blood vessels or in fact, a balloon, the relationship between pressure and volume changes is given by

$$C = \frac{dV}{dP} \tag{1.2}$$

where C is the *compliance*, a measure of the stiffness of the vessel and is a material property.* Equation 1.2 can be applied to a single distensible vessel such as an artery or to a system of distensible vessels such as the vascular system, the heart, or the lungs. It is also important to realize that vessels in the body are not homogeneous materials. Therefore, in general, C is not constant but is a function of pressure. As vessels are filled, the more compliant structural elements allow increased volume with relatively low changes in pressure. As vessels near their volumetric capacity, stiffer structural elements (less compliant) are engaged and more pressure (force/area) is required to increase volume further. Equation 1.2 can be rearranged to make this relationship more apparent:

$$dP = \left[\frac{1}{C(P)}\right]dV \qquad (1.2a)$$

Equation 1.2a illustrates that a decreased compliance at higher pressures means that changes in volume require higher changes in pressure than if the compliance were constant.

The compliance of the vascular system allows long-term regulation of BP by the kidney through vascular volume changes.

1.5 Regulation of Body Functions

The human body has thousands of control systems that act together to maintain homeostasis. The most complex of these are genetic controls that are present in all cells. These regulate intracellular processes for cell growth, metabolism, and motility, as well as all extracellular interactions. Cells can modify their extracellular matrix by secretion of proteins and remove protein channels in the cell membrane to control transport (e.g., aquaporins are channels that regulate water uptake and help control cell volume). Other systems operate to control individual parts of organs (renin–angiotensin–aldosterone regulate renal uptake of sodium in kidney collecting ducts, which controls water excretion by the kidney and blood volume). Other control systems operate throughout the body and control interactions between organs.

Example 1.1 Regulation of Oxygen and Carbon Dioxide Concentration in the Extracellular Fluid

The overall process is illustrated in Figures 1.4 and 1.5. Oxygen is essential for efficient energy production (ATP synthesis) by aerobic metabolism in the mitochondria of all cells. In aerobic metabolism, water and carbon dioxide are the

---

* For a discussion of the relationship between compliance and modulus of elasticity (Young's modulus), see Chapter 4.

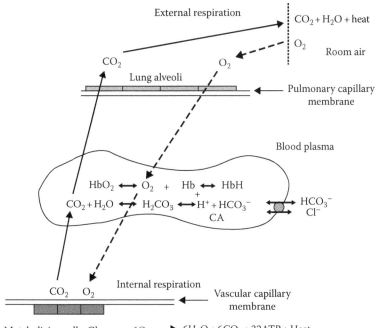

FIGURE 1.4

At the pulmonary capillaries, oxygen is transported by diffusion down its partial pressure gradient from the pulmonary alveoli (P_{AO_2} = 100 mm Hg) across the pulmonary capillary membrane and into the venous blood (P_{VO_2} = 40 mm Hg). From the blood plasma, oxygen diffuses into the red blood cell where it is taken up by a specialized carrier protein called hemoglobin (Hb). Each molecule of Hb can carry up to four oxygen molecules when fully saturated. Normal blood contains 15 g Hb/100 mL blood. When fully saturated (full capacity), Hb can bind 1.34 mL O_2/g Hb at body pressure and temperature.* Oxy-hemoglobin (HbO$_2$) then travels through the arterial system to the capillaries of organs and tissues. At the metabolizing cells, partial pressure of oxygen is low compared with the P_{AO_2} in the red blood cells. The HbO$_2$ in the red blood cells releases O_2 in exchange for CO_2. Most of the CO_2 in the red blood cell is rapidly converted to HCO_3^-, by the reversible relationship (Equation 1.3a) (~70%). Increasing concentration of CO_2 in the red blood cell drives the reaction (Equation 1.3a) to the right. Some of the CO_2 (~23%) reversibly combines with the amine residues on the proteins in the plasma and on the Hb molecule in the form of carbamino compounds. The remainder (~7%) exists as dissolved CO_2 in the blood fluids. The venous blood carries the CO_2 back to the lungs where the cycle repeats itself.

---

* An additional 0.003 mL O_2/P_{O_2} is dissolved in plasma at body temperature according to Henry's law of solubility. This means that in arterial blood at a P_{O_2} = 100 mm Hg, an additional 0.3 mL O_2 is dissolved in plasma (100 mm Hg P_{O_2} ×0.003 ml mL O_2/P_{O_2}). In venous blood at a P_{O_2} = 40 mm Hg, an additional 0.12 mL of O_2 is carried as dissolved gas in plasma.

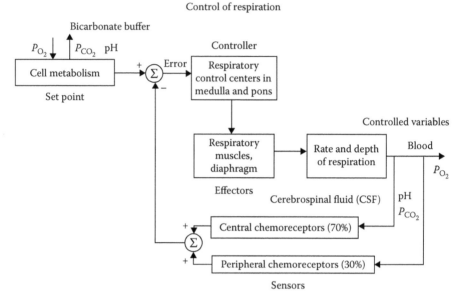

FIGURE 1.5

A schematic diagram of a negative feedback control system. Changes in P_{O_2} in arterial blood are sensed by peripheral chemoreceptors located in carotid and aortic bodies and send neural signals to the respiratory control center in the medulla. Changes in P_{CO_2} and H^+ in the CSF are sensed directly by the central chemoreceptors in the medulla. The respiratory control centers adjust the rate and depth of respiration to ensure proper P_{O_2}, P_{CO_2}, and pH in the tissues. The respiratory control and bicarbonate buffering systems are an example of *interacting control systems*. Note that mean arterial pressure also increases secondarily to respiration since increased sympathetic neural stimulation of the respiratory muscles also acts on the smooth muscle in the walls of arterioles to increase R_{TPR} as shown in Equation 1.1. The increased arterial pressure also helps by increasing blood flow and the gas transfer rates of O_2 and CO_2 (P_{O_2} and P_{CO_2}).

metabolic byproducts of the complete oxidation of glucose by the Krebs cycle and oxidative phosphorylation according to the following overall relationship:

$$C_6H_{12}O_6 + 6O_2 \rightarrow 6CO_2 + 6H_2O + 32ATP + heat$$
$$\text{(Glucose)}$$

$$CO_2 + H_2O \underset{\text{carbonic anhydrase catalyst}}{\longleftrightarrow} H_2CO_3^- \leftrightarrow H^+ + HCO_3^- \qquad (1.2a)$$

This reaction is normally very slow at normal body conditions of temperature and pressure. The catalyst, carbonic anhydrase (CA), present in the red blood cell but *not* in the plasma phase, speeds up this reaction by a factor of several thousand. The HCO_3^- produced in the red blood cell is rapidly exchanged for chloride ion (Cl$^-$) by a specialized exchanger on the red blood cell membrane (called the "chloride shift"). Therefore, much of the HCO_3^- produced by CO_2 transport is carried in the plasma phase where the conversion of HCO_3^- to CO_2 by Equation 1.3a is very slow. The presence of HCO_3^- in the plasma allows it to act as a buffer since it can easily scavenge any excess H^+ that may be present. In fact, the *bicarbonate buffer* system is the major *extracellular* buffer, keeping blood pH under very tight control.

In the lungs, CO_2 is lost from the red blood cell, driving Equation 1.3a to the left, bringing in HCO_3^- from the plasma and kicking out a Cl^- to maintain electrical neutrality.

1.6 Control of Respiration

- The only way that the body can remove the CO_2 produced during aerobic cellular respiration is by blowing it off into room air during breathing.

- The goal of respiration is to maintain proper partial pressures (concentrations) of O_2, CO_2, and H^+ in the tissues. Thus, the *rate and depth* of respiration is regulated by the rate of O_2 utilization and CO_2 production by cell metabolism as illustrated in Figure 1.5.

- The respiratory control center consists of three major groups of neurons in the medulla and pons.

  1. *The dorsal respiratory group in the medulla.* This group of neural cells receives input from *peripheral chemoreceptors* and several other receptors. They generate APs in an increasing ramp-like output to the respiratory muscles and diaphragm and are responsible for basic respiratory rhythm.

  2. *The pneumotactic center.* This group is located in the *pons* and helps control the rate and pattern of respiration. It transmits inhibitory signals to the dorsal respiratory group and limits the inspiratory phase of breathing. By so doing, it also increases breathing rate.

  3. *The ventral respiratory group in the medulla.* This group is not active in normal, quiet breathing but stimulates abdominal muscles when high rates and depth of respiration are necessary.* These muscles can be used to stimulate both inspiration and expiration depending on which neurons are stimulated. The abdominal muscles are auxiliary muscles called into play when high respiratory rates are required.

- The respiratory control center contains sensory neurons that are particularly sensitive to changes in H^+. However, H^+ and other *hydrophilic* (water soluble) molecules have difficulties crossing the

---

* At high lung volumes, nerve receptors in the walls of bronchi and bronchioles send signals to the dorsal respiratory group that inhibits the inspiratory signal and stops further filling. This is called the Hering–Breuer inflation reflex.

blood–brain (*blood–cerebrospinal fluid* [*CSF*]) *barrier.* Lipid-soluble molecules and CO_2 readily cross through the lipid bilayer cell membranes. Other important molecules such as glucose have specialized carriers that enable them to cross the barrier. Once in the CSF, CO_2 generates H^+ in the aqueous environment by Equation 1.3a.

- Excess CO_2 and H^+ in the CSF directly stimulate the respiratory center in the medulla (*the central chemoreceptors*) and causes increased breathing rate and depth.

- In contrast, oxygen acts on *peripheral chemoreceptors.* These are specialized cells located in the *carotid and aortic bodies* that transmit neural signals to the respiratory center. They are located in the path of the flowing arterial (oxygen rich) blood and are especially sensitive to low P_{O_2} (hypoxia) and to a lesser extent, to high H^+ (acidemia). These receptors are activated when arterial P_{O_2} falls to 60–70 mm Hg.

Example 1.2 Long-Term Regulation of Arterial BP
There are several systems that contribute to the regulation of arterial BP.

- *Short-term regulation* (beat-to-beat) is neutrally mediated through the baroreceptor system.
  - *Baroreceptors,* located in the walls of the carotid artery (carotid sinuses) and in the arch of the aorta, are specialized receptors that respond to stretch by changing the frequency of their AP signals to the CV control center in the medulla.
  - The CV control center then alters the sympathetic (and vagus nerve) neural output to the heart and vascular smooth muscle in the walls of arterioles to change HR, cardiac contractility (SV), and R_{TPR}, thereby adjusting $\overline{P}_{art}$, as shown in Equation 1.1.

- *Long-term regulation* is mediated through the kidney and the renin–angiotensin–aldosterone system and antidiuretic hormone (ADH; vasopressin). These processes regulate BP directly through chemical vasoconstriction (angiotensin II [AT-II]) of smooth muscle in the walls of arterioles and through blood volume by the kidney (see Equation 1.2a).
  - At low BP or following acute blood loss, cells of the *juxtaglomerular apparatus* (JGA) of the kidney release the enzyme *renin.* Renin acts on *angiotensinogen,* a protein secreted by the liver, to produce *angiotensin I* (*AT-I*). AT-I is not biologically active and is converted to *AT-II* by an enzyme called *angiotensin-converting enzyme* (*ACE*). ACE is released by the endothelial cells of several body tissues, most importantly the capillary endothelium of the lungs.
  - AT-II acts in several ways to control systemic BP and volume.

- AT-II
  - Is a potent vasoconstrictor and acts directly on arteriolar smooth muscle cells to increase R_{TPR}.
  - Stimulates reabsorption of Na^+ directly by acting on renal tubules and indirectly by stimulating the release of *aldosterone* from the **adrenal cortex**. Aldosterone also acts on renal tubules to increase Na^+ absorption. Since water follows Na^+ osmotically, both blood volume and pressure increase (see Equation 1.2).
  - Stimulates the hypothalamus to release ADH. ADH targets the kidney tubules to absorb more water and produce concentrated urine, increasing blood volume.
  - Decreases peritubular hydrostatic pressure allowing a net decrease in net filtration pressure and more water to be reabsorbed in the kidney.
  - Reduces the glomerular filtration rate (GFR) by contracting the filtration cells and reducing the surface area available for filtration.
  - **Of these five effects, the first two are the most important**.
- Atrial natriuretic hormone is released by the atrial muscle cells when BP increases. It blocks renin and aldosterone secretion and other mechanisms that affect water and Na^+ reabsorption. Its overall effect is to decrease BP by increasing urine flow and Na^+ concentration in urine.

Figure 1.6 is a block diagram of the negative feedback regulation of mean arterial pressure when a homeostatic imbalance is caused by a decrease in mean arterial pressure. The *inputs* are shown as dashed lines. Short-term effects are neural, acting through the baroreceptors and chemoreceptors. Long-term regulation is through the kidney and involves the renin–angiotensin–aldosterone system and ADH.

Baroreceptors are not important in long-term BP regulation since they reset to whatever pressure they are exposed to within 1–2 days. That is, their firing rate will initially increase or decrease with pressure changes but will return to normal levels within 1–2 days and then defend whatever the new pressure happens to be (e.g., idiopathic *hypertension*).

NOTE a: It has been demonstrated in animal experiments that in the absence of neural, beat-to-beat BP regulation, mean arterial BP remains approximately in the normal range over a long period of time. The standard deviation of the mean is, however, much larger than if beat-to-beat regulation is present.

NOTE b: While mean arterial and central venous pressure are very stable, the distribution of blood flow to organs and tissues according to their individual metabolic requirements is under local control (autoregulation) at

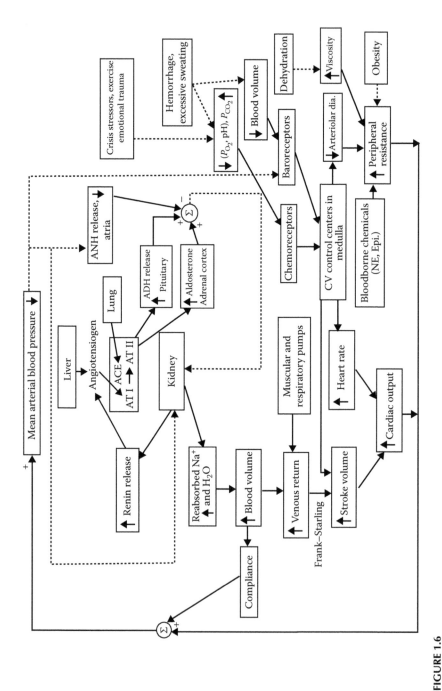

FIGURE 1.6
Short- and long-term regulation of mean arterial BP. The dashed lines show inputs to the system.

the cellular level. Tissues with high metabolic rates release local vasodila-
tors (e.g., nitric oxide) or vasoconstrictors (e.g., endothelin and leukotrienes)
that act on smooth muscle cells in feeding arterioles to alter vessel diameter
(hemodynamic resistance to flow) to increase or decrease local perfusion
rates. If there is a contention between systemic and local control, local
control always wins out.

1.7 Positive Feedback Systems

In general, *positive feedback* systems are present to increase the rate of a process. They are usually part of an *overall negative feedback* system that limits
their effects and returns the overall system to homeostasis. Previously, we
described the positive feedback system for mobilizing platelets to form a
platelet plug as part of the clotting mechanism. Below, we describe three
other positive feedback systems, one of which is *not* compensated by an
overall negative process.

Example 1.3 Generation of an AP

Axons communicate information over long distances by either transmitting
or not transmitting electrical signals called APs to other axons in the chain.
An AP is an all-or-nothing event. It either happens or does not. At rest, the
membrane potential of an excitable cell, such as a neuron, is about −70 mV,
with the inside of the cell membrane negatively charged with respect to
the outside. Two other factors are important here: First, the concentration
of sodium ions on the outside of the cell is higher than the inside (see Table
1.1). Second, the concentration of potassium on the inside of the cell is
higher than the outside. Therefore, there are concentration (electrochemical potential) gradients for driving Na^+ into the cell and K^+ out of the cell.
The cell membrane also contains water-filled pores or channels that allow
passive transport of many substances including ions like Na^+ and K^+. Some
of these channels are specific for Na^+, some for K^+, and some are nonspecific (i.e., they allow transport of either ions). The nonspecific ion channels
are called "leakage" channels. If there were no compensatory mechanism,
both sodium and potassium would "leak" down their electrochemical
potential gradients until ionic equilibrium between the inside and outside
of the cell was established and no further ion movement would occur. The
cell membrane potential would be about 0 mV and the neuron would no
longer be able to transmit electrical signals. However, there are also specialized active transport proteins on the cell membrane whose function
it is to transport Na^+ and K^+ against their electrochemical potential gradients, reversing the effects of leakage channels and stabilizing the resting membrane potential at −70 mV. These are called Na/K-ATPase pumps.

Since they were transporting ions against their electrochemical potential gradients, energy in the form of ATP is required. The pumps transport three Na^+ outside the cell for every two K^+ brought inside the cell to establish a stable resting membrane potential.

NOTE: When Na^+ channels are open, Na^+ lowers its electrochemical potential gradient from its high value **outside** the cell to its lower value **inside** the cell. The positive charges **entering** the cell *depolarize* the cell membrane (make it **more positive** than its resting membrane potential of $-70\,mV$). When K^+ channels are open, K^+ lowers its electrochemical potential gradient from its high value **inside** the cell to its lower value **outside** the cell. The positive charges **leaving** the cell *hyperdepolarize* the cell membrane (make it **more negative** than its resting membrane potential of $-70\,mV$).

1.7.1 How Is an Action Potential Generated?

Axons are connected to many other axons in series and/or parallel arrangements similar to electrical cables packed in a dense electrical distribution system. Each axon transmits information to the next axon in the chain in the form of low-level electrical signals called *graded* potentials. These potentials can either be excitatory (depolarizing the axon cell membrane) or inhibitory (hyperpolarizing the cell membrane). A graded potential, by itself, is not strong enough to depolarize an excitatory cell membrane to threshold. The cell membrane integrates (sums) all the graded inputs (excitatory and inhibitory) from neurons that it receives signals from, and when the cell membrane is depolarized to threshold value (about $-55\,mV$), the axon generates an AP. The AP travels the length of the neuron and, at the axon terminal, initiates a chain of events that sends a chemical or electrical signal to other axons in the chain.

While most axons use chemical signals to pass information on to other axons, muscles, or glands, a *very few* directly use electrical signals to communicate at the *synapse*.

At the synapse (a fluid-filled cavity between two neurons), *neurotransmitter* chemical is released by the presynaptic neuron, which diffuses across the space (*cleft*) between the presynaptic and postsynaptic neuron and binds to a *receptor* on the postsynaptic membrane. The receptor opens chemically gated ion channels in the postsynaptic cell membrane. Depending on the *neurotransmitter* and the *receptor type*, the opening of the ion channels can transport either positive ions (depolarization) or negative ions (hyperpolarization) into the postsynaptic cell. The positive ions are *excitatory* and the negative ions are *inhibitory* with respect to initiating an AP in the postsynaptic neuron. Graded potentials are low-level signals that rapidly decay in both time and position (space) from their entry point on the cell membrane.

The neurotransmitter only binds to the receptor for a short period of time and then either is released and diffuses back to the presynaptic terminal, where it is taken up and recycled, or is degraded by an enzyme on the post-synaptic membrane. Some of the degraded neurotransmitters then diffuse back and are taken up and reused to synthesize new neurotransmitters.

This electrochemical system ensures that there is only one-way transmission of information (electrical signal) and that each new piece of information is unique. That is, each new graded potential must have come from an AP on the presynaptic neuron.

The strengths of the graded potentials are **frequency and amplitude modulated**. That is, their signals can be made stronger by increasing both the frequency (rate) and the amplitude of the incoming signal. If more neurotransmitter is released by presynaptic neurons, it increases the number of ion channels that open on the postsynaptic membrane and therefore the signal strength.

1.7.2 Action Potential: A Positive Feedback Process

With respect to generation of APs, most of the changes observed can be attributed to the movement of sodium and potassium ions through voltage-gated ion channels.

The way it works is as follows:

1. Axons receive graded potential inputs from other axons in their information pathway. These can be excitatory (positively charged ions) or inhibitory (negatively charged ions). When the sum of all the inputs depolarizes the axon cell membrane to threshold ($\sim -55\,$mV), fast, voltage-gated* Na^+ ions start to open. This depolarizes the cell further, opening up more fast, voltage-gated Na^+ channels in **a positive feedback** mechanism that depolarizes the cell membrane from -55 to $+30\,$mV in about 5–10 ms.

2. The positive feedback process is **terminated** by two mechanisms. The fast, Na^+ channels have an *inactivation* gate that closes, cutting off the flow of Na^+ ions into the cell. At the same time, as the inactivation gates close, slow, voltage-gated K^+ channels open. This allows K^+ ions to leave the cell and repolarizes the cell.

3. The slow K^+ channels close after a period of time. However, they stay open a little longer than necessary, hyperpolarizing the cell below its resting membrane potential. Leakage channels then *bring Na^+ into the cell* and K^+ *leaks out of the cell* down their electrochemical potential gradients.

---

* Voltage-gated ion channels are protein channels, present in the cell membrane, that open in response to changes in cell membrane potential. When open, they provide a water-filled pathway for hydrophilic ions such as Na^+ and K^+ to enter or leave the cell through the lipid bilayer cell membrane. Normally, they stay open only for a brief period of time, then they close.

4. The Na/K-ATPase pumps on the cell membrane restore a stable resting membrane potential by simultaneously transporting three Na⁺ out of the cell (against its electrochemical potential gradient) and bringing two K⁺ into the cell (against its electrochemical potential gradient), thereby compensating for the effects of the leakage channels. These active transport pumps require energy in the form of ATP hydrolysis to adenosine diphosphate (ADP) and a PO_3^- ion. Each mole of hydrolyzed ATP yields about 7 kcal of energy.

5. Once initiated at the *axon hillock*, the AP is propagated in space along the axon by opening and closing these voltage-gated channels positioned along the axon.*

6. The AP is thus an *all-or-nothing* event and is *self-generating* as it moves down the axon from the cell body to the terminal. All APs look the same.

Example 1.4 Contractions of Uterus during Childbirth (Oxytocin)
The processes that take place during labor and childbirth are called *parturition*. At the end of the third trimester of pregnancy, the uterus begins a series of progressively stronger contractions that bring the baby out of the birth canal. This process is initiated by changes in the ratio of the hormones *estrogen* and *progesterone* and is facilitated by the increase in mechanical properties of the smooth muscle in the uterus.

- Estrogen and progesterone both act on the smooth muscle of the uterus in opposite ways. Estrogen *increases* smooth muscle contractility while progesterone *decreases* smooth muscle contractility.

- At the beginning of the third trimester, progesterone secretion stays constant, while estrogen secretion increases. The ratio of estrogen to progesterone continuously increases during the last trimester. This increases the sensitivity of the smooth muscle in the uterus to hormone stimulation.

- During the final weeks of pregnancy, the increased estrogen levels stimulate the *myomecium* cells of the uterus to increase the synthesis of oxytocin receptors on the smooth muscle cells in the walls of the uterus. This increases the *irritability* of the myometrium and the uterus exhibits slow, periodic contractions called "Braxton Hicks" contractions (sometimes leading to trips to the hospital and "false labor").

- As the third trimester begins to close, some fetal cells begin to produce *oxytocin*, which causes the placenta to produce *prostaglandins*. Both hormones elicit powerful contractions of the uterus.

---

* In myelinated axons, the voltage-gated channels are more concentrated at the *nodes of Ranvier* (spaces between *Schwann* cells). The AP seems to jump from node to node in a process called saltatory conduction. Myelinated axons have a higher conduction velocity than that of unmyelinated axons of the same diameter.

- At this time, the physical and emotional stress of the mother activate the mother's hypothalamus to also release oxytocin.
- The combined effects of the increased prostaglandin and oxytocin levels bring on the strong, periodic contractions of true labor.
- *Smooth muscle cells* are a major component of the walls of the uterus, and most hollow organs (except the heart) have certain mechanical properties that aid in *peristalsis* (movement of the contents through the organ). These are as follows:
  a. The muscle walls can sustain large amounts of stretch to accommodate changes in the volume of their contents without reflex contraction.
  b. When stimulated to contract, the force of contraction is relatively independent of the stretch.
  c. During contraction, the muscle can form actin–myosin "latch bridges." These are cross bridges that last for an extended period of time (force generation) with relatively low energy requirements.
- As the fetus grows, the uterus is continually stretched. Toward the end of pregnancy, the moving fetus further stretches the uterus and the cervix is distended by the growing fetus. Contractions that are initiated by the cervix spread upward through the uterus.
  a. Stretch of the cervix produces reflexes that further elicit release of oxytocin from the mother's pituitary gland.
- *Initiation of the onset of labor: a positive feedback process.*
  a. Once the hypothalamus is involved, greater contraction force causes the release of more oxytocin, which causes greater contraction force and more frequent contractions in a positive feedback mechanism.
  b. The positive feedback mechanism is terminated with the birth of the baby.

Figure 1.7 is a block diagram of the process. The dashed lines show the *two positive feedback processes*. At birth, the overall negative feedback loops shut down the positive feedback loops.

Example 1.5 Hypovolemic Shock

Hypovolemic shock is a life-threatening condition in which there is rapid, severe blood loss. When the volume of blood lost during trauma exceeds a certain amount, a positive feedback process takes place that, if left untreated, eventually shuts down the CV system.

The normal cardiac output of a 70 kg person is about 5 L/min. The same person also has about 5 L of total blood volume in their body. A rapid loss of a small volume of blood (up to 20% of total blood volume) is readily

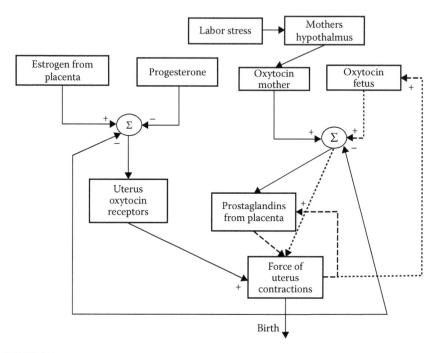

FIGURE 1.7
Parturition, showing the interaction of two positive feedback processes within an overall negative feedback system. The dashed lines show the *two positive feedback processes.*

compensated for by the same (mostly renal) mechanisms, as shown in Figure 1.6 (called *reversible* or *compensated* shock). BP will return to near normal values in a few hours.

In addition to the mechanisms shown in Figure 1.6, the lower blood volume will stimulate the thirst center in the hypothalamus.

- If the volume loss stems from trauma, the low red blood cell count will stimulate release of *erythropoietin* from the bone marrow to synthesize replacement red blood cells.
- If the volume loss stems from excessive sweating, diarrhea (acidosis and loss of K^+), or vomiting (alkylosis and electrolyte imbalance), other renal compensatory mechanisms that may involve *osmoreceptors*, which respond to low blood *osmolarity*, may also come into play and further enhance renal reabsorption of Na^+ and H_2O as well as H^+ and other ions.

With a rapid loss of 20% of a person's total blood volume (about 2 L), the normal compensatory mechanisms as shown in Figure 1.6 begin to deteriorate rapidly.

- The volume of blood remaining is not sufficient to support normal BP and perfusion of vital organs.
- The heart is one of the first organs to feel this effect. Low blood volume means low BP (Equation 1.2), low cardiac output, and low venous return (normally venous return = cardiac output).

  1. Low venous return means low ventricular filling during diastole, low end diastolic volume (preload), and poor perfusion of the coronary arteries.

  2. Low end diastolic volume (EDV) reduces the force of ventricular contraction by the Frank–Starling mechanism. Poor coronary artery perfusion reduces nutrient and O_2 transport to the cardiac myocytes (heart muscle cells), weakening of the heart muscle, and further reduction of cardiac output (CO) (and venous return).

  3. Steps 1–2 continue in a positive feedback process until the heart stops pumping and the patient dies.

1.8 Introduction to the Analysis of Block Diagrams

1.8.1 Input–Output Notation

Each of the blocks in Figures 1.1 through 1.7 has one or more inputs and one or more outputs. The block labels themselves are, in reality, a shorthand notation for the complex relationships between the inputs and outputs. We call these relationships "models." Models come in various flavors, depending on what you wish to accomplish and the degree to which you know the details of the relationships involved and their interactions. Some of these are as follows:

- *Descriptive.* A narrative that describes the mechanisms that operate to transform the inputs to outputs and the flow of information in the process. These can be superficial or very detailed depending on the author's objective and the knowledge base that was used. For example, the statement "… renin, released by the kidney, initiates a cascade of events that produces angiotensin II" leaves out much of the information about which intermediate compounds are involved, what their biological function may be, and where they are synthesized.

- *Graphical or empirical.* These models show the relationship between the inputs and outputs of blocks in a graphical form. These relationships can be linear or nonlinear, one-dimensional (1-D), two-dimensional (2-D), or three-dimensional (3-D) plots can be used, depending on the data set. Where no underlying mechanism is available, experimental data can often be fit by curve fitting them to an empirical mathematical model. These data are often fit using statistical models (see below).

- *Mathematical.* These models come in increasingly complex flavors.

  - *Algebraic.* Relationships between input and output can be described by an algebraic equation. The model may involve several coupled algebraic equations.

  - *Lumped parameter.* Relationships between inputs and outputs are modeled with ordinary differential equations with one independent variable (*time or space*). Again, several coupled sets of equations may be involved. These equations may be linear, nonlinear, or have time- or space-dependent parameters.

  - *Distributed parameter.* Relationships between inputs and outputs are modeled with partial differential equations, in time *and* space. Again, several coupled sets of equations may be involved. The space dependency can be 1-D (x, y, or z), 2-D (x, y), or 3-D (x, y, z). They may be linear or nonlinear, with time- and space-varying parameters, and several coupled sets of equations may be involved.

  - *Statistical.* These are usually used when the mechanism(s) that transforms the inputs to the outputs is uncertain or unknown. They are more common in the medicine and biology literature than in engineering where they are used to compare differences in outcomes among two or more groups.

    - The major concerns in using statistical models are proper design of experiments **before** taking data, choosing the proper statistics (parametric and nonparametric), the particular test or tests to use (t test, one-way or two-way ANOVA, correlation, and Kaplan–Meier), and the hypothesis to test in analyzing and interpreting the data. *An in-depth discussion of statistical methods is beyond the scope of this text.*

  - *Optimization models.* In these models, a *penalty function* is set up to achieve the best or optimum result. That is, to determine the operating conditions (trajectory of the manipulated variables) that maximize (or minimize) the penalty function, subject to mathematical constraints on the system inputs and variables (e.g., subject to conservation of mass, energy, and momentum). For example, the penalty function may weight excessive deviation from the set point in such a way that overshooting the homeostatic value or range is more costly than undershooting it. Specialized software is required to search for the optimum, subject to the constraints. Sometimes, the code for the iterative search algorithm (e.g., gradient) has to be written specifically for the problem at hand.

Methods for solving these models for various input and parameter values, in general, require computational software such as MATLAB®/Simulink®, LabVIEW, or Mathematica. In CV models, which involve fluid–vessel

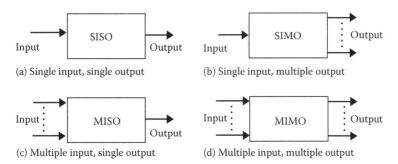

FIGURE 1.8
Various possible input–output blocks. Most of the information flow in physiological processes is MIMO.

interactions, computational fluid dynamics (CFD) software such as ADINA and COMSOL, which allow fluid–structure interactions, are necessary.

Information flow (input–output) in a block diagram can take several forms depending on the interactions between the inputs and outputs from each block. Figure 1.8 shows the four major types of input–output relationships:

a. Single input, single output (SISO)
b. Single input, multiple output (SIMO)
c. Multiple input, single output (MISO)
d. Multiple input, multiple output (MIMO)

1.8.2 Overall Process Inputs

The administration of drugs and biologics into humans and animals can only occur by a few routes. These are indicated in the following:

- *Injection.* In the literature, injections (intravenous [i.v.], intra-arterial, intramuscular, intraperitoneal, etc.) are usually described as a "bolus" input. That is, delivery of a fixed volume (or mass) of an injectate in a short period of time. Mathematically, these can usually be simulated by an impulse function or a probability distribution.
- *Intravenous drip.* This is a slow, i.v. infusion of a drug or biologic until a therapeutic level is achieved in the blood. Mathematically, this can be simulated by a step function.
- *Ingestion.* This is the usual way that pills and liquid formulations are taken. The pills dissolve in the stomach and diffuse into the blood through the GI tract along with food and liquids. These inputs to the blood compartment can often be simulated mathematically by a delayed step change.
- *Inhalation.* Inhalants are used as respiratory drugs and bronchodilators for asthmatics, although some newer proposals are looking at using

inhaled insulin for diabetics instead of injections. Inhalants usually get into the bloodstream very quickly due to the large surface area and efficient alveolar blood capillary transport in the lung. Mathematically, these can be simulated by a delayed step change.

- *Diffusion through skin (patches)*. Nicotine and other drug-dispensing patches are commonly used. They can also be mathematically simulated by a delayed step change in the blood concentration.

- *Circadian rhythms*. Many glands secrete hormones at rates that vary periodically with time. The period may be seconds, minutes, hours, days, or months. Some of these can be mathematically described by sine or cosine functions with the proper period or frequency.

- *Radiation, light, and other electromagnetic inputs*. These may be whole body or localized. Their effects on tissue are often hard to simulate and are specific for the type of input, dose, time, and where in the body the target tissue resides.

These are, of course, simple functions that can be made to fit more complex time-varying inputs by combining two or more of the simple functions.

- *CFD models*. These represent more complicated cases. Here, the inlet pressure is a complex, periodic function of time that may best be described by a graphical or empirical function. Also, the initial ($t = 0$) flow distribution in space may have to be specified.

Example 1.6 A SISO Model

A simple example of a SISO model is shown below. The relationship between the input and output of a block is given by a first-order linear differential equation with constant coefficients:

$$\tau \frac{dy}{dt} + y \qquad f(t) \longrightarrow \boxed{} \longrightarrow y(t)$$

a. If the input, $f(t)$, is an injection of an amount, A g (an impulse of size A), then the output* of the block (the information leaving the block) is

$$y(t) = Ae^{-t/\tau} \qquad\qquad (1.4)$$

---

* The solution of first-order linear differential equations with constant coefficients to impulse, step, and sine function inputs is available in standard mathematical textbooks and is presented here without details.

Here, we see that at $t = 0$, the output of the block is A and the output decays to zero after a long time. Since the solution depends on the exponential function, e, the decay will be 95%–98% complete between 3τ and 5τ time units (s, min, h, etc.). τ is called the *time constant* for the process described by the block. Equation 1.4 is called the *impulse response* of the process.

b. If the input, $f(t)$, is an i.v. drip with drug concentration A g/mL (a step change of size A), then the output of the block is

$$y(t) = A\left[1.0 - e^{-t/\tau}\right] \tag{1.5}$$

Here, we see that at $t = 0$, the output of the block will be zero and will exponentially approach A at long times. Again, the approach to 95%–98% of the final value, A, will take 3τ–5τ time units.

Equation 1.5 is called the *step response* of the process.

c. If the input, $f(t)$, is a simple circadian rhythm, it can be represented by a sine function of period T (s, h, days, etc.). Then, $f(t) = A \sin(\omega t)$. Here, ω is in rad/s and $\omega = 2\pi/T$. The output of the block is then

$$y(t) = A\left[\left(\frac{\tau\omega}{1+\tau^2\omega^2}\right)e^{-t/\omega} + \frac{1}{\sqrt{1+\tau^2\omega^2}}\sin(\omega t + \phi)\right] \tag{1.6}$$

where $\phi = \tan^{-1}(-\tau\omega)$. Equation 1.6 is called the *frequency response* of the process.

At long times, the first term (the transient contribution) decays to zero and the long-term solution is just

$$y(t \to \infty) = \frac{A}{\sqrt{1+\tau^2\omega^2}}\sin(\omega t + \phi) \tag{1.7}$$

where ϕ is the angle by which the output sine wave *lags* the input sine wave.

It is interesting to compare the ratio of the amplitudes of the output to the input sine waves, which is called the *gain*, G, of the process:

$$G = \frac{\text{amplitude out}}{\text{amplitude in}} = \frac{A/\sqrt{1+\tau^2\omega^2}}{A} = \frac{1}{\sqrt{1+\tau^2\omega^2}} \tag{1.8}$$

For this process, according to Equation 1.7, after a long time, the amplitude of the output sine wave will always be smaller than the input amplitude and also be a function of the frequency of the input, ω. Also, at steady state ($\omega \to 0$), the gain will approach 1 (output amplitude = input amplitude). This implies that the dynamics of the process are slow enough so that they can be ignored.

1.8.3 Interaction of Blocks in a Block Diagram

In block diagrams, information outputs from interconnected blocks are inputs to the next block in the loop. It is clear from Equations 1.4 through 1.6 that the time domain forcing functions become complex very quickly as one goes through the loop. Analytical solutions for the *overall output* of the controlled variable(s) to disturbances or set point changes are indeed complex, except for the simplest systems. One can use Laplace transforms to bring the differential equations in the time domain to algebraic equations in the Laplace domain and determine an overall relationship between the output of the closed-loop and the system input in the Laplace domain (called the *transfer function*). However, inverting the solution to the time domain can be a real challenge.

Fortunately, there are software such as MATLAB/Simulink and LabVIEW that allow relatively straightforward solutions of these coupled sets of equations with a minimum of programming. This is the approach we have taken in this textbook.

1.8.4 Constructing a Block Diagram from a Narrative Description of a Process

This section outlines a procedure for constructing a block diagram from the description of a feedback control process.

1. Determine the **controlled** variable(s).
2. What sort of **disturbances or inputs** would make the variables deviate from homeostasis? Where would they come from?
3. Determine the **sensor(s)** that sense the current value of the variable(s). There may be more than one variable and sensor.
4. If more than one variable and/or sensor, is there **more than one** feedback loop involved? If so, are any of the loops **positive feedback** loops?
   a. If there are positive feedback loops, how are they terminated within an **overall negative feedback** process?
   b. Are there **short-** and **long-term** controls? Identify them.
5. Where do these **sensors reside** (in the medulla, kidney, or vasculature)?
6. Is there a **set point** or is this a reflex process? If there is a set point, is it constant or can it be reset?
7. Where is the **controller located**? How do the sensors communicate with the controller(s)?
8. What are the **effector(s)** and how does the controller **communicate** with the effectors?

9. How do the effectors **change** the controlled variables when they **exceed** and when they are **less than** the set point? Is there **more than one organ** involved? If so, how do they **interact**?

10. What about inputs from **higher centers** of the brain and from **hormones** released from glands?

These steps are a guide to the considerations involved in constructing a block diagram. Example 1.7 takes you through the steps described above.

Example 1.7 A Block Diagram for Glucose Regulation

Glucose is one of the most tightly regulated substances in the body. Normal blood glucose levels are maintained between 90 and 110 mg/dL (1 dL = 100 mL). About 43% of the bond energy in the glucose molecule (686 kcal) is conserved by cells as ATP through glycolysis and the Krebs cycle plus oxidative phosporylation (38 mol of ATP are produced/mole of glucose). Cells and organs use the energy liberated by hydrolysis of ATP (to ADP + PO_4^{3-} + 7 kcal/mol of ATP) to drive *endergonic* (energy-consuming) synthesis reactions, active transport, and force generation by muscles.

The pancreas is the major sensor and controller of blood glucose levels. An anatomical structure in the pancreas called the islets of Langerhans contains four types of cells:

- *α-Cells.* These cells, which comprise ~25% of islet cells, respond to low levels of glucose by secreting *glucagon*, a *hyperglycemic* (glucose increasing) substance. Glucagon, in effect, reverses the effects of insulin.

- *β-Cells.* These cells, which comprise ~60% of the islet cells, are sensitive to high blood glucose levels and respond by secreting *insulin*, a *hypoglycemic* (glucose lowering) substance.

- *δ-Cells.* These cells secrete *somatostatin*. Somatostatin inhibits GI motility, secretion, and absorption. It is a potent inhibitor of both insulin and glucagon secretion. It delays transport of nutrients in the GI tract and utilization of absorbed substances by the liver and peripheral tissues.

- *Pancreatic polypeptide cells.* These cells secrete *pancreatic polypeptide* (PP). The physiological effects of PP are not firmly established but are thought to reverse the effects of somatostatin.

1.8.4.1 High Glucose Levels and Insulin Secretion

Both insulin and glucagon are secreted into the portal vein of the liver via the pancreatic vein. Thus, the liver has a higher concentration of these substances than do the peripheral tissues. After a meal, the complex hydrocarbons, proteins, and lipids (fats) in the ingested food are converted to simpler substances that cells can use. Much of the absorbed carbohydrates and lipids are ultimately converted to glucose and fatty acids. The resulting high blood glucose levels promote synthesis and secretion of insulin by the β-cells.

- In muscle, insulin promotes transport of glucose into muscle cells down their concentration gradient by upregulating glucose transporters in the cell membrane.
- Glucose uptake by cells is almost entirely dependent on insulin, so that glucose entry into muscle cells occurs mainly after a meal when insulin is secreted or *during exercise when glucose uptake is not insulin dependent.*
  - Glucose is stored in muscle cells as *glycogen*. Glycogen is a very large polymer of glucose of average MW ~ 5×10^6.
- In the liver, *insulin facilitates* glucose uptake while inhibiting glucose production. In the liver and in fat (adipose tissue), insulin
  - Increases the number of glucose transporters in the cell membranes, thus increasing *facilitated transport* of glucose into cells.
  - Increases glycogen synthesis (due to higher uptake of glucose).
  - Decreases the output of glucose by the liver by inhibiting *glycogenolysis* (production of glucose by breakdown of glycogen), *gluconeogenesis* (formation of new glucose molecules from amino acids and fat), and facilitated transport by removing glucose transporters from cell membranes.
  - Enhances synthesis of fatty acids that further inhibits gluconeogenesis.

1.8.4.2 Low Glucose Levels and Glucagon Secretion

- Glucagon secretion reduces blood glucose levels by several mechanisms. In the liver
  - It increases glycogenolysis and the secretion of glucose into the blood.
  - It inhibits glycolysis (the breakdown of glucose to form ATP by anaerobic metabolism) in all cells.
  - It stimulates gluconeogenesis.
- Amino acids, especially arginine and alanine, stimulate glucagon secretion.
  - After a protein meal, both insulin and glucagon secretion are stimulated. However, the glucagon response is attenuated if glucose is also ingested. Without the depressed glucagon effect, increased insulin secretion would induce *hypoglycemia* (low blood sugar levels).
- Fasting and exercise increase glucagon secretion. This effect may be mediated by sympathetic stimulation during exercise.

1.8.5 Metabolism of Insulin and Glucagon

Insulin and glucagon do not stay in the bloodstream very long. They are metabolized (enzymatically degraded) in the liver and kidney almost as soon as they are secreted. Approximately 50% of the insulin and glucagon entering the liver through the portal vein is metabolized in its first pass through the liver. Most of the remaining hormones are metabolized in the kidney. The half-life of these hormones is 5–10 min.

The half-life ($t_{1/2}$) is the time it takes for the concentration of a substance to decay to half of its original concentration. This is based on the assumption of a first-order exponential decay model as given in Equation 1.4. In Equation 1.4, y is the concentration of hormone at any time, t, τ is the time constant for the process, and A is the amount secreted at $t = 0$.

$$y = Ae^{-t/\tau} \tag{1.4}$$

At $t = t_{1/2}$, $y = A/2$, so that

$$y = \frac{A}{2} = Ae^{-t_{1/2}/\tau} \tag{1.9}$$

Dividing Equation 1.9 on both sides by A gives

$$\frac{1}{2} = e^{-t_{1/2}/\tau} \tag{1.10}$$

Taking natural ln's of (1.10)

$$\ln\left(\frac{1}{2}\right) = \frac{-t_{1/2}}{\tau} \tag{1.11}$$

By the properties of ln's

$$\ln(2) = \frac{t_{1/2}}{\tau} \tag{1.12}$$

Solving for τ

$$\tau = \frac{t_{1/2}}{\ln(2)} = \frac{t_{1/2}}{0.693} \tag{1.13}$$

Thus, knowing the half-life gives you the time constant for the process.

With an exponential decay process, we can predict that it takes 3–5 time constants (τ) for 95%–98% of the hormone to disappear after it has been secreted. With a half-life of 5–10 min, the range of the time constant is

$$\tau = \frac{5-10}{0.693} = 7.2 - 14.4 \text{ min}$$

Using the minimum value, 95% of the hormone will be metabolized within 3τ or between 21.6 and 43.2 min.

1.8.6 Diabetes Mellitus

Diabetes mellitus is the result of an impaired production of (Type I, also called insulin dependent or juvenile onset) and cellular response to (Type II, also called noninsulin dependent or adult onset) insulin.

- Type I diabetes is an *autoimmune* disease caused by destruction of the β-cells of the pancreas. It can also be caused by the loss of these cells due to a viral infection.
- Most of the pathophysiological manifestations of Type I diabetes are due to hyperglycemia, depletion of proteins and fats, and increased ketone production. These include the following:
  - High blood glucose levels saturate the glucose transporters in the kidney and lead to glucose excretion in urine (*glucosurea*), high blood osmolarity leading to *osmotic diuresis*, which leads to low blood volume (*hypovolemia*) and low BP (*hypotension*). The osmotic diuresis also leads to frequent urination (*polyuria*), dehydration, and excessive thirst (*polydipsia*). Although there is plenty of glucose present, it cannot be used, so there is excessive food consumption and hunger (*polyphagia*) with loss of weight and lack of energy. The last three signs are characteristic of diabetes mellitus.
  - Other symptoms are acidosis leading to diabetic coma if untreated, rapid, deep breathing, ketone breath, *hypercholesterolemia* (high blood cholesterol levels), atherosclerosis, and vascular disease.
  - Untreated diabetes eventually leads to end-organ damage (mainly heart and kidney failure), blindness, and neural degeneration.
- Type II diabetes is characterized by impaired ability of cells to utilize insulin. Insulin is produced by the β-cells but is poorly taken up by target cells. This is called *insulin resistance*.
  - Type II diabetes is much more prevalent than Type I. It is usually associated with obesity.

- Hyperglycemia still occurs, although insulin secretion may be higher than normal. However, increased lipid depletion and ketone formation are usually not present.
- Proper diet, exercise, and weight reduction usually improve insulin resistance.
- Blood glucose levels can be controlled by oral hypogylcemic drugs. However, in the later stages of the disease, insulin secretion is impaired and insulin injections are required.
- Tight control of blood glucose levels, postpones, but does not prevent, end-organ damage.

1.8.7 Creating a Block Diagram from the Narrative

Creating a block diagram from the description above requires us to specify an objective since there is much more information than we can effectively use.

Let us go through the 10-step process:

Step 1: The **controlled** variable is blood glucose level.

Step 2: The **disturbances are** *increased* blood glucose levels from a meal or disease (Type I or Type II diabetes) or *decreased* blood glucose levels from exercise.

Step 3: The **sensors** are the α- and β-cells of the pancreas. The variables are insulin and glucagon secretion.

Step 4: There are two feedback loops involved. None of the loops are **positive feedback** loops. Both the short- and long-term controls are the same.

Step 5: Where do these **sensors reside** (in the medulla, kidney, or vasculature)? Sensors reside in the α- and β-cells in the islets of Langerhans of the pancreas.

Step 6: Is there a **set point** or is this a reflex process? If there is a set point, is it constant or can it be reset? The set point is blood glucose levels between 90 and 110 mg/dL. The set point in nondiabetic people is constant. In Type I and Type II diabetes, the set point does not exist.

Step 7: Where is the **controller located**? How do the sensors communicate with the controller(s)? The controller(s) and set point are α- and β-cells in the islets of Langerhans of the pancreas.

Step 8: What are the **effector(s)** and how does the controller **communicate** with the effectors? The effectors are *insulin and glucagon* secreted by α- and β-cells in the islets of Langerhans of the pancreas.

Step 9: How do the effectors **change** the controlled variables when they **exceed** and when they are **less than** the set point? Is there **more than one organ** involved? If so, how do they **interact**? Insulin secretion in response to

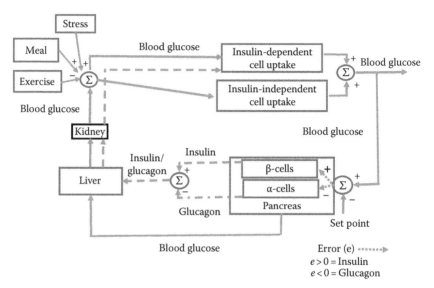

FIGURE 1.9
Block diagram for glucose regulation.

increased glucose levels in the blood increases glucose uptake and utilization by all cells and is *hypoglycemic*. Glucagon secretion in response to low blood glucose levels increases glucose availability in the liver and bloodstream and is *hyperglycemic*. Both insulin secretion and glucagon secretion are *negative* feedback responses.

Step 10: What about inputs from **higher centers** of the brain and from **hormones** released from glands? Fight or flight inputs from higher centers of the brain increase glucagon and insulin output from the pancreas. Type I and Type II diabetes modulate both insulin and glucagon effects in all cells. Exercise levels also modulate glucose levels.

Now, we can *construct a block diagram*. Since the controlled variable is blood glucose level, we start with all possible sources of glucose generation and loss from blood (see Figure 1.9):

- We then add the important organs (pancreas, liver, and kidney).
- The pancreas secretes either insulin or glucagon, depending on whether blood glucose levels are higher (insulin) or lower (glucagon) than the set point.
- The liver has two functions. It can create glucose and it can break down the hormones.
  - The liver creates glucose in two ways: by gluconeogenesis and by breaking down *glycogen* to form glucose.
  - The liver metabolizes both insulin and *glucagon*.

- Insulin increases uptake of glucose by all cells and inhibits the formation of glucose by the liver.
  - Some cells take up glucose by a non-insulin-dependent pathway.
- Any hormone left after a single pass through the liver is metabolized by the kidney.
- Increases in blood glucose levels (disturbances) are through food digestion (+), exercise (–), and stress (+).
- Figure 1.9 shows a block diagram of the process.
  - We have left out the contributions of skeletal muscle for two reasons. First, the effects of skeletal muscle on blood glucose levels are most significant during moderate to severe exercise. Normal daily movements can be simulated by the simple exercise block in Figure 1.9. Second, we have left the addition of an exercise contribution as a problem for students at the end of this chapter.

1.8.8 Analysis of Block Diagrams

To quantitatively analyze the effects of disturbances (or set point changes) on the controlled variable(s), we need *three* important pieces of information. First, we need *mathematical models* for each of the blocks in the closed-loop diagram. These models translate the inputs to each block to quantitative outputs to the next block. Second, we have to specify how the external inputs to the controlled system drive the system away from homeostasis (e.g., step change, injection, and circadian rhythm). Finally, we need a valid set of initial conditions and parameters for each of the block models and a set point for system homeostasis.

We discuss the creation of mathematical models in Chapter 3. Normally, the mathematical model for the closed-loop process consists of a coupled set of differential equations that must be solved simultaneously. For these processes, a closed form of analytical solution for the set of equations would be difficult, if not impossible. These models are best tackled with software such as MATLAB/Simulink.

1.8.9 Set Points

Table 1.1 lists the normal values and ranges for some important physiological variables that are under homeostatic control.

- Values outside the normal range are usually caused by illness.
- Note small range for most of these variables.
- The nonlethal limits for some of these parameters are very small. For example, an increase in core body temperature of 7°C above normal can lead to a positive feedback process of increased cellular metabolism that destroys cells. The lethal value for acid–base control is only 0.5 pH units outside of the normal value of 7.4.

TABLE 1.1

Normal Values and Ranges for Physiological Variables

Substance	Set Point	Normal Range (±)	Units
Breathing rate	12	3	BPM
Mean arterial pressure	100	40	mm Hg
Acidity	7.4	0.1	pH
Core temperature	37	0.22	°C
Blood glucose	100	10	mg/dL
Calcium	1.2	0.2	mmol/L
Sodium	142	4	mmol/L
Potassium	4.2	0.4	mmol/L
HCO_3^-	28	4	mmol/L
P_{O_2} (arterial)	100	10	mm Hg
P_{O_2} (venous)	45	5	mm Hg
P_{CO_2} (venous)	40	5	mm Hg

1.8.10 Laplace Transforms

For models that consist of coupled sets of linear differential equations with constant coefficients, an analytical solution for the closed-loop response may be possible using Laplace transforms.* The advantage of transforming time domain differential equations to the Laplace domain is that it replaces convolution integrals in the time domain with algebraic expressions in the Laplace domain when the output of one block acts as the input to another block. The price one pays is that the expressions in the Laplace domain become so complex that inverting them to the time domain requires great effort. However, MATLAB/Simulink software allows Laplace domain expressions to be used and inverts to the time domain directly.

1.8.11 Analysis of Closed-Loop Systems in the Laplace Domain

Consider Figure 1.3 that shows a simple closed-loop negative feedback control system. We first define a transfer function for each block, i, in the Laplace domain as follows:

$$G_i(s) = \frac{\text{output from block}}{\text{input to block}}$$

The transfer function is derived by taking the Laplace transform of the differential equation(s) that describes the mathematical model for that block.

---

* A discussion on the use of Laplace transforms is beyond the scope of this text. This section is included for completeness.

Note that the transfer function is not a function of the input to the block and only depends on the mass, energy, and momentum balances that occur in that space. For simplicity, we will omit the Laplace variable, s, from the notation.

The transfer functions have been identified for each block in Figure 1.3. There are only two rules to follow in this analysis of the closed loop:

1. The output from each block is its transfer function times the input to that block.
2. The output from a summation block is the algebraic sum of the inputs to the block.

We will start with the controlled variable, y, and go counterclockwise around the block, applying the two rules above, until we return to the starting point.

$$y = G_p e_d = G_p[d + e_m] = G_p d + G_p e_m = G_p d + G_p G_E e_c$$

$$y = G_p d + G_p G_E G_c e_o = G_p d + G_p G_E G_c[r_{sp} - e_s]$$

$$y = G_p d + G_p G_E G_c r_{sp} - G_p G_E G_c e_s = G_p d + G_p G_E G_c r_{sp} - G_p G_E G_c G_s y$$

Moving the y-dependent terms t, the left-hand side of the equation gives

$$y[1 + G_p G_E G_c G_s] = G_p d + G_p G_E G_c r_{sp} \tag{1.14}$$

Solving for the output of the controlled system, y:

$$y = \frac{G_p d}{1 + G_p G_E G_c G_s} + \frac{G_p G_E G_c r_{sp}}{1 + G_p G_E G_c G_s} \tag{1.15}$$

1.8.12 Two Transfer Functions

From Equation 1.15, we can define two transfer functions:

a. The closed-loop response of y to changes in set point, r_{sp}. In this case, we assume that no disturbances are entering the system so that $d = 0$.

$$H_{sp} = \frac{\text{output}}{\text{input}} = \frac{y}{r_{sp}} = \frac{G_p G_E G_c}{1 + G_p G_E G_c G_s} \tag{1.16}$$

b. The closed-loop response of y to disturbances entering the system, d. In this case, we assume that no set point changes take place, so $r_{sp} = 0$.

$$H_d = \frac{\text{output}}{\text{input}} = \frac{y}{d} = \frac{G_p}{1 + G_p G_E G_c G_s} \tag{1.17}$$

Our task then would be to select the appropriate input for either r_{sp} in Equation 1.16 or d in Equation 1.17, convert them to the Laplace domain, multiply thoroughly, and solve for y in the Laplace domain. The next and final task would be to invert the y values from the Laplace to the time domain.

In the next chapter, we take up some fundamental ways in which mathematical models are derived for simulation of the processes that make up the blocks in physiological systems.

Problems

1.1 The glomerular membrane of the kidney filters blood, rejecting the formed elements (red blood cells, white cells, and platelets). The glomerular membrane filters the blood producing a filtrate that contains all the ingredients that are present in blood plasma, except the large proteins. Most of the ingredients in the filtrate are subsequently reabsorbed (actively and passively) back into the blood as the filtrate passes through the kidney nephron. The remaining substances (urea, uric acid, fixed acids, and extra NaCl) are excreted in the urine.

Two of the major functions of the kidney are salt (NaCl) and water balance and long-term regulation of BP through the renin–angiotensin–aldosterone system. Each kidney nephron has a region known as the JGA that participates in both NaCl balance and BP regulation. Figure P1.1 is a schematic diagram of the glomerular membrane showing the location of the cells in the juxtaglomerular apparatus (JGA). The ascending limb of the

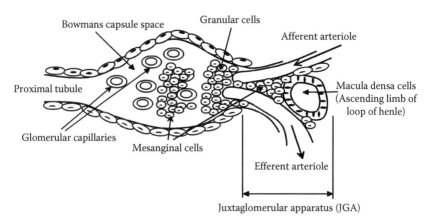

FIGURE P1.1

loop of Henle passes next to the afferent arteriole feeding blood to (and sometimes the efferent arteriole draining blood from) the glomerulus. Modified cells in both the ascending limb and in the afferent arteriole at the point of contact participate in the regulatory functions.

a. *Granular cells* in the arteriolar walls (also called *juxtaglomerular [JG] cells*) are enlarged smooth muscle cells that contain secretory granules of renin. The JG cells are *mechanoreceptors* that respond to changes in BP in the afferent arteriole.

b. *Macula densa cells* are closely packed cells in the loop of Henle that lie close to the JG cells. The macula densa cells are *chemoreceptors* that sense the NaCl content of the filtrate in the ascending limb. These cells constantly release a *vasoconstrictor chemical*, which maintains the tone of the smooth muscle cells in the walls of the afferent arteriole.

The approximately 2 million nephrons of the kidneys need a constant GFR to perform their job. The GFR is directly proportional to the systemic BP.

If the systemic BP falls, GFR decreases.

c. Decreased GFR decreases NaCl flow in the ascending limb of the loop of Henle. The decreased NaCl flow is sensed by the macula densa cells, and inhibits their release of the vasoconstrictor chemical. This acts to vasodilate the afferent arteriole and increases the GFR.

d. At the same time, the decreased GFR is sensed by the JG cells. The JG cells release their renin, which initiates the renin–angiotensin–aldosterone system and increases peripheral resistance and NaCl and water retention, increasing systemic blood volume and BP.

Draw a block diagram for this process. Identify the following:

a. Controlled variable(s)
b. Controller(s)
c. Manipulated variable(s)
d. Set point
e. Sensor(s)
f. Feedback pathway(s)

1.2 When osmolarity (plasma sodium concentration) increases above normal, the osmoreceptor-ADH feedback system operates as follows:

a. Increased extracellular fluid osmolarity stimulates osmoreceptor cells in the anterior hypothalamus to send AP signals that are relayed to the posterior pituitary gland.

b. These signals stimulate the release of ADH, which is stored in secretory granules in the nerve endings.

c. ADH is transported in the blood to the kidney where it increases water permeability of the late distal convoluted tubules, cortical collecting ducts, and medullary collecting ducts of the nephron.

d. Increased water permeability causes increased water reabsorption and therefore a small volume of concentrated urine. This causes dilution of solutes in the extracellular fluid.

The opposite sequence of events occurs when the extracellular fluid becomes too dilute (hypo-osmotic).

Draw a block diagram for this process. Identify the following:
a. Controlled variable(s)
b. Controller(s)
c. Manipulated variable(s)
d. Set point
e. Sensor(s)
f. Feedback pathway(s)

1.3 PTH is the single most important hormone controlling the calcium balance of the blood. PTH release is triggered by falling blood Ca^{2+} levels and inhibited by high levels of blood calcium (hypercalcemia). PTH increases calcium in the blood by stimulating three target organs: bones, which contain considerable amounts of calcium salts in their matrices, the kidneys, and the intestine.

PTH release (1) stimulates osteoclasts (bone-resorbing cells) to digest some of the bony matrix and release ionic calcium and phosphates into the blood, (2) enhances reabsorption of Ca^{2+} (and excretion of PO_4^{3-}) by the kidneys, and (3) increases absorption of Ca^{2+} by the intestinal mucosal cells. Calcium absorption by the intestine is enhanced indirectly by PTH's effect on vitamin D activation. Vitamin D is required for absorption of Ca^{2+} from ingested food, but the form in which the vitamin is ingested or produced by the skin is relatively inactive. For vitamin D to exert its physiological effects, it must first be converted by the kidneys to its active D_3 form (calcitrol), a transformation stimulated by PTH.

a. Draw a block diagram for the regulation of calcium ions in the blood by PTH.
b. Identify the controlled variable, the controller, the manipulated variable, the sensor, the set point, and the effector(s). Identify the feedback pathway.
c. PTH deficiency sometimes follows parathyroid trauma (hypoparathyroidism). What are some of the physiological effects that may occur with PTH deficiency?

2

Modeling of Physiological Processes

In this chapter, we introduce some of the basic concepts of model building that will set the stage for the more detailed treatment given in Chapters 3 through 5.

2.1 The Standard Person

The concept of individual variability is as old as human observation. Different people respond differently to the same dose of drug, to the same decibel noise level, or to the same temperature changes. Physiological parameters such as cardiac output (CO), diastolic blood pressure, and blood glucose level, to name just a few, all vary widely from person to person. To help us make sense of these variations and to provide a framework in which we can operate in a logical way, we usually think in terms of a *standard person*. Our standard person is a hypothetical statistical construct that allows us to use *average* values of physiological parameters in our models. When we use single values for physiological variables (i.e., 120/80 mm Hg for systolic/diastolic blood pressure), we implicitly recognize that we are using average values that only apply under a limited set of conditions. In fact, all physiological variables have a range of values when measured in clinical laboratories using similar techniques. These are called *normal* or *reference* values [1,2]. By convention, in the biomedical sciences, the normal range is reported as the 95% confidence interval [3] for a given variable. When there are variations in the values as measured in different laboratories, reference values are reported as measured in a particular laboratory (e.g., Stanford University and Massachusetts General). The mean value and the confidence interval for any variable may change with age, sex, weight, race, or one or more environmental, socioeconomic, or hereditary factors. Table Appendix A lists adult reference values for a number of routine laboratory tests performed in hospitals, clinics, and physicians' offices.

2.1.1 An Engineering Point of View

Throughout this chapter, our approach will always be that of an engineer. That is, we are prepared to view biological systems as *black boxes*. Where experimental data are unavailable or are insufficient or imprecise enough to allow a rigorous analysis of the process or to discriminate among alternative mechanisms, we will use an analysis that is appropriate for the quality and quantity of the data that have been reported in the literature. The philosophy is that, as engineers, we have to work with the data that are available. This does not mean, however, that we cannot use a more rigorous analysis to aid us in planning future experiments.

An engineering analysis of physiological processes can take place at several different levels. For example, from a thermodynamic point of view, a human body can be considered as an open system that is in quasi-steady state with its environment. Mass enters and leaves the system, and chemical reactions convert the entering reactants into products producing heat and mechanical work. The heat is ejected to the surrounding environment. Some of the available energy is irreversibly converted to heat, thereby increasing the entropy of the surroundings. Since thermodynamics does not care which paths are used to go from one thermodynamic state to another, the details of mass, energy, and momentum transport as well as the chemical kinetics would not be relevant for this type of analysis. On the other hand, if one were interested in the pharmacokinetics of an orally administered drug and its therapeutic levels in a target cell type or organ, a relevant engineering analysis would have to consider such things as blood–tissue transport in the gastrointestinal (GI) tract, uptake by the target organ or cell type, biotransformation (chemical conversion), and distribution of the drug in other body compartments in addition to the target regions and routes of elimination. On a smaller scale, one might be interested in the behavior of a leukocyte (white blood cell) that is rolling or sticking to the surface of the endothelium (the endothelium is the sheet of endothelial cells, which lines the walls of all blood vessels) in an arteriole. In that case, the details of the flow and pressure gradients, which determine the shear stresses at the cell–cell interface, along with estimates of the bond strengths formed by the adhesion molecules (integrins, selectins, and glycoproteins) on the surfaces of both cell types would be important. On yet a smaller scale, one might be interested in cell motility or shape changes in response to a chemical stimulus such as histamine. In that case, molecular events such as receptor–ligand interaction, second messenger signaling, and the cascade of intracellular events that lead to the contraction of the actin–myosin cytoskeleton of the cell would be important.

2.2 Mathematical Models

When we try to quantitatively describe complex processes, we often have to start with an incomplete or simplified description. This is true for a number of reasons. First, all the mechanisms may not have been fully identified, so the process cannot be entirely modeled. Second, the resulting set of equations may either be difficult or impossible to solve with the resources available (e.g., personal computer) or in the time frame allocated. Third, if the mathematical description is sufficiently detailed, parameters for one or more of the intermediate steps may not be available. Finally, available experimental data may limit the ability to discriminate among alternative models. One of the most difficult choices that we are faced with is that of determining the degree of detail that we should incorporate into any given model. A good rule of thumb is to use the simplest model consistent with established mechanisms that can best fit the data that are available using reasonable parameter values. Obviously, this still leaves a lot of room to be creative.

Mathematical models are all based on conservation laws. Conservation equations may be called by different names in different disciplines. *Ohm's law, Fick's laws of diffusion*, and *Newton's law of viscosity* are all well-known mathematical relationships that are statements of conservation of charge, mass, and momentum, respectively. These laws are also *mathematical models* that allow us to quantitatively describe the behavior of a particular process over a limited range. For example, a Newtonian fluid is defined as a fluid that can be described by Newton's law of viscosity at least over some defined range. That is, the shear stress exerted by the fluid is a linear function of the negative of its velocity gradient at any point in the fluid. Similarly, Fick's first law of diffusion allows us to define a binary diffusion coefficient for a pair of substances under conditions in which one of the substances is stationary and the mass flux (mass flow/unit area) of the diffusing substance is a linear function of its concentration gradient.

2.2.1 Mechanical and Electrical Analog Models

Many of the physiological models that appear in the literature are based on electrical and mechanical analogs of force, pressure, flow, resistance, and compliance (capacitance).

2.2.1.1 Electrical Analogs

Electrical analogs use resistors, capacitors, and inductors to represent the resistance to flow and storage of volume (capacitors) and kinetic energy (inertance or inductance). ΔE (voltage drop) is used as an analog of pressure drop and current (I) is used to simulate volumetric flow rate.

NOTE: There is no inductance concept in either thermal or chemical systems.

Figure 2.1 shows the basic units used in electrical analog models.

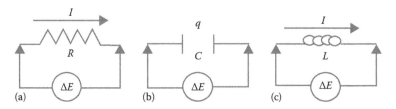

FIGURE 2.1
Basic electrical analog units: (a) resistance, (b) capacitance (compliance), and (c) inductance.

a. *Resistance.* The relationships in Figure 2.1a are given by Ohm's law:

$$\Delta E = IR$$

b. *Capacitance.* The relationships in Figure 2.1b are as follows:

$$\text{Capacitance} = \frac{\Delta \text{charge}}{\Delta \text{voltage}}$$
$$= \text{amount of electrical charge stored per unit voltage across capacitor}$$

$$C = \frac{q}{\Delta E} \tag{2.1}$$

where

$$q = \int_{0}^{t} I\, dt$$

Then,

$$q = C\Delta E = \int_{0}^{t} I\, dt \tag{2.2}$$

c. *Inductance.* Storage of kinetic energy is given by

$$\Delta E = L\frac{dI}{dt} \tag{2.3}$$

2.2.1.2 Mechanical Analogs

Mechanical analogs use springs and dashpots (viscous fluids) to simulate resistance to flow, force, and velocity, where

V is the velocity
F is the force
R_m is the resistance of fluid to flow (viscosity)

Figure 2.2 shows the basic mechanical elements commonly used:

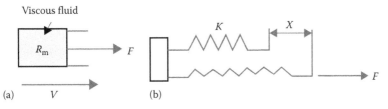

FIGURE 2.2
Basic mechanical analogs: (a) dashpot (shock absorber) and (b) spring.

a. *Dashpot*
   Here,

$$F = VR_m \tag{2.4}$$

where V is the velocity.

b. *Spring, stretched a distance, X*

$$F = KX \tag{2.5}$$

where K is the spring constant (stiffness).
Let

$$C_m = \frac{1.0}{K} \tag{2.5a}$$

where C_m is the capacitance of the spring.
Then,

$$F = \frac{X}{C_m} \tag{2.6}$$

and

$$X = C_m F \tag{2.7}$$

Both the basic electrical and mechanical analog units are used in series and parallel arrangements to simulate physiological processes. Table 2.1 shows the equivalence of these series and parallel arrangements.

TABLE 2.1

Resistances, Capacitances, Springs, and Dashpots in Series and Parallel Arrangements

Equivalent Series Arrangement	Equivalent Parallel Arrangement
Resistances and capacitances in series and parallel (electrical analogs)	
Resistances: $R_{eq} = R_1 + R_2$	Resistances: $R_{eq} = \left(\dfrac{1}{R_1} + \dfrac{1}{R_2} \right)^{-1}$
Capacitances: $C_{eq} = \left(\dfrac{1}{C_1} + \dfrac{1}{C_2} \right)^{-1}$	Capacitances: $C_{eq} = C_1 + C_2$
Springs and dashpots in series and parallel (mechanical analogs)	
Springs: $C_{eq} = C_{m1} + C_{m2}$	Springs: $C_{eq} = \left(\dfrac{1}{C_{m1}} + \dfrac{1}{C_{m2}} \right)^{-1}$
Dashpots: $R_{eq} = \left(\dfrac{1}{R_{m1}} + \dfrac{1}{R_{m2}} \right)^{-1}$	Dashpots: $R_{eq} = R_{m1} + R_{m2}$

Note: Dashpots = resistances to flow, springs = compliances.

In parallel arrangement of dashpots:

$$F = VR_{eq} = V\left(R_{m1} + R_{m2} \right)$$

where
   F is the total force
   V is the velocity

This is not the same as parallel resistances.
Be careful in converting from mechanical to electrical analog circuits.

The Three-Element Windkessel *Model*
The three-element windkessel model is an electrical analog model that is commonly used to model the relationship between flow, pressure drop, and resistance in the circulatory system.

Figure 2.3 is a schematic diagram of the three-element windkessel model.

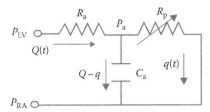

FIGURE 2.3
The three-element windkessel model of the circulation.

Here

R_a is the resistance to flow of the aortic valve
R_p is the peripheral resistance (mostly arteriolar) (variable)
C_a is the compliance of the arterial system
P_a is the aortic pressure
P_{RA} is the right atrial pressure
$Q(t)$ is the CO
$q(t)$ is the portion of CO that reaches the peripheral circulation
$Q - q$ is the portion of CO that is stored as volume due to the compliance of the aorta and bypasses the peripheral circulation

An electrical analog mathematical model of the circulatory system
From Figure 2.3, using Ohm's law, we can write $Q(t) = \Delta P/R_a = (P_{LV} - P_a)/R_a$, so that

$$P_a = P_{LV} - R_a Q(t) \tag{2.8}$$

Ohm's law again gives

$$P_a - P_{RA} = R_p q(t)$$

and

$$P_a = P_{RA} + R_p q(t) = R_p q(t) \tag{2.9}$$

Since the right atrial pressure can be assumed to be atmospheric (e.g., $P_{RA} \approx 0$). Now, by definition,

$$C_a = \frac{dV_a}{dP_a} \tag{2.10}$$

where
dP_a is the aortic pressure change
dV_a is the aortic volume change

Then,

$$dV_a = C_a dP_a \tag{2.10a}$$

For constant C_a, we can differentiate Equation 2.10a with respect to time to give

$$\frac{dV_a}{dt} = C_a \frac{dP_a}{dt} \tag{2.11}$$

From Figure 2.3,

$$\frac{dV_a}{dt} = Q(t) - q(t)$$

Substituting this into Equation 2.11,

$$C_a \frac{dP_a}{dt} = Q(t) - q(t) \tag{2.12}$$

Since we do not know $q(t)$, we have to eliminate it. From Equation 2.9, $P_a = R_p q(t)$ so that

$$q(t) = \frac{P_a}{R_p}$$

Substituting into Equation 2.12,

$$C_a \frac{dP_a}{dt} = Q(t) - \frac{P_a}{R_p}$$

Collecting terms and multiplying through by R_p gives

$$C_a R_p \frac{dP_a}{dt} + P_a = R_p Q(t) \tag{2.13}$$

Equation 2.1 shows, $P_a = P_{LV} - R_a Q(t)$, so that

$$\frac{dP_a}{dt} = \frac{dP_{LV}}{dt} - R_a \frac{dQ(t)}{dt} \qquad (2.14)$$

Substituting Equations 2.8 and 2.14 into Equation 2.13 and collecting terms

$$C_a R_p \frac{dP_{LV}}{dt} + P_{LV} = C_a R_p R_a \frac{dQ(t)}{dt} + \left(R_a + R_p\right) Q(t) \qquad (2.15)$$

Equation 2.15 is the starting point for many models of the circulation.

- In some models, R_a is called the *characteristic impedance* of the aorta to accommodate aortic compliance.
- P_{LV} is the left ventricular blood pressure. This is approximately the same as the arterial blood pressure during systole.

In order to use this model, we have to generate inputs for either CO [$Q(t)$] or P_{LV}. A block diagram for using Equation 2.15 in which we generate $Q(t)$ by a separate model is shown in Figure 2.4.

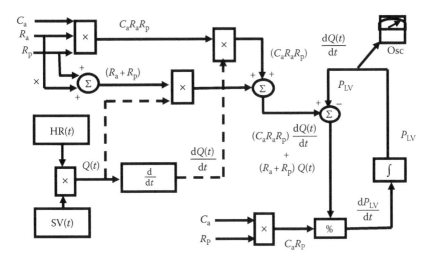

FIGURE 2.4
Block diagram of Equation 2.15a. The blocks are labeled according to their function. × indicates a multiplication of the inputs to the block, Σ indicates algebraic summation, ∫ indicates integration, % indicates division, and d*/dt indicates differentiation of the expression entering the block. The expression leaving each block is the result of the operation carried out in the block. Osc indicates an output device (oscilloscope).

In order to proceed, we first solve for the highest derivative of the unknown, P_{LV}. We then assume that we can recreate this derivative from the components on the right-hand side of the equation, so we integrate as many times as necessary until we get to the dependent variable. We then arrange the blocks to recreate the equation assuming, in this case, we have simulated $Q(t)$ with the product of HR and SV. The starting equation is then

$$\frac{dP_{LV}}{dt} = \frac{C_a R_p R_a \dfrac{dQ(t)}{dt} + \left(R_a + R_p\right)Q(t) - P_{LV}}{C_a R_p} \tag{2.15a}$$

Mechanical Analog Model
Mechanical analog models use springs and dashpots (shock absorbers) to model mechanics. In the following model, we look at a well-known mechanical analog model for skeletal muscle mechanics.

Over the physiological range (80%–120% of resting length), stretch of *skeletal* (and *cardiac*) muscle increases the force of contraction due to the better alignment of *actin–myosin cross-bridges*. That is, more cross-bridges are formed. Outside this range, the alignment is less and the force generated decreases. Figure 2.5 shows a typical force–length curve for the skeletal muscle. There are two components to the force generated when the muscle is stretched. These forces are called the *passive force* (dashed curve) and the *active force* (solid curve). The passive force occurs because of the passive resistance of the tendons, ligaments, and supporting structures surrounding the muscle and is similar to the resistance that one would feel when stretching a rubber band. The active force is generated by the series arrangement of muscle cells, parallel arrangement of muscle bundles, and the force generated by the sliding actin–myosin cross-bridges in each muscle cell. The cross-bridge formation and release requires calcium release by the cells in response to an action potential stimulus and ATP hydrolysis to provide the energy for the stroke and to actively reuptake the calcium at the end of the contraction (called a *twitch*).

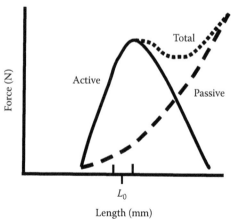

FIGURE 2.5
Length versus force generated by skeletal muscle showing the active and passive forces generated. L_0 is the resting muscle length. The two hash marks on either side of L_0 indicate the physiological range in which increasing stretch leads to greater *active* force of contraction.

The total force generated is the sum of the active and passive forces.

Figure 2.6 shows a standard mechanical model for muscle mechanics. It consists of a spring, which simulates the compliance of the elastic muscle components (e.g., tendons), in series with a parallel arrangement of a dashpot (which simulates the viscous nature of the tissue surrounding the contracting muscle) and an active force generator (the active force generated by the sliding cross-bridges). This arrangement is in parallel with a second spring that represents the compliance of the stretched muscle bundles (the passive force in Figure 2.5).

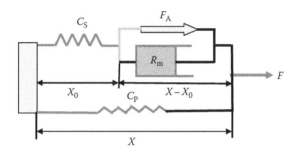

FIGURE 2.6
A mechanical analog model of skeletal muscle contraction. The muscle has been stretched an incremental distance, X, and then stimulated to contract. C_P represents the compliance of the parallel elastic element (passive stretch). C_S represents the compliance of the series elastic element. R_m represents the viscous damping of the tissue surrounding the muscle, F_A represents the active force generated by the contractile elements of the muscle, and F is the total force observed (the sum of the active and passive forces as shown in Figure 2.5).

A mathematical model for muscle contraction

1. We now stretch C_P by a length, X. The series-parallel arrangement of $C_S R_m$ and F_A is also stretched by a distance, X. However, C_S is not necessarily stretched an equal distance as the parallel arrangement F_A and R_m.
2. Since we do not know how far C_S is stretched, we assume that it is stretched a distance X_0. This means that the parallel arrangement F_A and R_m is stretched a distance $X - X_0$.
3. Force balance on the model in Figure 2.5 tells us the following:
   a. The force through the series elastic element, C_S, must be the same as the force through the dashpot-active force parallel component. So, using Equations 2.4 and 2.6,

$$\frac{X_0}{C_S} = F_A + R_m V = F_A + R_m\left(\frac{\mathrm{d}X}{\mathrm{d}t} - \frac{\mathrm{d}X_0}{\mathrm{d}t}\right) \tag{2.16}$$

where

$$V = \frac{\mathrm{d}}{\mathrm{d}t}(X - X_0) \tag{2.17}$$

b. The sum of the forces through the series elastic element and the parallel elastic element must equal the total force, F.

$$F = \frac{X}{C_P} + \frac{X_0}{C_S} \tag{2.18}$$

Since X_0 is unknown, we must eliminate it from Equations 2.16 and 2.18.

From Equation 2.18, we get

$$X_0 = C_S\left[F - \frac{X}{C_P}\right] \tag{2.19}$$

We also need $\mathrm{d}X_0/\mathrm{d}t$, so differentiating Equation 2.19 with respect to time gives

$$\frac{\mathrm{d}X_0}{\mathrm{d}t} = C_S\left[\frac{\mathrm{d}F}{\mathrm{d}t} - \left(\frac{1}{C_P}\right)\frac{\mathrm{d}X}{\mathrm{d}t}\right] \tag{2.20}$$

Substituting (2.19) and (2.20) back into (2.16) and collecting terms gives

$$\frac{\mathrm{d}F}{\mathrm{d}t} + \frac{F}{R_m C_S} = \frac{F_A}{R_m C_S} + \left(\frac{1}{C_S} + \frac{1}{C_P}\right)\frac{\mathrm{d}X}{\mathrm{d}t} + \frac{X}{R_m C_S C_P} \tag{2.21}$$

There are two simulations possible using this model:

1. Simulate the force, F, required to achieve a given velocity, $\mathrm{d}X/\mathrm{d}t$, with a given active force, F_A.

    In that case, we arrange Equation 2.21 to solve for $\mathrm{d}F/\mathrm{d}t$.

$$\frac{\mathrm{d}F}{\mathrm{d}t} = -\frac{F}{R_m C_S} + \frac{F_A}{R_m C_S} + \left(\frac{1}{C_S} + \frac{1}{C_P}\right)\frac{\mathrm{d}X}{\mathrm{d}t} + \frac{X}{R_m C_S C_P} \tag{2.21a}$$

2. Simulate the velocity $\mathrm{d}X/\mathrm{d}t$ achieved by the muscle with a given force, F, applied over a given time course, $\mathrm{d}F/\mathrm{d}t$.

The block diagram for simulation a, Equation 2.21a, is shown in Figure 2.7. We leave the block diagram for simulation b, as an exercise for the students.

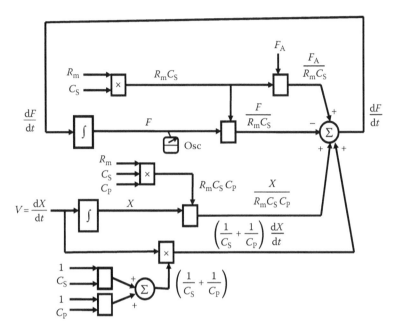

FIGURE 2.7
Block diagram of Equation 2.21a. *Note: Block diagrams are not necessarily unique and equivalent diagrams are acceptable.* Inputs to each block are denoted by arrows. Outputs from each block are the operations specified by the symbols in the block. ∫ indicates integration, Σ indicates summation, × indicates multiplication, and ÷ indicates division.

2.2.2 Conservation of Mass, Energy, and Momentum

Many models of physiological systems are concerned with transport of mass, energy, and momentum (flow). These tend to be constructed of coupled sets of partial differential equations, which may be multidimensional in space and time (e.g., x, y, z, t). *In all but the simplest cases, a closed form analytical solution is not possible.* Therefore, CFD software is required. For example, pulsatile flow in a distensible tube requires not only solution of the fluid velocity equation, but also the motion of the wall of the tube. The CFD software in that case would have to solve the equations of motion for the fluid and the wall as coupled equations. The CFD software would then have to contain fluid–structure interaction (FSI) capabilities.

When creating transport equations for models of fluid flow (momentum transport), diffusion (mass transfer), and energy (heat transfer), we know that these entities cannot be created or destroyed but can be transformed to different forms. For example, chemical reaction (metabolism and synthesis) can convert the mass of the reactants to those of the products, but the total mass must be conserved. Similarly, energy can be transferred to different forms (kinetic or potential) and work. Momentum transport is a statement of Newton's laws of motion according to which the sum of the forces acting on the system equals the mass times the acceleration.

Each basic conservation or *balance* equation can be expressed in words as

$$
\begin{matrix}
\text{Net} & & \text{Net} & & \text{Net} & & \text{Net} & & \text{Net} \\
\text{accumulation} & = & \text{transport} & - & \text{transport} & + & \text{generation} & - & \text{consumption} \\
\text{in system} & & \text{in through} & & \text{out through} & & \text{in system} & & \text{in system} \\
\text{volume} & & \text{system surface} & & \text{system surface} & & \text{volume} & & \text{volume}
\end{matrix}
$$

The entity that is accumulated, transported, generated, and/or consumed may be mass, energy, or momentum (fluid flow).

The balance equation is applied over each compartment or spatial region of interest for each entity and for each chemical species that is transported.

2.2.2.1 Fluxes and Rates

When constructing differential models, it is mathematically more convenient to start with the flux of a quantity such as mass or momentum rather than its total rate of transport. *The flux of a quantity is defined as its rate of transport per unit area.* For example, the mass flux, j_A, is the mass of component A per unit time per unit area at a given point in space at any time. Since the flux is defined as a rate per unit area, it represents a *point* or *intensive* quantity. The total rate of transport of the quantity is then its flux times the area perpendicular to its direction of motion (e.g., total rate of mass flow = mass flux × area).

Fluxes are either *vector* (mass and heat) or *tensor* (momentum) quantities and so have direction as well as magnitude. Some common fluxes and a consistent set of units for each are given in Table 2.2.

TABLE 2.2

Common Transport Fluxes

Name	Symbol	Units	Vector/Tensor
Mass flux	J	$g/s\ cm^2$	Three components (vector)
Heat flux	Q	$cal/s\ cm^2$	Three components (vector)
Momentum flux[a]	τ	$g\ cm/s^2\ cm^2$	Nine components (tensor)

[a] Momentum is the product of mass and velocity. The rate of momentum transport is the product of mass flow rate (g/s) and velocity (cm/s). Both mass flow and velocity are vector quantities (three dimensions). Thus, their *scalar product* is a tensor with nine components.

2.2.2.2 Linear Rate Laws

In general, fluxes are caused by spatial gradients in density of a given property. For example, mass will flow from a region of higher concentration (mass density) to a region of lower concentration. Similarly, heat will flow from a region of higher enthalpy to a region of lower enthalpy. The usual mathematical definition of a positive spatial gradient requires the quantity at position "2" to be greater than the quantity at position "1." Since transport takes place from the region of higher concentration or enthalpy to the region of lower concentration or enthalpy, it takes place in the direction of the negative gradient. We can now write a cause–effect definition of a flux:

$$\text{Flux of a quantity} = -f(\text{material parameter; gradient} * \text{in density of quantity})$$
(2.22)

This definition says that the flux of a quantity (the effect) is a function of (is caused by) two factors: the gradient in density of the quantity and a material parameter that characterizes the nature of the spatial domain within which the transport is taking place. The minus sign says that the flux takes place in the direction of the negative gradient. It is also important to realize that the flux is a vector or tensor quantity. Therefore, the material parameter may either be constant (*isotropic materials*) or vary with direction (*anisotropic materials*).

The simplest form for calculation of fluxes is to assume that the function in Equation 2.22 is linear. Examples of linear rate laws for isotropic materials at steady state are Fick's first law of diffusion, Fourier's law of heat conduction, and Newton's law of viscosity. Table 2.3 lists the 1D form of these laws.

TABLE 2.3

One-Dimensional Form of Some Linear Rate Laws

Law Equation	Quantity Transported	x-Component of Gradient (Units)	Material Parameter (Units)
Fick's first law of diffusion $j_{A,x} = -D_{AB}\dfrac{dC_A}{dx}$	Mass	$\dfrac{dC_A}{dx}$ concentration gradient (g/cm^3)	D_{AB} binary diffusion coefficient (cm^2/s)
Fourier's law of heat conduction $q_x = -k\dfrac{dT}{dx}$	Enthalpy	$\dfrac{dT}{dx}$ temperature (K)	K thermal conductivity (cal/s cm K)
Newton's law of viscosity[a] $\tau_{x,y} = -\mu\dfrac{dV_y}{dx}$	Momentum	$\dfrac{dV_y}{dx}$ y-component of velocity (cm/s)	μ viscosity (g/cm s) poise

[a] One can readily show that momentum flux has the same units as shear stress: (g/s) (cm/s) ($1/cm^2$) = (g cm/s^2) ($1/cm^2$) = (dynes/cm^2) = force/unit area = stress.

The linear rate laws can be used to model 1D isotropic, steady-state transport.

Example 2.1 Mass Transfer across a Biological Membrane

Use Fick's first law of diffusion to determine the steady-state mass flow of a solute, A, across a membrane barrier of cross-sectional area, S_A (cm²) and constant thickness Δl. The diffusion coefficient for solute A within the membrane is given as D_{Am}. Assume that the solution bathing one side of the membrane barrier has a steady-state concentration of solute A of C_{A1}, while the solution bathing the other side of the membrane barrier has a steady-state concentration of solute A of C_{A2}. There are no film resistances to diffusion (unstirred layers) on either side of the barrier.

Solution

A steady-state mass balance across a differential volume element of area S_A and thickness Δx, at some position, x, in the middle of the membrane (Figure E2.1) can be written as

$$\text{Rate of mass into volume} - \text{rate of mass out of volume} = 0$$

$$\text{Rate of mass in} = j_A|_x S_A$$

$$\text{Rate of mass out} = j_A|_{x+dx} S_A$$

Then,

$$j_A|_x S_A - j_A|_{x+dx} S_A = 0$$

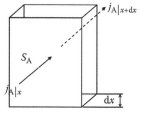

FIGURE E2.1
Rectangular exchange area for mass transfer across a membrane barrier.

Now divide by the volume of the control space, $V = S_A\, dx$ and take the limit as $dx \rightarrow 0$. Remember the definition of a derivative. This gives

$$-\frac{dj_A}{dx} = 0$$

Multiplying through by −1 and integrating,

$$j_A = C_1 \tag{2.23}$$

where C_1 is a constant of integration. Substituting Fick's first law of diffusion (Table 2.3) gives

$$-D_{Am}\frac{dC_A}{dx} = C_1$$

Rearranging, separating variables, and integrating again gives

$$C_A = -\frac{C_1 x}{D_{Am}} + C_2 \tag{2.24}$$

Here, C_2 is a second constant of integration. We now need two boundary conditions to evaluate C_1 and C_2. The boundary conditions given in the problem are

$$\text{at } x = 0, \quad C_A = C_{A1} \tag{2.25}$$

$$\text{at } x = Dl, \quad C_A = C_{A2} \tag{2.26}$$

Substituting (2.25) and (2.26) into (2.24) and solving the algebraic equations for C_1 and C_2 give $C_2 = C_{A1}$ and $C_1 = (C_{A1} - C_{A2})(D_{Am}/\Delta l)$, so that after canceling D_{Am} in the numerator and denominator, (2.24) becomes

$$C_A = \frac{(C_{A1} - C_{A2})x}{\Delta l} + C_{A1}$$

Rearranging, gives the final result

$$\frac{C_A - C_{A1}}{C_{A2} - C_{A1}} = \frac{x}{\Delta l} \tag{2.27}$$

This model gives us the concentration C_A at any position, x, between 0 and Δl. Notice that the steady-state concentration profile is linear with x and does not depend on the material constant, D_{Am}.

We can now evaluate the steady-state mass flux, j_{Ax}, by substituting for C_1 in (2.23). This gives us the integrated form of Fick's first law of diffusion as

$$j_A = C_1 = \left(\frac{D_{Am}}{\Delta l}\right)(C_{A1} - C_{A2}) \tag{2.28}$$

Equation 2.28 says that the steady-state mass flux across a membrane of thickness Δl, with constant concentrations C_{A1} and C_{A2} on either side is proportional to the macroscopic concentration gradient, $(C_{A1} - C_{A2})/\Delta l$, and a material parameter, D_{Am}.

Many times, the membrane thickness, Δl, is unknown. In that case, it is convenient to define a new material parameter, P, called the *permeability coefficient*:

$$P = \frac{D_{Am}}{\Delta l} \tag{2.29}$$

The integrated form of Fick's first law of diffusion then becomes

$$j_A = P(C_{A1} - C_{A2}) = P\Delta C \tag{2.30}$$

where ΔC is the concentration difference across the membrane $(C_{A1} - C_{A2})$. Since D_{Am} has units of cm²/s and Δl has units of cm, P has units of cm/s (the same as velocity).

The total integrated mass flow, J_A in g/s, is then the product of the integrated flux and the area:

$$J_A = j_A S_A = P S_A \Delta C \tag{2.31}$$

Equations 2.30 and 2.31 are the starting points for many 1D isotropic mass transfer problems.

The permeability of a solute across a membrane barrier is calculated from

$$PS_A = \frac{J_A}{\Delta C} \tag{2.31a}$$

The permeability–surface area product is the ratio of the measured mass flow of a solute across the membrane to the measured concentration difference across the membrane. Usually, the experiments are carried out using radioactively labeled solutes so that no osmotic gradients are established across the membrane. The radioactivity level is measured by either γ or β counters, depending on the radiolabel used.

There are many other macroscopic linear rate laws that appear in forms similar to Equations 2.30 and 2.31. Among these are Ohm's law in a resistive DC circuit and Hook's law of elasticity. The fluxes (current and strain rate) depend on material constants (resistance of the circuit and modulus of elasticity) and a macroscopic gradient (potential difference and change in stress).

2.3 Types of Models

In Chapter 1, we indicated that there were several ways by which we can classify mathematical models, the following notation will be used throughout this text:

2.3.1 Algebraic Models

These are models in which the relationships among the variables are governed by algebraic equations. In general, these relationships are nonlinear and involve transcendental and trigonometric functions. Algebraic relationships are commonly used to describe *equilibrium conditions, time averaged events, and those parts of a process that can be considered much faster than the rate-limiting steps.*

Example 2.2 Cardiac Output

CO in L/min is the product of heart rate (HR) in beats/min and stroke volume (SV) in L/beat.

$$CO = HR\,(beats/min) \times SV\,(L/beat)$$

$$= L/min$$

The CO as defined above is a time-averaged quantity. Both the HR and SV vary from beat to beat and are affected by such variables as age, sex, conditioning, exercise, levels of certain drugs, and hormones, and psychological factors such as stress. The normal range for adult HR is 55–100 beats/min at rest, with a mean of about 72 beats/min. The normal range for adult CO at rest is 4–6 L/min, with a mean of about 5 L/min. A mean value for resting SV can then be calculated as

$$SV = \frac{CO}{HR} = \frac{5\,L/min}{72\,beats/min} = 0.0694\,L/beat$$

We see that the SV in an average adult at rest is about 70 mL/beat.

It is important to point out at this time that while HR is relatively easy to measure (the pulse rate), the measurement of either CO or SV is not. The measurement of CO in patients (where it is clinically indicated) is an invasive procedure that requires the insertion of a Swan–Ganz catheter (a long, thin, tubular device that will be discussed in Example 2.4) into a large vein (usually in the leg or arm) and maneuvered into position in the right ventricle or pulmonary artery by a surgeon or anesthesiologist. These measurements are done in a cardiac catheterization laboratory. The SV is usually estimated from the CO and HR, although where more precise estimates are required, the SV can be estimated from the end diastolic and end systolic volumes (EDV and ESV, respectively) using imaging techniques.

Example 2.3 CO Using the Fick Principle

Since the entire CO passes through the lungs, a steady-state mass balance requires that the rate of uptake of oxygen by the blood must equal the rate of consumption of oxygen by the tissues. The rate of uptake of oxygen by the blood can be written as

$$Q(C_{art} - C_{ven})$$

where
Q is the volumetric flow rate of the bloodstream (e.g., the CO), L/min
C_{art} is the concentration of oxygen in arterial blood
C_{ven} the concentration of oxygen in mixed venous blood (to get a representative sample of mixed venous blood, the sample is usually taken from the right ventricle or the pulmonary artery)

Concentrations are expressed in mL O_2/dL of blood, where $1\,dL = 100\,mL$.
   The rate of consumption of oxygen by the tissues at rest is represented by $\dot{V}$, mL O_2/min. In an adult at rest, $\dot{V} \approx 300\,mL/min$.
   The Fick equation can then be written as

$$Q = (C_{art} - C_{ven}) = \dot{V}$$

Solving for Q (the CO) gives

$$Q = \frac{\dot{V}}{C_{art} - C_{ven}} \tag{2.32}$$

On knowing or measuring the oxygen consumption, the oxygen concentration in arterial and mixed venous blood samples, one can calculate the CO.

Example 2.4 Cardiac Output Using Thermal Dilution

CO using thermal dilution is an application of the Fick principle using temperature as the measured indicator. Instead of a mass balance, an enthalpy balance is required. The idea is that a small quantity of cold saline is injected into the bloodstream at a point where the entire CO passes (e.g., at the right ventricle or the pulmonary artery) and the drop in temperature versus time is recorded downstream after mixing of the cold saline with the warmer bloodstream. An enthalpy balance then allows calculation of the unknown CO from the measured temperature–time data, knowing also the amount and temperature of the injected cold saline.

Thermal dilution measurements are usually done in cardiac catheterization laboratories. The saline is injected and temperature measured using a *Swan–Ganz* catheter placed in the right atrium or pulmonary artery under anesthesia.

The Swan–Ganz Catheter

A typical Swan–Ganz catheter consists of a bundle containing several concentric flexible tubes, about 1 m long, arranged so that the annular regions of each tube are not connected (Figure E2.2). Each annular region is called a *lumen*. Ports at each end of each tube allow for introduction or withdrawal of blood, an electrical connection to a thermistor for temperature measurement, or a hydraulic connection to an external pressure gauge. An inflatable balloon at the end is connected to one of the tubes. Air from a syringe at the other end is used to inflate and deflate the balloon. Once the catheter is inserted, the balloon is inflated. This allows the catheter to *float* in the middle of the flowing bloodstream for easier placement. A second lumen is filled with *heparinized* (anticlotting agent) saline and is connected to an external pressure gauge by a fluid-filled tube. The pressure recordings allow for exact placement of the catheter in the atria, ventricles, pulmonary artery, or vena cava.

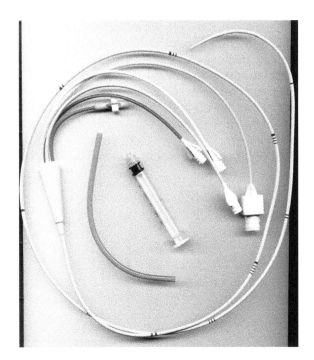

FIGURE E2.2
A Swan–Ganz catheter, showing computer connection and four luminal connections.

Swan–Ganz catheters come in various outer diameters to accommodate blood vessels of different sizes. These sizes are indicated by a number followed by the letter F (e.g., 7.5F). The number indicates the size (diameter) in French units. For adults, sizes 7.5F to 8F are typical. To convert the outer diameter of the catheter to millimeters, multiply the French value by π.

For thermal dilution experiments, a third lumen is used to inject cold (0°C) saline. A thermistor placed at the outlet of this lumen measures the temperature of the injected saline as it enters the bloodstream. A fourth lumen contains a second thermistor located several centimeters downstream from the exit port of the injectate. This thermistor measures the temperature of the mixed blood plus saline.

In a typical measurement of CO by thermal dilution, in an adult, a 7.5F quadruple-lumen, balloon-tipped, flow-directed Swan–Ganz catheter is used. Typically, 10 mL of sterile, ice cold (0°C) isotonic (0.9% NaCl) saline is injected through one of the lumen. To measure CO, the catheter is introduced through a slightly larger (e.g., 8F) introducer sheath that has been previously inserted into the right or left jugular vein. A heparinized (to prevent clotting), saline-filled line is connected to a pressure gauge for recording intravascular pressure. The thermistor electronics are connected to a flow computer. The balloon is inflated and the catheter is advanced (floats) into the right atrium, through the atrioventricular (AV) valve into the right ventricle and, if required, through the pulmonic valve into the pulmonary artery. The pressure readings in each of the structures are sufficiently different so that the physician knows exactly where the catheter is at any time. It is important to place the catheter in the pulmonary artery (or pulmonary outflow track) so that there is complete mixing of the cold saline with the warm blood to ensure an accurate determination of CO.

Once the catheter is placed, the saline is injected and the computer calculates the CO from measurements of the blood temperature versus time curve recorded in the right ventricle or pulmonary artery. Two or three determinations of CO can be made very quickly.

A typical thermal dilution curve is shown in Figure E2.3.

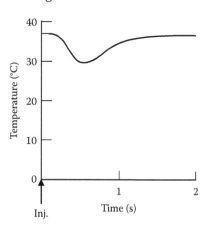

FIGURE E2.3
Typical thermal dilution curve.

The calculation of CO (L/min) depends on an enthalpy balance around the ventricle (or pulmonary artery) segment. A simplified schematic diagram for the control region is shown in Figure E2.4.

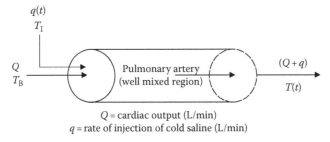

$$Q = \text{cardiac output (L/min)}$$
$$q = \text{rate of injection of cold saline (L/min)}$$

FIGURE E2.4
Control volume for thermal dilution enthalpy balance.

An enthalpy balance over the control volume for a small time increment, Δt, is

$$\text{Enthalpy of leaving stream} = \text{enthalpy of entering streams}$$

$$\rho C_p (Q+q)[T(t) - T_r]\mathrm{d}t = \rho C_p q[T_I - T_r]\mathrm{d}t + \rho C_p Q[T_B - T_r]\mathrm{d}t$$

To simplify the mathematics, we subtract an equal amount of enthalpy, $\rho C_p q[T_B - T_r]\mathrm{d}t$, from both sides of the equation:

$$\rho C_p (Q+q)\left[T(t) - T_r\right]\mathrm{d}t - \rho C_p q\left[T_B - T_r\right]\mathrm{d}t$$
$$= \rho C_p q\left[T_I - T_r\right]\mathrm{d}t + \rho C_p Q\left[T_B - T_r\right]\mathrm{d}t - \rho C_p q\left[T_B - T_r\right]\mathrm{d}t$$

Collecting terms and dividing by ρC_p

$$(Q+q)\left[T(t) - T_r - (T_B - T_r)\right]\mathrm{d}t = q\left[T_I - T_r - (T_B - T_r)\right]\mathrm{d}t$$

Simplifying

$$(Q+q)[T(t) - T_B]\mathrm{d}t = q[T_I - T_B]\mathrm{d}t$$

Dividing both sides by $(T_I - T_B)$

$$\frac{(q+Q)[T(t) - T_B]\mathrm{d}t}{[T_I - T_B]} = q\,\mathrm{d}t$$

For mathematical simplicity, we assume that the cold saline is injected instantaneously as an impulse,* so that $q = V_0\delta(t)$, where V_0 is the amount of cold saline injected (mL) and $\delta(t)$ is the unit impulse function. We then integrate the equation from 0 to ∞ to arrive at the total enthalpy of the entering and leaving streams:

$$\int_0^\infty \frac{(q+Q)[T(t)-T_B]}{[T_I-T_B]}\,dt = \int_0^\infty q\,dt = \int_0^\infty V_0\delta(t)\,dt = V_0 \int_0^\infty \delta(t)\,dt = V_0$$

Now, we let $Q + q$, the flow rate of the leaving stream be denoted by F. Assuming that $q \ll Q$, F is then (approximately) the CO. This then gives

$$F\int_0^\infty \frac{[T(t)-T_B]}{[T_I-T_B]}\,dt = V_0$$

Solving for the CO,

$$F = \frac{V_0}{\int_0^\infty \frac{[T(t)-T_B]}{[T_I-T_B]}\,dt} \tag{2.33}$$

If we were to plot $[T(t) - T_B]/[T_I - T_B]$ against time, we obtain a curve that looks like that in Figure E2.5.

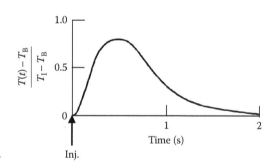

FIGURE E2.5
Normalized thermal dilution curve.

We recognize that the integral in the denominator of Equation 2.33 is just the area under the curve. The computer can readily calculate this numerically. Equation 2.33 tells us that knowing the volume injected, the temperature of the cold saline, and the blood temperature, and measuring the temperature change with time at the tip of the Swan–Ganz catheter, we can calculate the CO.

The Esophageal Doppler

Because of the invasive nature of the Swan–Ganz catheter, various methods have been introduced over the years to overcome this limitation. The procedure that has emerged as an acceptable alternative to thermal dilution involves the *esophageal Doppler*. The ascending *aorta* runs very close to the *esophagus*. A Doppler ultrasound probe placed appropriately in a patient's esophagus and pointed to the aorta will easily measure the flow velocity in the aorta. Since the centerline velocity is the maximum velocity in a tube, the average velocity (compare Equations 4.20 and 4.22) in the ascending aorta can be readily calculated using the esophageal Doppler. Estimates of the aortic diameter can then be used to continuously calculate the CO in an anesthetized patient *noninvasively*.

Compartmental or "Lumped Parameter" Models

A compartment is defined as a region of interest for the purposes of the model. Compartments may or may not have a precise anatomical identity. For example, the *blood compartment* encompasses all blood components (plasma, red cells, white cells, and platelets) that are present in all blood vessels without regard to their particular anatomical location. Similarly, the *extracellular compartment* includes blood vessels, lymphatics, and the interstitial space: again, without regard to their exact anatomical location. One of the assumptions in a compartmental model is that the compartment is spatially homogeneous. That is, concentrations of substances, pressure, etc. are uniform throughout the compartmental volume. In these models, the relationships among the variables are described by sets of ordinary differential equations and/or mixed sets of algebraic and ordinary differential equations. In general, these sets of equations are nonlinear and are coupled in some way. Time is the usual independent variable. However, in some models, spatial changes may be of interest and so distance along some defined axis would be the independent variable.

Pharmacokinetic Models

Pharmacokinetic models are lumped parameter models that use the idea of a compartment to follow the kinetics of a drug or anesthetic agent at various sites in the body as the drug is metabolized or converted to another form (biotransformation) in the body. The models are generally used to assess the short-term delivery of drugs. They are used to assess the consequences of various drug delivery modalities and to evaluate the possibility of under- and overdosing when these issues are important. Pharmacokinetic safety and efficacy studies are an important part of phase I clinical trials that are required in a new drug application (NDA) to the U.S. Food and Drug Administration (FDA).

If a drug is administered as a bolus injection and the plasma concentration of the drug is plotted on a log scale against time, the data typically appear as in Figure E2.6a and b.

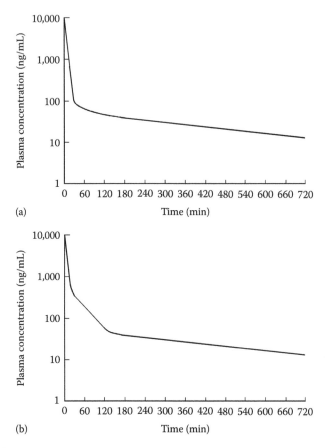

FIGURE E2.6
(a) Typical log-concentration versus time graph obtained for a bolus injection and a two-compartment model. (b) Typical log-concentration versus time graph for a bolus injection and a three-compartment model.

In Figure E2.6a, there is a rapid fall in concentration, followed by a prolonged steady fall that lasts for the duration of the study. In Figure E2.6b, there is an initial rapid fall in concentration, followed by a slower fall that lasts much longer and a final prolonged steady fall that lasts for the duration of the study.

In the first case, the phases correspond to a rapid distribution phase and a slow elimination phase (either through renal elimination, liver metabolism, or biotransformation). In the second case, the phases correspond to a rapid distribution phase, a slow distribution phase, and a slow elimination phase.

A schematic diagram for a three-compartment pharmacokinetic model for drug distribution is illustrated in Figure E2.7.

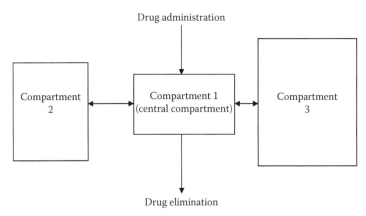

FIGURE E2.7
Schematic diagram of a three-compartment drug distribution model.

In this model, the drug is administered as a bolus injection into the central compartment (compartment 1). This is usually the plasma compartment. We assume that the drug is eliminated from the central compartment. The central compartment equilibrates with two other compartments. One is a rapidly equilibrating compartment (compartment 2) and one is a slowly equilibrating compartment (compartment 3). The steady-state volume of distribution of the drug is the sum of the volumes of the three compartments.

Qualitatively, the time course of action of the drug concentration in compartment 1 can be described as follows:

- The drug is administered as a bolus into compartment 1. The concentration in this compartment rises rapidly to a peak and then falls as the drug moves into compartments 2 and 3. This corresponds to the rapid distribution phase.

- The amount of drug in compartment 2 (the rapid compartment) rises to a peak and then falls as the drug is distributed back to compartment 1. During this phase, there is still net movement of drug from compartment 1 to 3. However, the rate of fall in concentration in compartment 1 slows because of the return of drug from compartment 2. This produces the slow distribution phase.

- The amount of drug in compartment 3 rises to a peak and then falls as the drug returns to compartment 1. The fall in drug concentration in compartment 1 slows further.

- While these redistribution processes are going on, there is irreversible loss of drug from compartment 1. After a while, the concentrations in each compartment will be virtually the same and the slow elimination of the drug from compartment 1 will continue to deplete the concentration in that compartment. This corresponds to the final slow elimination phase.

In reality, only the plasma compartment (compartment 1) has an anatomical basis. This compartment represents the physical volume of distribution of the drug and is a function of the physicochemical properties of the drug (e.g., solubility, affinity for proteins, and oil/water distribution). The other compartments are mathematical constructs that help to simulate the data and, in general, do not correspond to real physical or anatomical spaces.

We can derive the mathematical equations for this model using a mass balance around each compartment. Figure E2.8 shows the nomenclature.

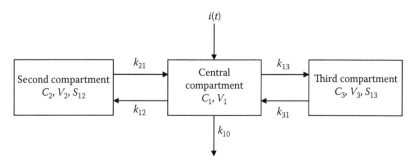

FIGURE E2.8
Three-compartment drug distribution model showing nomenclature.

Mass balance around compartment 1

$$V_1 \frac{dC_1}{dt} = k_{21}S_{12}C_2 + k_{31}S_{13}C_3 - \left(k_{10}S_{10} + k_{12}S_{12} + k_{13}S_{13}\right)C_1 + i(t) \qquad (2.34)$$

Mass balance around compartment 2

$$V_2 \frac{dC_2}{dt} = k_{12}S_{12}C_1 - k_{21}S_{12}C_2 \qquad (2.35)$$

Mass balance around compartment 3

$$V_3 \frac{dC_3}{dt} = k_{13}S_{13}C_1 - k_{31}S_{13}C_3 \qquad (2.36)$$

where
V_I are the compartment volumes (mL)
C_I are the concentrations of drug in each compartment (g/mL)
k_{ij} are the kinetic rate constants (cm/s) for transport of drug between compartments I and j
S_{ij} are the surface areas (cm²) available for transport of drug between compartments I and j
$i(t)$ is the injection rate for the drug (g/s)

Since the $k_{ij}S_{ij}$ always appear together, the mass balances are often divided by their respective volumes and presented as follows:

Mass balance over compartment 1

$$\frac{dC_1}{dt} = K_{21}C_2 + K_{31}C_3 - (K_{10} + K_{12} + K_{13})C_1 + I(t) \qquad (2.37)$$

Mass balance over compartment 2

$$\frac{dC_2}{dt} = K_{12}C_1 - K_{21}C_2 \qquad (2.38)$$

Mass balance over compartment 3

$$\frac{dC_3}{dt} = K_{13}C_1 - K_{31}C_3 \qquad (2.39)$$

where $K_{ij} = k_{ij}S_{ij}/V_i$ (s^{-1}) and $I(t) = i(t)/V_1$ (g/s mL).

Notice that the equations are coupled. That is, they must be solved simultaneously. In general, analytical solutions are difficult to obtain for more than two simultaneous equations, and numerical integration is required for three or more sets of simultaneous equations. In order to integrate these mass balances, we need to supply values for the rate constants (K_{ij}), a set of initial conditions, and an expression for the input dose $i(t)$.

A set of initial conditions that applies before any drug is administered is

$$\text{at } t = 0, \quad C_1 = C_2 = C_3 = 0 \qquad (2.40)$$

There are two ways of administering a drug:

1. *A bolus injection.* This is a fast injection of the drug into a vein using a syringe. A bolus injection can be mathematically approximated by an impulse, so that one possible representation for $I(t)$ is

$$I(t) = A\delta(t) \qquad (2.41)$$

   where
   $\delta(t)$ is the unit impulse function
   A is the amount of drug injected

The unit impulse function is defined by

$$\int_0^\infty \delta(t)dt = 1, \quad \delta(t) = \begin{cases} 0, & t \neq 0 \\ \infty, & t = 0 \end{cases} \tag{2.42}$$

2. *A constant infusion*. This is done using a syringe pump (or sometimes an intravenous [i.v.] drip). In this case, a constant rate of infusion of drug occurs. A constant infusion can be approximated mathematically by a step function, so that a possible representation of $I(t)$ is

$$I(t) = AU(t) \tag{2.43}$$

where
  A is the rate of injection
  $U(t)$ is the unit step function

The unit step function is defined as

$$U(t) = \begin{cases} 0, & t < 0 \\ 1, & t > 0 \end{cases} \tag{2.44}$$

Note that if the equations are being solved numerically, the bolus and step changes can be real functions of time and, in fact, cannot occur instantaneously.

Example 2.5 Two-Compartment Drug Distribution Model
A drug is being administered intravenously at a constant rate of d g/min. The drug is distributed between the blood and tissue compartments whose apparent volumes are V_b and V_T mL, respectively. In the tissue compartment, the drug is being metabolized according to a first-order reaction rate mechanism with a reaction velocity constant of k_r s^{-1}. Transport between the two compartments occurs according to first-order rate mechanisms with rate constants k_{12} and k_{21} cm/s. The surface area available for mass transfer between the compartments is S m^2. The process is shown schematically in Figure E2.9.

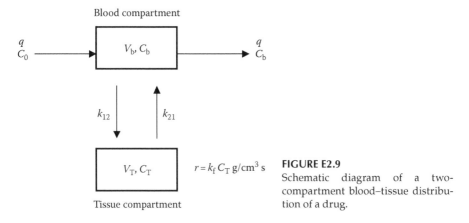

Blood compartment

$$V_b, C_b$$

$$\frac{q}{C_0}$$

$$\frac{q}{C_b}$$

$$k_{12}$$ $$k_{21}$$

$$V_T, C_T$$ $$r = k_f C_T \text{ g/cm}^3 \text{ s}$$

Tissue compartment

FIGURE E2.9
Schematic diagram of a two-compartment blood–tissue distribution of a drug.

Using mass balances over both the blood and tissue compartments, develop a model for the concentration of drug in each compartment with time after the start of the infusion. Assume that there is no drug present in either compartment prior to the start of infusion and that the CO is q L/min.

Solution
We apply the general conservation equation presented earlier to each compartment in the form of a mass balance on the infused drug since the entity being conserved is the mass of the infused drug.

Blood compartment

$$\text{Net accumulation} = \text{net input of drug} - \text{net output of drug} \qquad (2.45)$$

Net accumulation:

$$V_b \frac{dC_b}{dt} = \text{mL}\,(\text{g/mL/s}) = \text{g/s} \qquad (2.46)$$

Net input*:

$$\left(\frac{1000}{60}\right) q C_0 + (10^4) k_{21} S C_T$$

$$= (\text{mL/L/s/min})(\text{L/min})(\text{g/mL}) + (\text{cm}^2/\text{m}^2)(\text{cm/s})(\text{m}^2)(\text{g/mL})$$

$$= \text{g/s} + \text{g/s} = \text{g/s} \qquad (2.47)$$

Note that $(1000/60)\,qC_0 = d/60$ allowing calculation of C_0 if q is known.

Net output:

$$\left(\frac{1000}{60}\right) q C_b + (10^4) k_{12} S C_b = \text{g/s} \qquad (2.48)$$

Substituting (2.33), (2.34), and (2.35) back into (2.32) and collecting terms gives us the mass balance for the blood compartment:

$$V_b \frac{dC_b}{dt} = (1000/60)q(C_0 - C_b) + (10^4)S(k_{21}C_T - k_{12}C_b)$$

(2.49)

$$g/s = g/s + g/s$$

Tissue compartment

$$\text{Net accumulation} = \text{Net input} - \text{Net output} - \text{Net consumption} \quad (2.50)$$

$$\text{Net accumulation} = V_T \frac{dC_T}{dt} \quad (2.51)$$

$$\text{Net input} = (10^4)k_{12}SC_b \quad (2.52)$$

$$\text{Net output} = (10^4)k_{21}SC_T \quad (2.53)$$

$$\text{Net consumption in tissue volume} = k_f V_T C_T \,(\text{s})^{-1}(\text{mL})(g/\text{mL}) = g/s \quad (2.54)$$

Substituting (2.51) through (2.54) back into (2.50) and collecting terms gives us the mass balance for the tissue compartment.

$$V_T \frac{dC_T}{dt} = (10^4)S(k_{12}C_b - k_{21}C_T) - k_f V_T C_T \quad (2.55)$$

Equations 2.36 and 2.42 constitute a compartmental model for the process. Note that there is only one independent variable, time. Therefore, Equations 2.49 and 2.55 are ordinary differential equations. The equations are coupled, since they both contain C_b and C_T. This means that they must be solved simultaneously.

To solve this set of equations, we must apply numerical values for the parameters V_b, V_T, S, k_{12}, k_{21}, k_f, q, and d along with a set of initial conditions. For this problem, a set of initial conditions are

$$C_b(0) = C_T(0) = 0 \quad (2.56)$$

There are many ways of solving the set of Equations 2.49, 2.55, 2.56, for any consistent set of parameter values and a drug infusion sequence $d(t)$. For a review of standard techniques, see [4].

Loading doses
Clinically, when anesthesia is given intravenously by *infusion* (a constant rate), a *loading dose* is required. The loading dose consists of a bolus injection. This is then followed by the infusion at a constant rate. Loading doses are used for the following reasons:

- To reach an effective plasma drug concentration in a reasonable period of time
- To increase the level of anesthesia because the patient is responding to the surgical stimulus

Extending the pharmacokinetic model to include an effect site
For most anesthetics to take effect, they must leave the plasma compartment and reach a particular site of action such as a neuromuscular junction for a neuromuscular blocking agent or a specific brain area for a hypnotic drug. We call the site of action of a particular drug *the effect site*. In humans, it is virtually impossible to measure the concentration of drug at an effect site. However, we can model the effect site as a compartment connected to the central (plasma) compartment. This means that, at equilibrium, the concentration of drug at the effect site will be the same as that of the central compartment. Since we cannot measure the volume of the effect site either, we use the mass balances in which the compartment volumes have been combined with the surface areas and permeability coefficients (Figure E2.10).

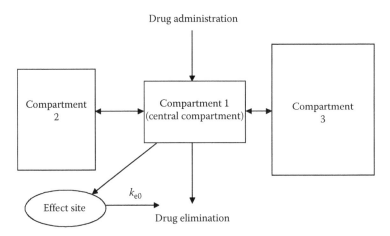

FIGURE E2.10
Three-compartment model with effect site. The dashed lines indicate different possibilities (see text).

There are several possible assumptions that have been used to derive models for this case. We will address two possibilities:

- There is no elimination of drug from the effect site ($k_{e0} = 0$) and the site equilibrates with the main compartment ($k_{41} \neq 0$). Mass balances over the four compartments give the following:

Mass balance over compartment 1

$$\frac{dC_1}{dt} = K_{21}C_2 + K_{31}C_3 + K_{41}C_4 - (K_{10} + K_{12} + K_{13} + K_{14})C_1 + I(t) \qquad (2.57)$$

Mass balance over compartment 2

$$\frac{dC_2}{dt} = K_{12}C_1 - K_{21}C_2 \qquad (2.58)$$

Mass balance over compartment 3

$$\frac{dC_3}{dt} = K_{13}C_1 - K_{31}C_3 \qquad (2.59)$$

Mass balance over compartment 4

$$\frac{dC_4}{dt} = K_{14}C_1 - K_{41}C_4 \qquad (2.60)$$

- The second assumption is that drug is eliminated from the effect site ($k_{e0} \neq 0$) and the effect compartment does not equilibrate with the main compartment ($k_{41} = 0$). The mass balances for this case would be as follows:

Mass balance over compartment 1

$$\frac{dC_1}{dt} = K_{21}C_2 + K_{31}C_3 - (K_{10} + K_{12} + K_{13} + K_{14})C_1 + I(t) \qquad (2.61)$$

The mass balances for compartments 2 (2.45) and 3 (2.46) would remain unchanged.

Mass balance over compartment 4

$$\frac{dC_4}{dt} = K_{14}C_1 - K_{e0}C_4 \qquad (2.62)$$

The K_{ij} is defined as above.

Coupling pharmacokinetic and hemodynamic models
Patients with cardiovascular problems are often given drugs to enhance their cardiovascular function during surgery. The effects of these drugs on cardiovascular hemodynamics (blood pressure, HR, and SV) can be modeled

by coupling a hemodynamic model to a pharmacokinetic model. This coupling requires a correlation (either graphical, as a table lookup or as an equation) of the effects of drug concentration at the effect site with hemodynamic parameters (e.g., vascular resistance and HR). A schematic representation of this coupling is shown in Figure E2.11.

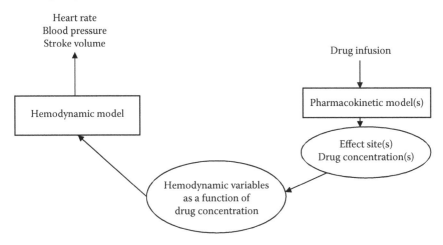

FIGURE E2.11
Coupling of hemodynamic and pharmacokinetic models.

2.3.2 Distributed Parameter Models

Compartmental models always have a finite volume associated with each compartment within which the contents are assumed to be spatially homogeneous and conditions are allowed to change with time. In a distributed parameter model, we allow conditions to change from point to point as well as with time. Obviously, this is a more complex arrangement. In general, distributed parameter models consist of sets of coupled partial differential equations.

Example 2.6 Interstitial Diffusion of Macromolecules
Nugent and Jain [5,6] have studied the interstitial diffusion of macromolecules in the rabbit ear chamber preparation in normal and neoplastic tissue using intravital fluorescence microscopy. They report interstitial diffusion coefficients for sodium fluorescein (NaF, a fluorescent molecule), fluorescein isothiocyanate (FITC)-labeled bovine serum albumin (BSA, average molecular weight ~67,000 Da), and a series of FITC-labeled dextran molecules of various molecular sizes (FITC-Dx, average molecular weight = 20,000–70,000 Da).

The following distributed parameter model was used to fit the experimental concentration versus time and distance data.

$$\frac{\partial C}{\partial t} = D_{AB}\frac{\partial^2 C}{\partial X^2} \tag{2.63}$$

The independent variables for this model are now the spatial variable, X, and time (t). The model is second order in distance (X) and first order in time. In general, to solve a partial differential equation, one must specify one boundary condition for each order for each spatial variable as well as one initial condition for each order of time. For this model, two boundary conditions are required on X and one initial condition on t.

The boundary and initial conditions that were used to approximate the experimental conditions used by Nugent and Jain [5,6] were

$$C_{X,0} = 0 \tag{2.64}$$

$$C_{0,t} = f(t) \tag{2.65}$$

$$C_{\infty,t} = 0 \tag{2.66}$$

In these equations, C represents concentration of the test molecule at any position, X, and at any time, t. The position variable, X, is measured relative to a specified origin (e.g., a position just outside a vessel wall). D_{AB} is the effective binary diffusion coefficient of the test molecule under these experimental conditions.

Equation 2.64 represents the initial condition that the test molecules were not present in the extravascular space prior to the experiment.

Equation 2.65 represents the boundary condition at $X = 0$ (just outside the vessel wall).

Equation 2.66 represents the boundary condition at $X = \infty$. Since intercapillary distances in this preparation were of the order of $100\,\mu m$, the position $X = \infty$ was chosen at $50\,\mu m$ as a conservative estimate.

A solution to this set of equations, which is applicable to short penetration distances (in order to minimize capillary interactions), is readily available [7,8]. For a step change in concentration at $X = 0$, Equation 2.65 becomes

$$f(t) = C_0 U(t) \tag{2.65a}$$

where $U(t)$ is the unit step function. $C_{X,t}$ is then given by

$$C_{X,t} = 2\pi^{-\frac{1}{2}} \int_{\lambda} f\left[\frac{t - X^2}{4D_{AB}\mu^2}\right] e^{-\mu^2} \, d\mu \tag{2.67}$$

where
$\lambda = X/(4D_{AB}t)^{1/2}$
μ is a dummy variable for integration

Experimentally measured values of $C_{X,t}$ for each of the test molecules are then plotted on the same coordinate axis as values of $C_{X,t}$ calculated from Equation 2.48 using different values of D_{AB}. The best estimates of D_{AB} for each of the test molecules is then determined by a least squares fit of the experimental data by the model.

2.4 Validation of Models

2.4.1 Parameter Estimation

Since a model is only an abstraction of reality, we must demonstrate that a proposed model is able to represent real data, at least over some limited range. This is called *validating the model*. There are two parts to a validation process. First, we must demonstrate that the model output fits experimental data using *physiologically realistic* values of the parameters. This is often referred to as the *parameter estimation* problem. The experimental data are often data that have been taken in the laboratory of the investigator who is doing the modeling. Parameter values (e.g., diffusion coefficients and kinetic rate constants) are estimated using the *best fit* of the model output to the experimental data. These values are then compared with generally accepted values of the same parameters or with values reported by other investigators, typically using different techniques. When comparing parameter values reported by different investigators, it is important to take into account that these values may be derived from data taken under different experimental conditions. Often, apparent conflicts in reported values can be resolved by considering the differences in experimental conditions.

If original experimental data are not available, an alternative approach is to compare the model output with appropriate experimental data reported in the literature. The disadvantage of this approach is that available literature data may not correspond to the assumptions used in deriving the model. In that case, the assumptions used in deriving the model would have to be altered to correspond as closely as possible to the reported experimental conditions.

The second part of model validation is called a *sensitivity analysis*. The question we are asking here is *how sensitive is the model output to changes in each of the parameters?* The approach here is to start with a set of parameters, initial conditions, and boundary conditions in which the model output matches a given set of experimental data. Each of the parameters, in turn, is systematically varied over a wide range (typically by ± a factor of 10). Changes in the model output are then compared with changes in the parameter values to arrive at estimates of the sensitivity of the model to changes in parameter values.

Example 2.7 Estimation of Capillary Permeability
The multiple indicator dilution technique [9–13] has been used to estimate the permeability of the capillary wall to small solutes and water. The basic experimental procedure can be briefly explained with the aid of Figure E2.12.

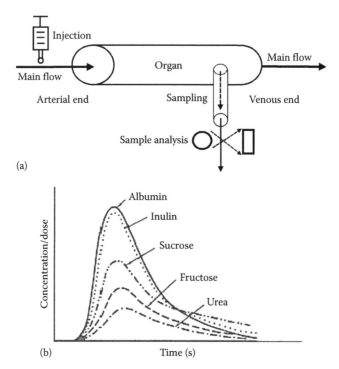

(a)

(b)   Time (s)

FIGURE E2.12
Multiple indicator dilution schematic: (a) experimental setup and (b) typical outflow curves for
injection of tracer amounts of several small hydrophilic solutes using albumin as the vascular
reference indicator.

At $t = 0$, a cocktail consisting of tracer* quantities of at least two solutes
(a nonpermeable *reference* solute and at least one permeable *test* solute) are
injected as a bolus (sudden injection or impulse) into the arterial side of a
capillary bed, which is perfusing an organ (e.g., lung, heart, and liver) or
tissue (e.g., gracilis muscle) of interest. Simultaneously, blood samples are
collected from a vein that is draining the organ or tissue of interest. This is
shown schematically in Figure E2.12a. The outflow samples are then analyzed for the concentration of each of the solutes injected using appropriate
analytical techniques (e.g., spectroscopy and β- or γ-counting of radioactive
labels). A graph of outflow concentration divided by amount injected versus
time is shown schematically in Figure E2.12b for several solutes.

Figure E2.12b demonstrates that when various indicators are injected
into the arterial inlet of the skeletal muscle, the venous outflow curves
all have their peak value practically simultaneously. The test substances,
however, all have lower peak heights than the vascular reference indicator (albumin). These data suggest two things. First, that a fraction of the
test molecules has left the vascular space and second, that the barrier to
the passage of molecules from the blood to the interstitial space is able to

discriminate among molecules based on their molecular size. Since the test molecules are present in tracer quantities, these data have been interpreted to mean that (in some tissues) transport of small molecules from the blood to the interstitial space is significantly impeded by a barrier of low permeability. That is, the rate of transport of these molecules through this barrier is several orders of magnitude lower than it would be through a film of water of equal thickness. The question then is how can these data be interpreted quantitatively in terms of the permeability of the blood–interstitial space barrier?

The Bohr–Kety–Renkin–Crone–Perl Model
The interpretation of the indicator outflow curves in terms of a permeability barrier can be illustrated using Figure E2.13. We assume that the indicators travel together through the nonexchange portion of the vasculature until they reach the (capillary) exchange region. At the exchange region, the permeable indicators are transported between the blood and tissue regions according to a first-order diffusive mechanism. The vascular reference indicator does not cross the barrier and stays in the blood compartment. Since we are dealing with tracer quantities, we can ignore osmotic effects.

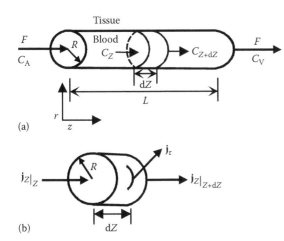

FIGURE E2.13
Schematic of the blood–tissue exchange region: (a) schematic diagram of the microvascular exchange unit and (b) a differential volume element of radius R and length dZ at position Z somewhere between $Z = 0$ and L.

Here
  F is the volumetric flow rate in mL/s
  C_A is the mass concentration of any solute at the *arterial end* of the exchange region in g/mL
  C_V is the concentration of any solute at the *venous end* of the exchange unit in g/mL
  C_Z is the concentration of any solute at position Z *within the exchange unit*
  C_{Z+dZ} is the concentration of any solute at position $Z + dZ$ *within the exchange unit*

In Figure E2.13b, $j_{Z|Z}$ represents the Z component of the **mass flux vector**, j, of any solute evaluated at position Z in $g(s)^{-1}(cm^2)^{-1}$. The rate of mass flow at position Z is then the mass flux at position Z times the area available for mass flow (e.g., $g(s)^{-1}(cm^2)^{-1}$ times $cm^2 = g(s)^{-1}$). In the Z direction, this is just πR^2 for a cylindrical vessel. Similarly, $j_{Z|Z+dZ}$ represents the Z component of the mass flux vector evaluated at position $Z + dZ$ and j_r represents the radial component of the mass flux vector. The rate of blood–tissue transport of any solute through the walls of the exchange vessels is then the radial mass flux times the blood–tissue exchange area. In the radial direction, the exchange area is just the surface area of the differential cylinder $(2\pi R\, dZ)$.

Basic assumptions of the model:

1. Single input, single output.
2. Steady state (accumulation of mass is zero).
3. Constant permeability of the vessel wall and constant exchange area (not functions of distance along the exchange unit or time).
4. Barrier limited diffusible solutes. Permeation by Fick's first law of diffusion at each position along the exchange length. That is, $j_r = P[C(Z) - C_T]$. Where P is the permeability coefficient of the vessel wall in cm/s and C_T the local concentration of solute in the tissue space.
5. No axial diffusion. That is, $j_Z = V_Z [C(Z)]$. Here, V_Z is the (constant) plug flow velocity of blood in the axial direction.
6. The inlet and exit *nonexchange* regions have no effect on the measurements of C_A and C_V.
7. The input function C_A not a function of time.
8. Instantaneous radial equilibration, so that C represents the average radial concentration at any position, Z, in the exchange unit. That is,

$$C(Z) = \int_{0}^{R} C(r, Z)\, dr$$

With these assumptions, the equation for conservation of mass becomes

Rate of mass into differential volume – rate of mass out – rate of permeation = 0

$$\text{Rate of mass in} = \pi R^2 j_Z \big|_{Z}$$

$$\text{Rate of mass out} = \pi R^2 j_Z \big|_{Z+dZ}$$

$$\text{Rate of permeation} = 2\pi R\, dZ\, j_r$$

Substituting these into the mass balance equation gives

$$\pi R^2 j_Z\big|_Z - \pi R^2 j_Z\big|_{Z+dZ} - 2\pi R\, dZ\, j_r = 0$$

Dividing by $\pi R^2 dZ$ and taking the limit as dZ approaches zero gives a steady-state differential mass balance for the exchange unit as

$$-\frac{d}{dZ}(j_Z) = \frac{2}{R} j_r$$

Using assumptions 3 (Fick's first law) and 4 (no diffusion in the axial direction) to replace j_Z and j_r in terms of C, the concentration of diffusible solute at any position, Z, on the blood side of the barrier gives

$$-\frac{d}{dZ}(V_Z C) = \frac{2P}{R}(C - C_T)$$

If V_Z is not a function of Z, this becomes

$$V_Z \frac{dC}{dZ} = -\frac{2P}{R}(C - C_T) \tag{2.68}$$

Rearranging and dividing by V_Z, gives

$$\frac{dC}{dZ} + \left(\frac{2P}{V_Z R}\right) C = \left(\frac{2P}{V_Z R}\right) C_T \tag{2.69}$$

Equation 2.56 is a first-order ordinary differential equation with constant coefficients. For constant C_T, its solution is given by

$$C(L) = C(0) e^{-\left(\frac{2P}{V_Z R}\right) L} + C_T \left[1 - e^{-\left(\frac{2P}{V_Z R}\right) L}\right] \tag{2.70}$$

where
  $C(L)$ is the concentration of tracer solute measured at $Z = L$
  $C(0)$ the concentration of tracer solute measured at $Z = 0$

These are represented by C_V and C_A, respectively. Equation 2.57 then becomes

$$C_V = C_A e^{-\left(\frac{2P}{V_Z R}\right)L} + C_T \left[1 - e^{-\left(\frac{2P}{V_Z R}\right)L} \right] \tag{2.71}$$

If we now assume an infinite sink in the tissue space, $C_T = 0$, then Equation 2.71 becomes

$$C_V = C_A e^{-\left(\frac{2P}{V_Z R}\right)L} \tag{2.72}$$

Using the relationships, $F_C = V_Z A_C = V_Z \pi R^2$, so that

$$V_Z = \frac{F_C}{\pi R^2}$$

Equation 2.72 becomes

$$C_V = C_A e^{-\left(\frac{2\pi RLP}{F_C}\right)} = C_A e^{-\left(\frac{PS_C}{F_C}\right)} \tag{2.73}$$

where $S_C = 2\pi RL$, the surface area available for transport in the radial direction. Rearranging Equation 2.73 and taking the ln of both sides gives

$$\ln\left(\frac{C_V}{C_A}\right) = -\frac{PS_C}{F_C} \tag{2.74}$$

Recall now that our vascular reference indicator is chosen to be a substance that will not leave the blood (vascular) compartment during a single pass through the organ. Generally speaking, albumin with an average molecular weight of 67,000 Da is an adequate vascular reference. This means that the concentration of the vascular reference solute will not change as it passes through the exchange region. We can then use the measured outflow concentration of the vascular reference, C_R, as an estimate of C_A. On the other hand, the concentration of diffusible indicator will decrease as substances travel through the exchange region. We assume that there is no loss of indicator in the nonexchange region between the exit of the exchange area and the sampling point. The concentrations of diffusible indicators as measured at the outflow, C_D, are then an estimate of C_V. We then define an extraction, E, as the ratio of the concentrations of diffusible indicator to reference indicator as measured at the sampling point:

$$E = \frac{C_D}{C_R} = \frac{C_V}{C_A} \tag{2.75}$$

Using (2.74) and (2.75), we can now solve for the permeability–surface area product, PS_C as

$$PS_C = -F_C \ln(1.0 - E) \tag{2.76}$$

where
    S_C is the (capillary) surface area available for blood–tissue transport of the diffusible substance
    F_C is the blood flow through the (capillary) exchange region

If we have an estimate of the surface area available for exchange, S_C, then Equation 2.76 allows quantitative interpretation of the experimental measurements in terms of a phenomenological permeability coefficient, P. P is a measure of the velocity of *extravasation* of a solute across a given biological barrier.

Discussion
If we think for a moment about the assumptions that went into the derivation of Equation 2.76, we realize that in order for the result to be true, the quantity $(1.0 - E)$ **cannot** be a function of time. However, if we look again at the data presented in Figure E2.12b, we realize that the ratio of diffusible to reference concentration (E) changes with time. In fact, at the tail of the curves, the concentrations of most of the diffusible indicators are higher than the reference concentration. This is because the tissue space is not an infinite sink as assumed in the derivation. Therefore, as the tissue concentration increases, we have back diffusion of the diffusible solutes into the vascular space. Since E is not constant with time, the question becomes what value of E should we use in Equation 2.76? While several interpretations are possible, the idea is to use the early values of E, before there is significant back diffusion. Some investigators use only those early values of E that are constant, while most investigators just average the values of E from appearance time to the peak of the vascular reference curve. A good rule of thumb is that back diffusion can be ignored and Equation 2.76 can provide a reasonable estimate of permeability – surface area product if $E < 0.2 - 0.3$.

Recognizing the limitations of the upslope model, several investigators have derived models for interpretation of multiple indicator dilution experiments which take into account finite tissue concentration [14,15] and also dispersion of the solute in the exchange region [16,17].

Table E2.1 lists permeability coefficients for several solutes in different organs which were estimated using the multiple indicator dilution technique.

TABLE E2.1

Permeability Coefficients for Several Solutes

Solute	Stokes–Einstein Radius (cm) $\times 10^8$	Organ	P (cm/s) $\times 10^6$	References
Na	1.4	Dog lung	0.3–0.5	[18]
Urea	1.6	Dog lung	25	[18]
Glucose	3.7	Dog lung	10	[18]
Sucrose	4.4	Dog hind limbs	7.4	[11]
Raffinose	5.6	Dog lung	6.5	[18]
Inulin	15.2	Dog hind limb	2.6	[11]
		Rabbit heart	5.1	[19]
Dextran 70[a]	60	Dog paw	0.0005–0.0002	[20,21]

[a] Weight average molecular weight of 70,000 Da.

Overview

In this chapter, we introduced the concepts of electrical and mechanical analogs to model physiological systems. These models use electrical circuit concepts and force balances to derive mathematical models for biomechanics and for flow in the circulatory system. We followed that by defining forces, fluxes, and material parameters and derived general conservation equations for creating models for transport of mass, energy, and momentum. We presented a way of classifying models according to their degree of mathematical complexity. We presented the general form for a balance equation and showed how this is used to generate simple models for calculating CO using the Fick principle and thermal dilution. We introduced the concept of a compartment and presented several examples of pharmacokinetic models. We showed schematically how pharmacokinetic models can be coupled to hemodynamic models using an effect site. Finally, we discussed the need for validation of models and we illustrated how models can be used to interpret experimental data in multiple indicator dilution experiments.

Problems

2.1 In the section on *pharmacokinetic models*, we set up a three-compartment model with an effect site. Let us consider a three-compartment drug distribution model without an effect site, where the main compartment is the blood compartment and there are two parallel compartments on either side. One of the compartments is a lipid soluble compartment and the other is a muscle tissue compartment. Assume that the drug of interest is introduced into the blood compartment. The drug is mainly metabolized in the muscle tissue compartment by a first-order reaction with rate constant k_e. Drug is introduced into the blood compartment at a rate of $AC_0(t)$. Assume that the concentrations in each compartment at any time can be represented by C_1, C_2, C_3, the volumes by V_1, V_2, V_3, and the transport coefficients by $k_{i,j}$, where $i, j = 1, 2; 2, 1; 1, 3; 3, 1$.
   a. Do an unsteady-state mass balance around each compartment and present the balance equations in terms of the parameters indicated. Use the initial conditions that the drug concentrations in all compartments are zero at $t = 0$.
   b. From the mass balance equations set up a Simulink®-type diagram to calculate C_1, C_2, and C_3 as a function of time for various inputs, $AC_0(t)$.
   c. **(Optional).** Use the following parameter values: $k_e = 0.0005$, $k_{12} = 0.001$, $k_{13} = 0.0001$, $k_{21} = 0.0001$, $k_{31} = 0.0001$, $V_1 = 5$, $V_2 = 2$, $V_3 = 3$, and $A = 25$. Assume two types of input:
      1. *An injection.* In that case, $C_0(t) = \delta(t)$, where $\delta(t)$ is the unit impulse function.
      2. *An i.v. drip.* An i.v. drip can be simulated by $U(t)$, where $U(t)$ is the unit step function.

Plot: C_1, C_2, and C_3 as a function of time. Be sure to carry out the simulation for a long enough time for the concentrations in each vessel to reach their steady-state values.

2.2 An obvious extension of the windkessel model is the respiratory system. During breathing, the lungs and chest wall move together. In normal individuals, except at very high and very low lung volumes, the compliances of these two systems are approximately the same. A simple model of the respiratory system consists of a resistance in series (R_S) with two compliances (lung and chest wall) in series. The series resistance represents the resistance to flow of the airways (mostly the medium-sized bronchioles).
   a. Set up a block diagram that will simulate the air flow during inspiration as a function of the pressure drop between the alveoli, P_A, and the atmosphere, P_B.
   b. Write down the mathematical relationship between the pressure drop across the respiratory system and the inspiratory flow using an electrical analog model.

c. Which parameters would change for a person with asthma: for a person with asbestos lung disease or for a person with scoliosis? Why?

d. (**Optional**). Use the following parameter values: $P_A = AP_A(t)$, $A = -4$ cm H_2O, $PB = 1$ atm $= 0$ mm Hg, $R_S = 8$ cm $H_2O/L/min$. $C_L = C_W = 0.3$ L/cm H_2O. Plot inspiratory flow, Q_i, versus $P_A(t)$. We can simulate $P_A(t)$ using a unit step change $U(t)$.

2.3 During exercise, the cardiac output, Q (L/min), is redistributed between the working skeletal muscles and the gastrointestinal (GI) track, to provide a greater proportion of the total flow to the working skeletal muscles and less to the GI. Alternatively, after a meal, the cardiac output is redistributed to provide more flow to the GI track than to the relaxed skeletal muscles. The flow redistribution can be modeled by a five element windkessel model as shown schematically in Figure P2.3.

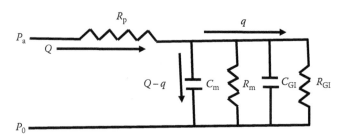

FIGURE P2.3
Five element wind kessel model.

Q = cardiac output (L/min). Normal is 4.5–5.0 in a young, fit person (depends on age, weight, height, and sex).

P_a = mean arterial pressure (mm Hg). Normal is 90–100 mm Hg in a young, fit person.

P_0 = atmospheric pressure = 0 mm Hg.

q = portion of the cardiac output, Q, that reaches the skeletal muscles and GI track (L/min). This is then distributed into q_1, q_2, q_3, and q_4, through the two resistances and two compliances in parallel. The sum of these four flows must equal Q, the total flow through the circuit.

C_m = compliance of the working skeletal muscles (L/mm Hg)

C_{GI} = compliance of the GI track (L/mm Hg).

R_p = flow resistance of the large arteries, not including the skeletal muscles and GI track (mm Hg/L/min).

R_m = flow resistance of the working skeletal muscles (mm Hg/L/min).

R_{GI} = flow resistance of the GI circuit (mm Hg/L/min).

a. Analyze the circuit in Figure P2.3 to determine the flows $Q(t)$, and $P_1(t)-P_4(t)$ as a function of arterial pressure changes, $P_a(t)$.

b. Set up the Simulink® block diagram to calculate the flow responses to a step change of size A in P_a.

c. During vigorous exercise, P_a can reach 160–180 mm Hg in a young, fit person and cardiac output can more than double. Discuss the changes you can make to the model variables that will allow you to simulate this effect.

d. After a heavy meal, the GI system can receive a larger share of the cardiac output, while the skeletal muscles, at rest, decrease their share of the cardiac output. How would the model variables have to change to simulate this effect?

2.4 Figure P2.4 is a schematic diagram of a kidney dialysis unit. A membrane that is permeable to small solutes such as urea separates the blood and dialysate. Let P_D be the overall permeability coefficient (cm/s) between the two fluids, S be the surface area (cm²) available for mass transfer between the compartments, and Q_D and Q_B be the volumetric flow rates (mL/s) of the dialysate and the blood, respectively. Let the concentrations of solute (g/mL) in the dialysate and blood be represented by C_D and C_B.

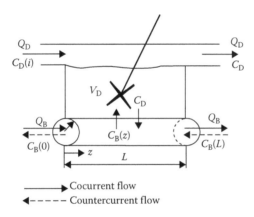

FIGURE P2.4
Schematic diagram for transport in kidney dialyzer unit.

a. There are two possible configurations for the flow of the blood and the dialysate, co-current and countercurrent. For each of these possibilities, use steady-state mass balances around each compartment to derive a steady-state model for solute transport between blood and dialysate. Be sure to use appropriate boundary conditions for each case.

b. Solve these models for the steady-state concentration profiles in each compartment.

c. Extend the countercurrent model to the nonsteady-state case. *Note:* This means a separate mass balance across the membrane. For a step change in solute concentration in the blood, what are the appropriate initial and boundary conditions?

2.5 In arterial blood at 37°C, P_{CO_2} = 40 mm Hg, pH = 7.4, and a P_{O_2} of 100 mm Hg, normal hemoglobin is 98% saturated with oxygen. Normal blood contains 15 g of hemoglobin/dL of blood (1 dL = 0.1 L = 100 mL). The hemoglobin molecule is a large protein that contains four individual polypeptide chains (two α chains and two β chains). Each of these chains contains a heme (iron containing) protein that can bind an oxygen molecule to iron in its ferrous (Fe^{+2}) form. The molecular weight of hemoglobin is 64,485 g/g mol. The uptake of oxygen by hemoglobin can be described by the following stoichiometry:

$$4O_2 + Hb \Leftrightarrow HbO_8$$

Oxygen + deoxyhemoglobin $\Leftrightarrow$ oxyhemoglobin

a. Use a mass balance and the ideal gas law to determine the theoretical maximum oxygen carrying capacity of hemoglobin (mL O_2/100 mL blood) in normal arterial blood.

b. Experimentally, it has been determined that 1 g of Hb will combine with 1.39 mL of oxygen under ideal conditions. How does this value compare with the stoichiometric determination? Comment on the possible reasons for any difference.

2.6 Henry's law tells us that the solubility of a gas in a liquid is proportional to the partial pressure of the gas above the liquid. Arterial blood is exposed to an oxygen partial pressure of 100 mm Hg in the pulmonary alveoli. The Henry's law constant for oxygen dissolved in plasma at 37°C is 0.003 mL O_2/(100 mL plasma-mm Hg oxygen partial pressure). At rest, the normal oxygen consumption for an adult is 300 mL O_2/min (basal metabolic rate).

a. If there were no red cells to carry oxygen and the tissues could extract all the oxygen dissolved in plasma, what would the CO have to be to supply the oxygen for basal metabolism? Compare this with the normal CO.

b. Normal blood has a hematocrit of 45% (volume of red cells/(volume of red cells + volume of plasma) × 100). There are 15 gHb/dL of blood. Mixed venous blood at a P_{O_2} of 40 mm Hg is 75% saturated. Based on a normal CO of 5 L/min, how much oxygen (mL) does each gram of Hb carry?

2.7 To measure the total body red cell volume, 100 μC_I (microcuries) of radiolabeled red cells were injected into a patient. After 10 min, a 10 mL sample of the whole blood was withdrawn from the patient and was found to have a concentration of 0.45 μC_I/mL. The patient's hematocrit (volume of red cells/volume of whole blood) × 100 was 48%.

a. Calculate the red cell volume for this patient.

b. What is this patient's plasma volume?

2.8 Given the following schematic diagram of the human placenta. The fetal and maternal sides of the placenta are separated by a membrane which is permeable to solutes (Figure P2.8).

a. Derive an expression for the steady-state concentration distribution of the transported substance C_M and C_F on the maternal and fetal sides, respectively, for the countercurrent model. List all assumptions. What are appropriate boundary and initial conditions for this model?

b. Suppose that the maternal circulation was so large as compared with the fetal circulation that, as a first approximation, the concentration of solute in the maternal circulation could be considered constant (e.g., not a function of Z). Solve the countercurrent model for the steady-state solute distribution in the fetal circulation.

c. Repeat parts a and b for the co-current (parallel) flow model. List the appropriate boundary and initial conditions for this model.

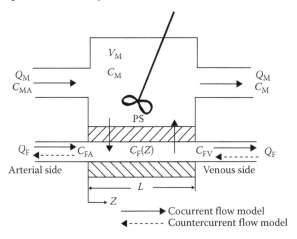

FIGURE P2.8
Schematic diagram for umbilical cord transport of solutes and gases between mother and fetus.

2.9 The drug halothane is a short-acting anesthetic, which is inhaled and distributed into the circulation from the lungs. It is not metabolized and is eliminated from the kidneys by a first-order mechanism. The anesthetic is inhaled into the lungs and then transported into the circulation from where it is carried to the kidneys.

a. Draw a schematic diagram for a three-compartment model for this drug, using the circulation as the central compartment.

b. Using mass balances, develop a pharmacokinetic model for the distribution of this drug. Use a bolus input for the drug. What are the boundary and initial conditions that you would use to solve this model? Which parameters would have to be specified before this model can be solved?

2.10 A one-compartment pharmacokinetic model is often used to determine elimination rates for a particular drug in phase 1 and 2 clinical trials. The one-compartment pharmacokinetic model is shown schematically in Figure P2.10:

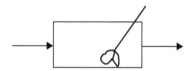

FIGURE P2.10
Schematic for one compartment pharmacokinetic model.

Assume that you are interested in determining the value of the first-order elimination constant, K, whose unit is L/h.

a. Start with an unsteady-state mass balance around the compartment.
b. What values would you need to specify in order to solve the model for K?
c. Suppose you can assume that the amount of drug is injected as a bolus (which can be approximated mathematically by an impulse of size A), what data would you have to collect, and how would you use these data to evaluate K?
d. Suppose that all the drug is accumulated in the bladder and eliminated in the urine. The average urine output is about 1.5 L/24 h. How would this information help you solve for K?

2.11 The circulation of a newborn child with a patent (open) ductus arteriosus (a shunt between the aorta and the pulmonary artery that normally closes at birth) is shown in Figure P2.11:

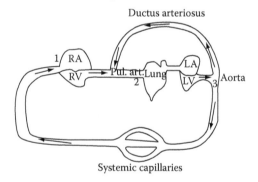

FIGURE P2.11
Flow pathway for a patent ductus arteriosis.

Well-mixed blood samples were obtained by cardiac catheterization prior to the repair of the shunt. The O_2 content of the samples was as follows:

Site of Sampling	O_2 Content (mL O_2/100 mL Blood)
(1) Right atrium	13
(2) Pulmonary artery	16
(3) Aorta	19

Oxygen consumption is 240 mL/min.

Using the Fick principle and conservation of mass,
a. What is the flow through the systemic circulation?
b. What is the flow through the lungs?
c. What is the flow through the ductus arteriosus?

2.12 The splanchnic circulation consists of the blood supply to the gastrointestinal tract, liver, spleen, and pancreas. The intestines and pancreas are supplied by a series of parallel circulations via the branches of the superior and
inferior mesenteric arteries. The stomach, spleen, and liver are supplied by
branches of the celiac artery and hepatic artery. The liver is also supplied
by the portal vein that receives outflow from the intestines, pancreas, and
spleen. Figure P2.12 is a schematic diagram of the circulation. The hepatic
and portal circulations both drain into the hepatic sinusoids. The sinusoids
are large, low pressure collecting ducts that allow blood from terminal
hepatic arterioles and terminal portal venules to mix and flow through to
the hepatic collecting veins and then to the larger hepatic veins.

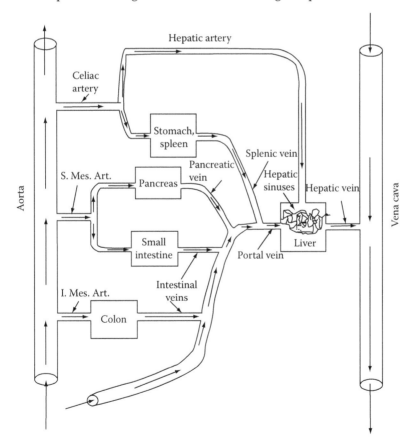

FIGURE P2.12
Flow pathway for hepatic circulation.

The splanchnic organs receive about 30% of the CO at rest.
Given the following information on flows and pressures

Position	Resting, Fasting Flow (mL/min)	Exercise Flow (mL/min)
Hepatic artery	500	125
Stomach, spleen	200	50
Pancreas	250	62.5
Small intestine	250	62.5
Colon	300	75

Position	Resting Pressure (mm Hg)	Exercise (mm Hg)
Arteries	90	95
Arterioles	88–35	90–20
Capillaries	35–18	20–8
Venules	18–10	8–5
Portal vein	10–8	5–3
Liver sinusoids	8–2	3–1
Vena cava	2–1	1–0

a. Calculate the hemodynamic resistance of the organs listed in the first table under resting and exercise conditions. Prepare a table summarizing your calculations. State any assumptions that you make.
b. Calculate the blood flow through the portal system under resting and exercise conditions.
c. Estimate the hemodynamic resistance of the portal system under resting and exercise conditions.
d. The splanchnic circulation normally contains 20%–30% of the total blood volume. Discuss how this volume may change during exercise.

References

1. Hardman, J.G., Limbird, L.E., Malinoff, P.B., and Ruddon, R.W. (Eds.). *Goodman & Gilman's The Pharmacological Basis of Therapeutics* (9th edn.). McGraw-Hill, New York, Chapter 2, 1996.
2. Speicher, C.E. *The Right Test. A Physicians Guide to Laboratory Medicine* (2nd edn.). W. B. Saunders Co., Philadelphia, PA, pp. 241–243, 1993.
3. Freund, J.E. and Simon, G.A. *Modern Elementary Statistics*. Prentice Hall, Englewood Cliffs, NJ, Chapter 11, 1992.
4. Zill, D.G. and Cullen, M.R. *Advanced Engineering Mathematics*. PWS-Kent Publishers, Boston, MA, Chapter 1, 1992.

5. Nugent, L.J. and Jain, R.K. Plasma pharmacokinetics and interstitial diffusion of macromolecules in a capillary bed. *Am. J. Physiol.* 246 (*Heart Circ. Physiol.* 15): H129–H137, 1984.

6. Nugent, L.J. and Jain, R.K. Extravascular diffusion in normal and neoplastic tissues. *Cancer Res.* 44: 238–244, 1984.

7. Crank, J. *The Mathematics of Diffusion.* Oxford University Press, London, U.K., 1957.

8. Carslaw, H.S. and Jaeger, J.C. *Conduction of Heat in Solids.* Oxford University Press, London, U.K., 1959.

9. Chinard, F.P., Vosburgh, G.J., and Enns, T. Transcapillary exchange of water and of other substances in certain organs of the dog. *Am. J. Physiol.* 183: 221–234, 1955.

10. Zierler, K.L. Theory of use of indicators to measure flow and extracellular volume and calculation of transcapillary movement of tracers. *Circ. Res.* 12: 464–473, 1963.

11. Crone, C. The permeability of capillaries in various organs as determined by the use of the "indicator diffusion" method. *Acta Physiol. Scand.* 54: 292–295, 1963.

12. Perl, W., Silverman, F., Delea, A.C., and Chinard, F.P. Permeability of dog lung endothelium to sodium, diols, amides and water. *Am. J. Physiol.* 230: 1708–1721, 1976.

13. Renkin, E.M. Multiple pathways of capillary permeability. *Circ. Res.* 41: 735–743, 1977.

14. Goresky, C.A., Ziegler, W.H., and Bach, G.G. Capillary exchange modeling: Barrier-limited and flow-limited distribution. *Circ. Res.* 27: 739–764, 1970.

15. Johnson, J.A. and Wilson, T.A. A model for capillary exchange. *Am. J. Physiol.* 210: 1299–1303, 1966.

16. Bassingthwaigthe, J.B. A concurrent model for extraction during transcapillary passage. *Circ. Res.* 35: 483–503, 1974.

17. Harris, T.R. and Newman, E.V. A comparative analysis of the proposed mathematical models of circulatory indicator–dilution curves. *J. Appl. Physiol.* 28: 840–850, 1970.

18. Chinard, F.P. The alveolar-capillary barrier: some data and speculations. *Microvasc. Res.* 19: 1–17, 1980.

19. Wittmers Jr., L.E., Bartlett, M., and Johnson, J.A. Estimation of the capillary permeability coefficients of inulin in various tissues of the rabbit. *Microvasc. Res.* 11: 67–78, 1976.

20. Garlick, D.G. and Renkin, E.M. Transport of large molecules from plasma to interstitial fluid and lymph in dogs. *Am. J. Physiol.* 219: 1595–1605, 1970.

21. Arthurson, G. and Granath, K.A. Dextrans as test molecules in studies of the functional ultrastructure of biological membranes. *Clin. Chim. Acta* 37: 309–322, 1972.

Further References on Theoretical and Experimental Work on Capillary Permeability

Alverez, O.A. and Yudilevich, D.L. Heart capillary permeability to lipid-insoluble molecules. *J. Physiol. (Lond.)* 202: 45–53, 1969.

Chinard, F.P., Enns, T., Goresky, C.A., and Nolan, M.F. Renal transit times and distribution volumes of T-1824, creatinine, and water. *Am. J. Physiol.* 209: 243–247, 1965.

Chinard, F.P., Thaw, C.N., Delea, A.C., and Perl, W. Intrarenal volumes of distribution and relative diffusion coefficients of monohydric alcohols. *Circ. Res.* 25: 343–357, 1969.

Kety, S.S. Theory and applications of the exchange of inert gas at the lungs and tissues. *Pharmacol. Rev.* 3: 1–15, 1951.

Martin, P. and Yudilevich, D. Theory for the quantification of transcapillary exchange by tracer–dilution curves. *Am. J. Physiol.* 207: 162–170, 1964.

Sangren, W.C. and Sheppard, C.W. Mathematical derivation of the exchange of a labeled substance between a liquid flowing in a vessel and an external compartment. *Bull. Math. Biophys.* 15: 387–394, 1953.

Tepper, R.S., Lee, H.L., and Lightfoot, E.N. Transient convective mass transfer in Krogh tissue cylinders. *Ann. Biomed. Eng.* 6: 506–530, 1978.

3

Cell Physiology and Transport

3.1 Introduction

Most of the regulatory activities of the human body are carried out at the cellular level. Cells replicate themselves as part of the normal process of degeneration and repair of organs and tissues; specialized cells synthesize receptors, proteins, hormones, and peptides, which may be released into the bloodstream to regulate the function of other cells and also act as local feedback inhibitors to the overproduction of the regulatory molecule. All cells are surrounded by a *plasma membrane* that keeps the cell contents localized, allows for the spatial separation of charge, and acts as a barrier to the external environment. The cell membrane is a *phospholipid bilayer* with a thickness of about 50 Å.* Specialized protein structures that span the membrane connect the exterior environment with the cell interior and function as *pumps, channels,* and *receptors*. These receptors, pumps, and channels help to maintain the cells' resting membrane potential, allow cell-to-cell signaling (e.g., propagation of the action potential in nerves and muscle), and provide mechanisms by which molecules and ions may enter and leave the cell. Remember that the internal part of the phospholipid bilayer that forms the cell membrane contains the *hydrophobic* (water insoluble) portion of the lipid molecules. Most ions and small molecules such as glucose that are essential for the functioning of the cell are *hydrophilic* (water soluble). Without the specialized transporters and fluid-filled protein channels, hydrophilic molecules could only enter and leave the cell very, very slowly (if at all) through the highly hydrophobic interior of the bilayer.

In addition, *intracellular* membrane structures such as those that surround the *mitochondria, sarcoplasmic reticulum* (SR), and other *organelles* compartmentalize the cell contents and aid in the synthesis, transport, and metabolic functions of the cell.

The plasma membrane of mammalian cells separates two aqueous solutions of very different composition. Table 3.1 shows the intracellular and extracellular concentrations of some important ions.

---

* 1 Å = 10^{-10} m.

TABLE 3.1

Intracellular and Extracellular Environments

Substance	Extracellular	Intracellular
Na$^+$ (mM)[a]	140	5–15
K$^+$ (mM)	4	120
Cl$^-$ (mM)	110	~10
HCO$_3^-$	20	~20
Ca^{2+} (mM)	2.5	1×10^{-4}
Osmolarity (mOsm/L)[b]	295	295
pH	7.4	7.0–7.2

a   mM = millimolar = 10^{-3} Molar.
b   mOsm/L = milliOsmoles/L.

Note that the K$^+$ concentration is much greater inside the cell than outside. Also note that the reverse is true for Na$^+$. Moreover, the concentration of Ca^{2+} is very low inside the cell as compared with the external environment. In addition, in most cells, the *electrical potential* of the interior of the cell is *negative* (–70 mV) with respect to the exterior. These ion differences and the electrical potential difference are generated and maintained by *active transport* systems for these ions.

In addition to the concentration differences between the inside and outside of the cell, *intracellular organelles* may contain high concentrations of ions such as Ca^{2+} (the endoplasmic reticulum [ER]) with respect to the *cytoplasm* of the cell. These compartments maintain their ionic concentration gradients by means of specific transport systems within the organelle membrane.

The cell is not a static structure in which the ion gradients are maintained ad infinitum. Cellular signaling mechanisms call for rapid changes in the membrane potential and/or changes in free [Ca].* These changes take place when specific channels within the plasma membrane are activated, which then allow ions to rapidly diffuse down their concentration gradient and alter the ionic composition of the cell interior. This mechanism forms the basis for muscle contraction, neural activity, and many other *signal transduction* mechanisms in different kinds of cells.

3.2 How Do Ions and Molecules Get into and out of Cells?

The four major mechanisms by which molecules move into and out of cells are *passive diffusion* (uncharged molecules, electrolytes, water); *convection*, which includes *filtration and osmotic flow* (water and dissolved solutes); *facilitated transport* (rate of transport greater than that predicted by passive

---

* The square brackets surrounding a chemical symbol will be used to denote concentration. That is, [Ca] means "the concentration of calcium."

diffusion; requires specialized transport systems such as hemoglobin); and *active transport* (transport against an electrochemical potential gradient; requires expenditure of metabolic energy).

3.2.1 Diffusion

Diffusion is a molecular mechanism involving the collision of molecules in random motion. The driving force for diffusion of uncharged molecules from one spatial point to another is the gradient in *chemical potential* (approximated by the concentration gradient in dilute solutions) of the molecule of interest between the two points. A simple quantitative relationship between concentration gradient and solute flux is given by Fick's first law of diffusion (see Table 2.2 and Example 2.1). A more complete discussion of diffusion and equilibrium in terms of *chemical potential* is given in Section 3.3.

Example 3.1 Diffusion of Weak Acids and Bases

A weak acid or base only partially dissociates at physiological pH. We can express this as

$$HA \Leftrightarrow H^+ + A^- \quad K_a = \frac{[H][A]}{[HA]}$$

$$BH^+ \Leftrightarrow H^+ + B,$$

where K_a is the dissociation constant for the weak acid. It turns out that the charged species (A^-, BH^+) are much less permeable than the uncharged species (HA, B). The various species establish equilibrium very rapidly so that, as soon as the uncharged species diffuses across the membrane, it rapidly dissociates ($H^+ + A^-$) or reassociates with a proton (BH^+). This allows, for example, weak bases to accumulate to high concentrations within acidic organelles (e.g., lysosomes) because the base will associate with a proton, acquire a net positive charge, and be "trapped" within the organelle. The same property allows the use of organic dyes such as *Fura-2* to be used for the measurement of intracellular pH.

Dissolved CO_2 is very permeable across biological membranes but exists in very low concentrations as the dissolved gas. Remember that gases such as CO_2 and O_2 are only sparingly soluble in aqueous media (i.e., plasma). Much of the CO_2 produced by the tissues as a result of metabolism exists in the blood as bicarbonate (HCO_3^-) and is transported to the lungs in this form. The conversion of CO_2 to HCO_3^- is a relatively slow process. The rate-limiting step is the hydration of CO_2 to H_2CO_3 (carbonic acid) that immediately dissociates into H^+ and HCO_3^-. The hydrolysis is catalyzed by the enzyme *carbonic anhydrase*, which exists in high concentrations in red blood cells:

$$H_2O + CO_2 \Leftrightarrow H_2CO_3 \Leftrightarrow H^+ + HCO_3^-$$

Major shifts of HCO_3^- and Cl^- occur in red blood cells as they pass through the tissues and the lungs. The rate of this process depends on the enzyme carbonic anhydrase, the existence of a Cl^-/HCO_3^- exchanger/carrier in the red cell membrane and the high permeability of the red cell membrane to O_2 and CO_2.

3.2.2 Diffusion of Ions

Definitions: **Electrical potential, electric field,** and **the electroneutrality principle**. An *electrical potential*, Ψ, represents the magnitude and distribution of electrical charges at a certain point in the spatial domain. An *electric field* is given by the spatial gradient of the electrical potential. The one-dimensional (1-D) representation of an electric field is $d\Psi/dx$. The *electrical force* exerted on a charged particle is proportional to the magnitude of the electric field.

The electroneutrality principle states that there must always be equal numbers of positive and negative charges in any *macroscopic* (i.e., *measurable*) *volume* of a solution of an electrolyte. Any charge separation (inequality of positive and negative charges) can occur only within a microscopic volume of molecular dimensions. On the microscopic scale, charge separation is always associated with the formation of an electric field. If, for example, one ionic species diffuses faster than another, the separation of charges, which would occur at the diffusion boundary, creates an electric field in such a way as to accelerate the slower ion and slow down the faster ion. In the steady state, both the slower and faster ions would diffuse down their electrochemical potential gradient with the same velocity, where any differences in rate due to molecular motion would be compensated for by a difference in electrical potential across the diffusion boundary. This potential difference is called a *diffusion potential*.

3.2.3 Membrane Potentials

If a membrane separates two electrolyte solutions of differing composition, and one ionic species crosses the membrane more rapidly than another, an electrical potential will be developed across the membrane. This is created by a separation of charge between the two membrane surfaces. Remember that the electric field only exists within the molecular dimensions of the membrane and not in the bulk solutions on either side (the electroneutrality principle).

3.3 Free Energy and Chemical Potential

At constant pressure, the total amount of work that can be obtained by a process in going from some initial state, 0, to some final state, 1, is given by the change in the *Gibbs free energy* for the process $\Delta G = G_1 - G_0$. If the free energy is negative (<0), the process has gone from a higher state of free energy to a lower state and we call this a *spontaneous process*. If $\Delta G > 0$, the process has acquired free energy during its transition from state 0 to state 1 (from its surroundings) and we say that this type of process *does not occur spontaneously*.

Finally, if $\Delta G = 0$ for the process, we say that the process is in *equilibrium*. Another way of saying this is that $\Delta G = 0$ represents a *necessary and sufficient* condition for equilibrium in a constant pressure process. The *chemical potential* of a substance, μ, is defined as the change in Gibbs free energy of the system as a result of adding an *infinitesimal* amount of the substance to the process at constant temperature, pressure, and amount of all other substances present:

$$\mu = \left(\frac{\partial G}{\partial n_i} \right)_{T,P,n_j} \tag{3.1}$$

where
  n_i is the number of moles of solute i added at constant temperature, pressure
  n_j is the number of moles of all other substances in the process

When there is no difference in chemical potential of a solute on either side of a membrane, we say that the substance is in equilibrium across the membrane. That is, $\Delta\mu = \mu_2 - \mu_1 = 0$. For a membrane separating two solutions at constant temperature, the difference in chemical potential can be written as

$$\Delta\mu = RT \ln \frac{C_i}{C_o} + \bar{V}\Delta P \tag{3.2}$$

where
  R is the gas law constant
  T is the absolute temperature
  $\bar{V}$ is the partial molar volume of the solute (i.e., $18\,\text{mL/mol}$ for water)
  ΔP is the pressure difference across the membrane
  C_i and C_o represent the concentration of solute on either side of the membrane in mol/L

At constant P, the ΔP disappears and we have simple diffusion.
  Because ions are charged, we expand this definition to include an electrical term and the free energy change per mole of solute added becomes known as the *electrochemical potential*, $\tilde{\mu}$. For a membrane separating two solutions with different concentrations of a *diffusible ion*, the electrochemical potential gradient can be written as

$$\Delta\tilde{\mu} = \Delta\mu + zF\Delta\psi \tag{3.3}$$

where
  z is valence of the ion (i.e., +1 for Na, −2 for SO_4^{-2}, etc.)
  F is Faraday's constant (the amount of charge carried by a gram equivalent of a monovalent ion $96,500\,\text{C/g}$ equivalent)
  ψ is electrical potential

These equations will be the starting point for developing expressions for calculating equilibrium potentials (the *Nernst equation*) and for osmotic pressure.

3.3.1 Equilibrium Potentials for Ions

When an ion is in equilibrium across a cell membrane, its electrochemical potential gradient is zero. From Equations 3.2 and 3.3, we can determine an *equilibrium potential*, $\Delta\psi_{eq}$, from the ions' measured concentration gradient. This is the electrical potential that would be necessary to establish thermodynamic equilibrium at the observed concentration gradient. Setting $\Delta\tilde{\mu} = 0$ in Equation 3.3 and substituting Equation 3.2 for $\Delta\mu$ with $\Delta P = 0$ gives

$$\Delta\psi_{eq} = -\frac{RT}{zF}\ln\frac{C_i}{C_o} \tag{3.4}$$

Equation 3.4 is called the *Nernst equation*.

Example 3.2 Equilibrium Potentials for Na and K

Using the intracellular and extracellular concentrations for Na and K given in Table 3.1 at 25°C, calculate the equilibrium potentials for Na and K.

Solution

a. For Na^+, $C_i = 14$ mM, $C_o = 140$ mM, $R = 1.99$ cal/K mol, $F = 23{,}070$ cal/V mol, and $T = 298$ K, $z = +1$. Equation 3.4 then gives

$$\Delta\psi_{eq}(Na) = -\frac{(1.99)(298)}{(+1)(23{,}070)}\frac{(cal/K\,mol)(K)}{cal/V\,mol}\ln\frac{14}{140}$$

$$\Delta\psi_{eq}(Na) = -(0.0257)(-2.30) = +0.0592\,V = +59.2\,mV$$

We interpret this to mean that, for Na^+ to be in equilibrium, there must be a +59.2 mV potential gradient across the cell membrane with the inside of the cell positive with respect to the outside of the cell. The measured resting membrane potential for a cell with these ion distributions would be about −70 mV (with the inside of the cell negative with respect to the outside of the cell). This means that Na^+ is far from thermodynamic equilibrium under these conditions.

b. For K^+, $C_i = 120$ mM, $C_o = 4$ mM. Then, Equation 3.4 becomes

$$\Delta\psi_{eq}(K) = -0.0257\ln\frac{120}{4} = -0.0874\,V = -87.4\,mV$$

We interpret this result to mean that K^+ is also not at thermodynamic equilibrium under these conditions but is closer to its equilibrium potential than is Na^+ to its equilibrium potential.

Nonequilibrium ion gradients are maintained by metabolic energy through active transport processes.

The difference between the measured membrane potential and the equilibrium potential for a particular ion ($\Delta\psi - \Delta\psi_{eq}$) can be interpreted as an electrical driving force for transport of that ion across the membrane. If $\Delta\psi$ is more positive than $\Delta\psi_{eq}$ as is the case for K^+, the difference ($\Delta\psi - \Delta\psi_{eq}$) = $-70 - (-87.4) = +18.4\,mV$ represents a positive force on the K^+. Since K^+ are positively charged, this force promotes transport of K^+ *out* of the cell. On the other hand, for Na^+, $\Delta\psi$ is much more negative than $\Delta\psi_{eq}$ and the difference ($\Delta\psi - \Delta\psi_{eq}$) = $-70 - (+59.2) = -129.2\,mV$ represents a large negative driving force for Na^+ transport *into* the cell.

3.3.2 Water Transport

There are two routes of water transport across biological membranes:

1. Diffusion through the lipid bilayer.
2. Passage through aqueous pores or channels. *Aquaporins* are a family of proteins that form water channels when inserted into membranes. The aquaporin content of membranes is regulated by a cyclic adenosine monophosphate (c-AMP)-dependent mechanism in renal collecting ducts. The aquaporins are stored in vesicles in the cell cytoplasm, just below the cell membrane. When *vasopressin*, also known as *antidiuretic hormone* (a peptide released by the kidney, which regulates the water permeability of the kidney collecting ducts), is released, intracellular vesicles containing aquaporin proteins fuse with the plasma membrane, inserting water channels and increasing the permeability of the renal collecting ducts to water.

3.3.3 Osmotic Pressure

To begin our discussion of osmotic pressure, we use the observation that the presence of a dissolved solute *lowers* the chemical potential for water in an aqueous solution. Consider Figure 3.1 in which aqueous solutions of a solute, i, in two closed compartments (**A** and **B**) are separated by a membrane. In the *top panel*, the membrane is freely permeable to both the solute and water.

At $t = 0$, an amount of solute is added to compartment **A**. Since the membrane is freely permeable to the solute, it will diffuse down its chemical potential gradient and, some time later, equilibrium will be established and

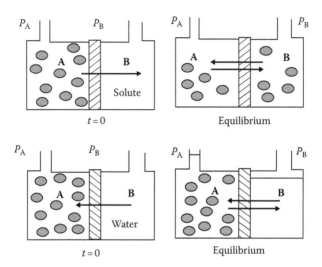

FIGURE 3.1
Demonstration of osmotic pressure.

the solute concentration will be equal on either side of the membrane. In the *bottom panel*, we replace the membrane with a membrane that allows the passage of water but does not allow passage of the solute. At $t = 0$, an amount of solute is again added to compartment **A**. In this case, when the solute tries to establish equilibrium by diffusing down its gradient in chemical potential, it cannot cross the membrane barrier. However, water, which can freely cross the membrane, will move down its gradient in chemical potential from compartment **B** to compartment **A** in an attempt to establish equilibrium. This flow of water down its gradient in chemical potential is called *osmotic flow*. Since the solute cannot leave compartment **A**, there is no way in which the concentrations can equilibrate on both sides of the membrane. Therefore, another mechanism comes into play to equilibrate the chemical potential for water on both sides of the membrane. As water flows from compartment **B** to compartment **A** by osmotic flow, the additional volume in compartment **A** increases the hydrostatic pressure head in that compartment. This establishes a pressure gradient between compartments **A** and **B**, which moves water by *filtration* in the opposite direction to the osmotic flow. These two competing processes continue until the pressure difference between compartments **A** and **B** is such that the rate of water flow by filtration just balances the rate of water flow by osmotic flow. The hydrostatic pressure difference that is required to establish equilibrium is called the *osmotic pressure*. We can write an expression for the change in chemical potential for water that describes these two competing processes as

$$\Delta\mu_w = \bar{V}_w \Delta P + RT \ln\left(\frac{n_w}{n_w + n_s}\right) \tag{3.5}$$

where

$\Delta\mu_w = \mu_{wA} - \mu_{wB}$ is the change in chemical potential between compartments **A** and **B**

$\overline{V}_w$ is the partial molar volume of water = 18 mL/g mol

$\Delta P = P_A - P_B$ is the hydrostatic pressure difference between compartments **A** and **B**

R is the gas law constant

T is the absolute temperature [273 + ($T°C$)]

ln ( ) is the natural log of argument

$n_w/(n_w + n_s)$ is the mole fraction of water. Here, n_w is the number of moles of water in the solution in compartment **A** and n_s is the number of moles of the solute in solution in compartment **A**

We now note that at equilibrium $\Delta\mu_w$ must equal zero. We can then solve for ΔP, the pressure difference at equilibrium as

$$\Delta P = -\frac{RT}{\overline{V}_w}\ln\left(\frac{n_w}{n_w + n_s}\right) \tag{3.6}$$

This pressure difference is given the symbol Π for osmotic pressure.
Now, by the property of logarithms,

$$-\frac{RT}{\overline{V}_w}\ln\left(\frac{n_w}{n_w + n_s}\right) = \frac{RT}{\overline{V}_w}\ln\left(\frac{n_w + n_s}{n_w}\right) = \frac{RT}{\overline{V}_w}\ln\left(1 + \frac{n_s}{n_w}\right)$$

For dilute solutions, $n_s/n_w \ll 1$ and from the property of logarithms, we can write $\ln(1 + a) \cong a$ if $a \ll 1$ so that Equation 3.6 can be approximated as

$$\Pi = RTC_s \tag{3.7}$$

where C_s is the molar concentration of the dissolved solute. Equation 3.7 is called the *van't Hoff* relationship and only applies to dilute solutions under conditions where the solute cannot cross the membrane. It should be noted that osmotic flow is in the *opposite* direction to the osmotic pressure gradient. Water will flow from a region of higher chemical potential (lower solute concentration) to a region of lower chemical potential (higher solute concentration). That is from a region of lower osmotic pressure to a region of higher osmotic pressure.

When there is solute on both sides of the membrane, Equation 3.7 can be written for the change in osmotic pressure as

$$\Delta\Pi = \Pi_A - \Pi_B = RT\left(C_{sA} - C_{sB}\right) = RT\Delta C_s \tag{3.8}$$

3.3.4 Permeant Solutes

Solutes that are permeable across a biological membrane also induce water flow by the competitive processes of filtration and osmotic flow. The net volumetric flow (of water) is given by the nonequilibrium thermodynamic transport equation as

$$j_v = L_p[\Delta P - \sigma \Delta \Pi] \tag{3.9}$$

where
  j_v is the volumetric flux of water across the membrane (mL/cm^2 s)
  L_p is the hydraulic conductivity of the membrane (cm/s mm Hg)
  ΔP is the hydrostatic pressure difference across the membrane (mm Hg)
  $\Delta \Pi$ is defined by Equation 3.8
  σ is called the *osmotic reflection* coefficient

Equation 3.9 was first derived using the principles of *nonequilibrium thermodynamics* and is often referred to as the *Kedem–Katchalsky* equation.

- The contribution to the total water flux by filtration is given by the product $L_p \Delta P$.
- The contribution to the total water flux by osmotic flow is given by the product $L_p \sigma \Delta \Pi$.
- L_p, the hydraulic conductivity, is a measure of the velocity of water transport across the membrane and is determined by the structure of the pores or channels in the membrane and the frictional resistance to water molecules as they flow through the channels.
- σ, the osmotic reflection coefficient, is a measure of relative size of the channels or pores in the membrane as compared with the hydraulic radius of the solute. If the pore or channel diameters are large compared with the solute radius, the membrane structure does not retard the velocity of the solute with respect to the velocity of the water molecules. The solute and water move through the aqueous channel together. There is no osmotic gradient. In that case, $\sigma = 0$ and there is no osmotic flow. Small, uncharged solutes such as glucose have an osmotic reflection coefficient approaching zero. On the other hand, if the pores or aqueous channels in the membrane have radii that are small compared to the solute hydraulic radius, the solute molecules will not be able to enter the channels and be *reflected off the wall* of the membrane. In that case, only water will move down its electrochemical potential gradient, setting up an osmotic gradient. In that case, $\sigma = 1$, and we have osmotic flow across an ideal semipermeable membrane. Large molecules such as albumin have osmotic reflection coefficients that approach 1.0.

Total water (volumetric) flow is given by the product of j_v and S, the surface area (cm²) available for water transport across the membrane.

$$J_v = j_v S = j_v S[\Delta P - \sigma \Delta \Pi] \qquad (3.10)$$

where J_v is the total volumetric flow rate (mL/s).

3.3.5 Osmolarity

Osmolarity of a dissolved solute is a concept that was introduced to allow quantitative discussion of the chemical potential for water. Unlike osmotic pressure, the osmolarity of a solution is a quantity that is *measured* either using a membrane osmometer or by the freezing point depression of the pure solvent (water). Osmolarity measurements take into account the nonideal behavior of real solutions. Osmolarity can be related to molar concentration of a dissolved solute as follows:

$$\text{Osmolarity} = \text{Osmoles/L} = g\Sigma v C_s \qquad (3.11)$$

where
  C_s is the molar concentration of the solute g mol/L
  v is the number of ions formed on complete dissociation of the solute
  g is an experimentally determined osmotic coefficient that accounts for nonlinear solution behavior

Since an Osmole is a rather large quantity for biological solutions, we usually report osmotic activity of solutions as milliOsmoles/L, abbreviated mOsm/L or mOsM.

For membranes that separate solutions of different osmolarities,

$$\Delta \Pi = RT \Delta \text{Osmolarity}$$

Example 3.3 Osmolarity and Osmotic Pressure

  a. What is the osmolarity of a 0.9% by weight solution of NaCl if the osmotic coefficient for NaCl is estimated as 0.85?

Solution

Using Equation 3.9 and the dissociation relationship:

$$\text{NaCl} \Leftrightarrow \text{Na}^+ + \text{Cl}^-$$

$v = 2$, $g = 0.85$.

$$C_s = \frac{(0.9\,\text{g NaCl}/100\,\text{g sol})(1.0\,\text{g sol}/\text{mL sol})(1000\,\text{mL}/\text{L})}{58.5\,\text{g NaCl}/\text{g mol NaCl}}$$

$C_s = 0.15\,\mathrm{g\ mol\ NaCl/L} = 0.15\,\mathrm{M} = 150\,\mathrm{mM}$

$$\mathrm{Osmolarity} = 2(0.85)(150) = 255\,\mathrm{mOsm/L\ NaCl}$$

b. A membrane separates two solutions. One side (compartment **A**) contains 0.3 M NaCl, the other side (compartment **B**) contains 0.4 M sucrose. What is the difference in osmotic pressure between the two solutions and in which direction will osmotic flow of water take place? Assume that the osmotic coefficient for sucrose is 1.0.

Solution

Osmolarity of sucrose solution = 0.4 Osm/L, Osmolarity of NaCl solution = 0.85(2)(0.3) = 0.51 Osm/L.

$\Delta\Pi = (\Pi_B - \Pi_A) = RT\Delta\mathrm{Osm} = (0.082\,\mathrm{L\,atm/g\,mol\ K})(298\,\mathrm{K})(0.40 - 0.51)\,(\mathrm{Osm/L}).$

$\Delta\Pi = -2.69\,\mathrm{atm}$

It is interesting to note that the osmotic pressure difference, −2.69 atm, is quite large. Since this difference represents the driving force for water movement, it indicates that osmotic forces can play substantial roles in biological systems.

Since we used the NaCl compartment as compartment **A** in this calculation, the negative sign for $\Delta\Pi$ indicates that osmotic flow will occur from compartment **B** to **A**.

Because of the definition of osmotic pressure, it should be noted that water will flow *in the direction of* the **highest** osmotic pressure.

3.3.6 Osmotic Relations in Cells

The osmolarity of the inside of most cells and biological tissue is approximately that of *blood plasma*, about 300 mOsm/L. If a red blood cell is suspended in a 300 mOsm/L solution, where the dissolved solute is *impermeant*, there will be no net water movement between the red cell and the bathing solution and the volume of the cell will be stable. We call such a solution an *isotonic* solution. If we suspend a red blood cell in a solution, whose osmolarity is >300 mOsm/L, there will be a net movement of water out of the cell and it will shrink in volume. Since the solute is impermeant, no stable osmotic equilibrium will be reached. Such a solution is called *hypertonic*. If we suspend a red cell in a solution whose osmolarity is <300 mOsm/L, there will be a net movement of water into the red cell and it will swell. Again, since the solute is impermeant, no stable osmotic equilibrium will be reached and the red cell will eventually burst. Such a solution is called *hypotonic*.

Example 3.4 Osmotic Effects of Permeable Solutes

  a. Red blood cells are suspended in a solution of 0.3 M urea. Cell membranes are permeable to urea. Describe the sequence of events that takes place.

Solution

Assuming an osmotic coefficient of 1.0 for urea, the osmolarity of the solution is 300 mOsM. The solution is therefore isotonic with respect to the interior of the red cells so that initially no volume change will occur. However, since urea is not normally present in the interior of red cells and the cell membrane is permeable to the solute, it will diffuse into the cell down its chemical potential gradient. This will raise the osmolarity of the cell interior causing water to flow down its chemical potential gradient into the cell. Therefore, after a short isovolumetric period, the cell will swell. Osmotic equilibrium can never be established since the proteins present in the cell interior cannot get out. Therefore, the cell will continue to swell until it bursts.

  b. Red blood cells are suspended in a solution of 0.6 M urea.

Solution

With the same assumptions as above, the osmolarity of the solution is 600 mOsM. Therefore, the solution is hypertonic with respect to the cell interior. Initially, water will flow down its chemical potential gradient from the cell interior to the bathing solution and the cell will shrink. At the same time, urea will be transported into the cell down its chemical potential gradient causing an increase in the osmolarity of the cell interior. If urea were to equilibrate, its concentration would be 0.3 M on either side of the cell membrane. However, the interior of the cell would then be 600 mOsM, while the bathing solution would be 300 mOsM. Now the cell interior would be hypertonic with respect to the bathing solution and water would flow into the cell down its chemical potential gradient. The cell would then swell. Since the cell proteins could not leave the cell, an osmotic equilibrium is not possible and the cell would continue to swell until it bursts.

  The problem here is the presence of an impermeant solute(s) in the cell interior (proteins and hemoglobin). In order to establish an osmotic equilibrium, we also need an impermeant solute in the bathing solution at isotonic concentration. Under most circumstances, 0.15 M Na (0.9% by weight of NaCl) can be used as the impermeant ion. Because of the action of the Na–K–ATPase pump (located in the cell membrane) the Na concentration inside the cell is maintained at a very low level. When Na diffuses into the cell down its concentration gradient, it is immediately exchanged for an extracellular K (three Na for two K), by the pump. Therefore, as far as the cell is concerned, Na acts as an impermeant ion as long as the pump is intact. Other solutes such as sucrose and inulin do not get into cells easily and can be used to establish osmotic equilibrium in the presence of a permeable solute. For example, if 0.9% NaCl were added to the bathing solution with the 0.6 M urea in part b above the red cells would initially shrink and then swell again. However, under these conditions, the cells would only swell to their original volume since, at equilibrium, the osmolarity of the solutions on both sides of the membrane would be the same. *Please note that mature red blood cells do not have Na–K–APase pumps on their cell membranes.*

3.4 Gibbs–Donnan Equilibrium

The Gibbs–Donnan equilibrium can occur whenever a membrane separates a protein solution from an electrolyte solution that contains little or no protein. Proteins have complex three-dimensional (3-D) structures that allow them to carry a net ionic charge. These net charges may either be positive or negative. If, for example, the protein contains a given net number of negative charges per protein molecule, in solution, these negative charges must be balanced by an equal number of positive charges to maintain electrical neutrality. This situation occurs quite often in cellular physiology and in the laboratory when proteins are purified by *dialysis*.

If all ions present are free to equilibrate across the membrane, equilibrium will be established when the membrane potential equals the equilibrium potential for each ion. That is, at equilibrium, the distribution of ions on either side of the membrane will be determined by Equation 2.4 under conditions such that $\Delta\psi_{eq}$ is the same for each diffusible ion. Because of the presence of the impermeant protein, there will also be an osmotic gradient across the membrane.

Example 3.5 Gibbs–Donnan Equilibrium

A protein contains six net negative charges per protein molecule. For simplicity, we will say that these negative charges are balanced by Na^+ for electrical neutrality. An amount of this protein is then dissolved in a 15 mM NaCl solution. The concentration of protein in solution is 6 mM. The protein solution is then placed in a dialysis membrane, tied off, and placed in a bath containing a 15 mM NaCl solution. The membrane is freely permeable to Na^+, Cl^-, and water but not protein. Assume that the volumes are initially equal on both sides of the membrane. What is the equilibrium potential for this system? What are the concentrations of Na^+ and Cl^- on either side of the membrane at equilibrium? What is the osmolarity difference across the membrane at equilibrium?

Solution

In order to solve this problem exactly, we would have to know the *hydraulic conductivity* of the membrane, the effective membrane surface area available for water flow, and the time of the experiment. Lacking this information, we can obtain an approximate solution to the problem by assuming that the

osmotic flow is small enough so that the volumes on either side of the membrane remain constant. With this assumption:
  Initial concentrations *inside* the membrane:

  Protein: 6 mM

  Total $[Na]_i$: (six net negative charges/protein molecule) (one Na^+/negative charge) (6 mM protein) = 36 mM Na^+ + 15 mM from NaCl in solution = 51 mM

  $[Cl]_i$: 15 mM from NaCl solution

  Initial concentrations *outside* the membrane:

  $[Na]_o$: 15 mM from NaCl in solution
  $[Cl]_o$: 15 mM from NaCl in solution

Because $[Na]_i$ is initially $>[Na]_o$, Na^+ will diffuse down their concentration gradient from inside to outside the membrane, creating a diffusion potential (positive outside the membrane), which will retard further diffusion of Na^+. The same diffusion potential will accelerate the transport of Cl^- from inside to outside the cell. The result is that both Na^+ and Cl^- will diffuse (in equal amounts to maintain overall electrical neutrality) from the inside to the outside until equilibrium is established. At equilibrium, $\Delta\psi = \Delta\psi_{eq}(Na^+) = \Delta\psi_{eq}(Cl^-)$. Then from Equation 3.4, we can write

$$\Delta\psi = \Delta\psi_{eq}(Na^+) = -\frac{RT}{zF}\ln\frac{[Na]_i}{[Na]_o} = \Delta\psi_{eq}(Cl^-) = -\frac{RT}{zF}\ln\frac{[Cl]_i}{[Cl]_o}$$

Since $z = +1$ for Na^+ and $z = -1$ for Cl^-, we can divide by RT/F, use the properties of logarithms, and take antilogarithms of both sides to get

$$\frac{[Na]_o}{[Na]_i} = \frac{[Cl]_i}{[Cl]_o} \tag{3.12}$$

This result is perfectly general. It can be shown that all ionic species that are permeable across the membrane will be distributed according to the Donnan ratio, r, where

$$r = \left(\frac{[C]_o}{[C]_i}\right)^{1/z} \tag{3.13}$$

where
  $[C]$ is the concentration of ion in question
  z is its valence

Note that this equilibrium is maintained by a stable membrane potential $\Delta\psi$, which is the same for all the permeable ions. That is,

$$\Delta\psi = \frac{RT}{F}\ln r \qquad (3.14)$$

where r is given by Equation 3.13. $\Delta\psi$ will be negative inside if the net charge on the *impermeable* ion(s) is negative ($r < 1$) and positive on the inside if the net charge on the impermeable ion is positive ($r > 1$). In this example, the net charge on the impermeable ion (the protein) is negative, so that at equilibrium, $\Delta\psi$ will be negative inside.

We can now calculate the equilibrium concentrations and the membrane potential.

Let x be the number of moles of Na^+ transported down its electrochemical gradient, from inside to outside. These are accompanied by an equal amount of Cl^-. If the volumes are equal and constant, we can use Equation 3.12, at equilibrium:

$[Na]_o[Cl]_o = [Na]_i[Cl]_i$

$[Na]_o = 15 + x$ mM; $[Cl]_o = 15 + x$ mM; $[Na]_i = 51 - x$ mM; $[Cl]_i = 15 - x$ mM

Substituting into Equation 3.12 and solving for x gives

$$x = 5.625\,\text{mM}$$

The concentrations at equilibrium are then

$[Na]_o = [Cl]_o = 20.625\,\text{mM}$
$[Na]_i = 45.375\,\text{mM}$
$[Cl]_i = 9.375\,\text{mM}$

The Donnan ratio, r, is

$$r = \frac{[Na]_o}{[Na]_i} = \frac{[Cl]_i}{[Cl]_o} = \frac{20.165}{45.375} = \frac{9.375}{20.625} = 0.4545$$

The equilibrium potential at 25°C from Equation 3.12 is

$$\Delta\psi = \frac{(1.99\,\text{cal/K mol})(298\,\text{K})}{23,070\,\text{cal/V mol}}\ln(0.4545) = -0.0193\,\text{V} = -19.3\,\text{mV}$$

The osmotic pressure difference across the membrane (assume that the osmotic coefficient for the protein is 1.0) is

Osmolarity inside: protein + Na + Cl = 6.0 + 45.375 + 9.375 = 60.75 mOsM
Osmolarity outside: Na + Cl = 20.675 + 20.675 = 41.25 mOsM
ΔOsmolarity = inside − outside = 60.75 − 41.25 = 19.5 mOsM

The positive sign indicates that the osmotic pressure inside the bag is greater than the osmotic pressure outside the bag and water will move by osmotic flow from *outside* to *inside*.

The osmotic pressure difference across the bag can be calculated from Equation 3.8 as

$$\Delta\Pi = 0.477 \, \text{atm}$$

We see here that the difference in osmolarity is equal to the concentration of the protein plus an additional amount due to the presence of the counterions associated with the protein. The additional osmotic contribution of the counterions is called the *Donnan effect*. In cells, the osmotic effect of the impermeable ions inside the cell is balanced by the effective impermeability of the cell to extracellular Na ions. The Na–K–ATPase active transport system "pumps" Na out of the cell and, among other functions, maintains the osmotic stability of mammalian cells.

3.4.1 Cellular Volume Regulation

The red cell is rather unique. A mature red cell has no nucleus and thus can do little to regulate its internal environment. It is primarily a membrane, which contains a solution of hemoglobin. Most other cells respond to osmotically induced changes in cell volume by changing the osmotic content of their cytosol to restore the original cell volume. They do this by opening and closing K and Cl channels to provide a regulatory volume decrease (RVD) in response to a hypotonic environment. They will activate a Na–K–2Cl active transporter to obtain a regulatory volume increase (RVI) in response to the effects of a hypertonic environment.

3.5 Carrier-Mediated Transport

Most of the cells have a number of *electrogenic pumps, carrier-mediated* transport systems, and *ion channels* embedded in their membranes, which allow them to transport ions and molecules into and out of the cell interior in response to an appropriate signal. The signal may be electrical in nature, such as the action potential, but most often it involves the *binding* of a *regulatory molecule* to a specific *receptor* on the surface of a cell membrane. Figure 3.2 is a schematic diagram of the most important electrogenic pumps and ion channels that form an integral part of just about every mammalian cell membrane. In addition to the transporters shown in Figure 3.2, cells in the liver, lung, kidneys, and other organs have many additional specialized transporter systems, depending on cell function.

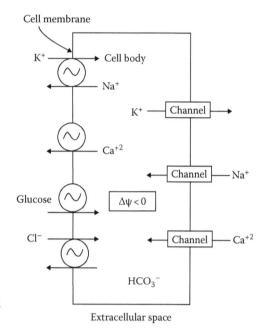

FIGURE 3.2
Pumps and channels in a typical cell membrane.

3.5.1 Characteristics of Carrier-Mediated Transport Processes

Carrier-mediated transport processes differ from passive diffusion in several important ways.

1. Carrier-mediated transport kinetics are similar to enzyme kinetics. They show a nonlinear (hyperbolic) dependence of transport rate with increasing substrate concentration. The carriers become saturated at high substrate concentration (J_{max}). Figure 3.3 shows a typical plot of the transmembrane flux (J_i mol/s cm^2) of a species, i, against its concentration on one side of a membrane, C_i.

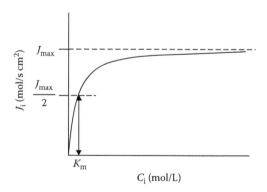

FIGURE 3.3
Kinetics of carrier-mediated transport.

The hyperbolic nature of the response allows the kinetics to be modeled by Michaelis–Menten-like relationships, which take the following form:

$$J_i = \frac{J_{max}C_i}{K_m + C_i} \qquad (3.15)$$

Equation 3.15 is the equation of a hyperbola. J_{max} is the maximum obtainable flux at high substrate concentrations (the saturation region) and K_m, the Michaelis constant, is related to the forward and reverse dissociation constants for the substrate–carrier complex.

At low concentrations, $K_m \gg C_i$ and $J_i = KC_i$, where $K = J_{max}/K_m$. There is a linear relationship between J_i and C_i. At high concentrations, $C_i \gg K_m$ and $J_i = J_{max}$, so that J_i is independent of concentration. Also, at $J_i = J_{max}/2$, $K_m = C_i$. These concepts are illustrated in Figure 3.3.

2. Carrier-mediated transport systems show structural specificity for the transported substrate. For example, D-glucose is transported by the glucose transporter but not L-glucose. That also means that structurally related substances often *competitively* inhibit the transporter. For example, galactose competitively inhibits glucose transport in many cells.

3. Some transport systems are *noncompetitively* inhibited by heavy metals at low concentrations. For example, Hg at low concentrations inhibits glucose transport.

4. The rate of transport is often faster than would be predicted by simple diffusion of the molecule across a lipid bilayer. Glucose transport is much faster than predicted by glucose diffusion across a lipid bilayer, since lipid membranes are impermeable to glucose.

5. In the case of *active transport*, the transport system depends on a source of metabolic energy. An example is the Na–K–ATPase pump, which depends on the hydrolysis of ATP.

3.5.2 Classification of Carrier-Mediated Transport

1. *Facilitated diffusion.* In these processes, transport occurs down an electrochemical gradient and does not depend on a source of metabolic energy. For nonelectrolytes, this means transport *from a region of higher chemical potential (concentration) to a region of lower chemical potential (concentration)*. An example is glucose transport in liver cells by the glucose transporter.

2. *Active transport.* In these processes, net transport of the substance occurs against an electrochemical gradient. For nonelectrolytes, this means *transport from a region of higher concentration to a region of lower concentration*.

a. *Primary active transport.* Transport is directly coupled with a chemical reaction such as ATP hydrolysis. Most, but not all, primary active transport systems are ATPases.

b. *Secondary active transport.* Transport of a substance against an electrochemical gradient is coupled with the movement of a second substance down its electrochemical gradient.

– *Cotransport* (also called *symport*). Both transported substances move across the membrane in the same direction. An example is Na-dependent glucose uptake in intestinal epithelial cells. Transport of glucose against a concentration gradient into the cells is coupled with the movement of Na^+ into the cells down an electrical and concentration gradient.

– *Exchange* (also called *countertransport* and *antiport*). The transported molecules move in opposite directions across the membrane. An example is the Na–Ca exchanger in which Ca movement out of the cell against its electrochemical gradient is coupled with the inward movement of three Na^+ for each Ca^{2+}.

– *Complex.* Transport involves the movement of more than one substance in more than one direction. An example is the Na–HCO_3/Cl exchanger. Here, Na^+ and HCO_3^- are transported into the cell in exchange for outward movement of Cl^-.

3.5.3 Molecular Characteristics of Transport Carriers

Transport carriers consist of hydrophobic membrane-bound proteins that cross the membrane 10–12 times. (In contrast, *receptors* tend to have fewer transmembrane segments. Both the dopamine D_3 receptor and the β-adrenoreceptor have seven transmembrane segments.) The transmembrane segments consist of about 20–30 hydrophobic amino acids. These transmembrane segments are thought to contain the *binding sites* for the transported substances. Many transporters also contain large *hydrophilic* domains.

3.5.3.1 ATPases

For transport ATPases, the hydrophilic domains contain the ATP binding sites. In the case of one type of ATPase (the P-type), the hydrophilic domains also contain the machinery for the formation of the *phosphorylated intermediate*. There are several different types of ATPases that have been identified. They are synthesized in different cells and even in different cell organelles. They differ in their mechanisms for binding and phosphorylating ATP and in the types of molecules that they transport. A discussion of their mechanisms of action, similarities (*homology*), and differences is beyond

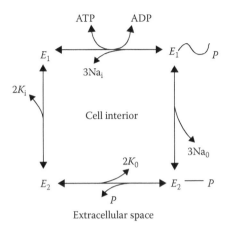

FIGURE 3.4
Reaction cycle: Na–K–ATPase.

the scope of this text. A model for the *reaction cycle* of the Na–K–ATPase is shown schematically in Figure 3.4. The ATPase enzyme (E) can exist in two conformational states, E_1 and E_2, which are in dynamic equilibrium with each other. In the E_1 conformation, the enzyme binds three Na^+ and one ATP inside of the cell. A phosphate group is transferred from ATP to an aspartate residue in a hydrophilic portion of the enzyme to form the E_1–P intermediate. Here P stands for the PO_4^{-2} group. A conformational change results in *occlusion* and translocation of three Na^+ as the E_1–P conformation is converted to the E_2–P conformation. The term *occlusion* refers to the loss of accessibility of the Na^+ as they are buried within the membrane during the transport process. The E_2–P conformation gives up three Na^+ at the external membrane surface and binds two K^+, leading to the hydrolysis of the *phosphodiester bond* to generate the free E_2 conformation of the ATPase. The E_2 conformation then spontaneously converts to the E_1 conformation with occlusion, translocation, and release of the K^+ at the internal membrane surface. The cycle then repeats itself. The turnover rate of the Na–K–ATPase enzyme is $\approx 100\,s^{-1}$.

In other transporters, the hydrophilic domains are the sites for regulation of cell behavior through *second messengers* or interaction with the *cytoskeleton* of the cell.

Figure 3.5 is a schematic representation of the secondary structure of the Na-dependent glucose transporter named *SGLT1*.

Example 3.6 Calcium Homeostasis

The resting intracellular free Ca concentration is about 100 nM in most cells, while the Ca concentration in the extracellular fluid is about 2.5 mM. Thus, there is a very large ($\cong 2.5 \times 10^4$ M) concentration gradient for Ca to enter the cell. Since the resting membrane potential in most cells is negative inside, there is also an electrical potential force driving Ca^{2+} into the cell. Since high concentrations of free intracellular Ca^{2+} are toxic to cells, intracellular concentrations of free Ca must be maintained at low resting levels.

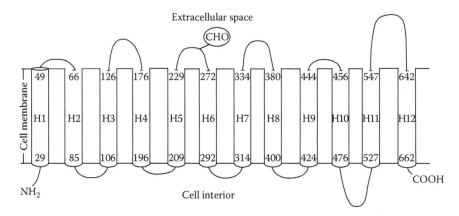

FIGURE 3.5
Secondary structure of the Na-dependent glucose transporter, SGLT1.

Free intracellular Ca concentration, $[Ca]_i$, is also important, since changes in $[Ca]_i$ act as a *second messenger* in signal transduction in response to interaction of hormones and regulatory molecules with their receptors on the plasma membrane. Furthermore, $[Ca]_i$ plays a crucial role in the formation and turnover of *actin–myosin crossbridges* in the contractile mechanism of cells. $[Ca]_i$ is maintained at low levels by two mechanisms: *sequestration of calcium by intracellular organelles* (the mitochondria and the ER) and two *carrier-mediated transport systems* in the cell membrane that transport Ca out of the cell. These are illustrated in Figure 3.6.

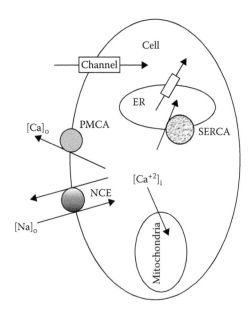

FIGURE 3.6
Transport mechanisms involved in Ca homeostasis.

3.5.3.2 Organelle Sequestration of Ca

Ca accumulation in the ER is due to the action of one of the specialized ATPase active transport pumps (the SERCA ATPase pump; SERCA comes from sarcol(*endo*)plasmic *reticulum* Ca ATPase). When certain hormones or *first messengers* (e.g., norepinephrine, NE) interact with their receptors on cell membranes, the interaction initiates a cascade if *intracellular* events including the formation of *inositol 1,4,5-triphosphate* (IP$_3$) or *diacyl glyceride* (DAG) from membrane phospholipids. These intermediates stimulate the release of second messengers such as Ca from the ER. In muscle cells, the ER has been specialized to initiate muscle contraction by releasing Ca. *In muscle cells, the ER is called SR.* Reaccumulation of Ca by the SR is important in bringing about the relaxation of contracted muscle cells.

In mitochondria, there is a transport mechanism in the inner membrane that allows Ca to enter in response to a negative mitochondrial membrane potential. In the mitochondria, Ca is important in regulation of pyruvate dehydrogenase, a key enzyme of the tricarboxylic acid (TCA) or Krebs cycle in aerobic cells. Under resting conditions, Ca accumulation by the mitochondria is a relatively low level process. However, if [Ca]$_i$ increases, the mitochondria can accumulate large amounts of Ca and form Ca–phosphate precipitates within the mitochondria matrix. This is a sign of Ca overload and poor physiological condition of the cell.

3.5.3.3 Cell Membrane

Here, two carrier-mediated transport mechanisms operate to transport Ca out of the cell. The first involves one of the ATPases (the PMCA ATPase; PMCA comes from *p*lasma *m*embrane Ca ATPase). This pump is found in the plasma membrane of most cells. The second is a secondary active transport system that is found in some but not all cells. It is an antiport Na/Ca exchanger (NCE). This system exchanges three Na down its electrochemical gradient into the cell for one Ca against its electrochemical gradient out of the cell. The downhill movement of three Na into the cell and the negative resting membrane potential of the cell ($\Delta\Psi \approx -70\,mV$) provide the energy necessary for transport of the Ca against its electrochemical gradient. The NCE is abundant in cardiac muscle, where it is the primary mechanism for Ca transport out of cardiac myocytes.

Since the thermodynamic driving force for Na/Ca exchange depends on [Na]$_i$, very small changes in intracellular sodium concentration can have very large effects on Ca transport. *Cardiac glycosides* used in the treatment of heart failure partially inhibit the Na–K–ATPase active transport pump. This causes an increase in [Na]$_i$, which reduces Ca transport out of the cell. Therefore, more Ca is accumulated by the SR during relaxation, leading to an increase in free intracellular Ca and an *increased force of contraction*.

3.5.4 Ion Channels and Membrane Potential

3.5.4.1 Resting Membrane Potential

In most cells, there is a difference in electrical potential across the cell membrane at rest called the *resting membrane potential*. This potential is mainly caused by the diffusion potentials generated by ions, as they diffuse across the membrane down their gradients in electrochemical potential. These electrochemical potential gradients are established and maintained by the actions of the ion transport ATPases.

The contribution of a particular ion gradient to the resting membrane potential depends on two factors: the magnitude of the *concentration gradient* for that ion and the *permeability of the membrane* for that ion. The two most important ions for the resting potential in *excitable tissue*, that is, tissue like nerve and muscle that produce *action potentials* are Na^+ and K^+. The relationships between resting membrane potential and the concentrations and permeabilities of Na^+ and K^+ are described mathematically by the *Goldman–Hodgkin–Katz* equation (also called the *constant field* equation).

3.5.4.2 The Goldman–Hodgkin–Katz Equation

The driving force for transport of an ionic species is the negative of the gradient in electrochemical potential for the ion across the membrane.

The electrochemical potential, $\tilde{\mu}$, for any ion, i, at constant pressure is defined as (see Equations 3.16 and 3.17):

$$\tilde{\mu}_i - \tilde{\mu}_{i0} = RT \ln\left[C_i\, e^{zF\psi/RT} \right] = RT \ln C_i + zF\psi \tag{3.16}$$

where $\tilde{\mu}_{i0}$ is the electrochemical potential in an appropriate reference state. For a single dimension, x, the negative of the gradient in chemical potential can be written as

$$-\frac{d\tilde{\mu}_i}{dx} = -\left[RT\frac{d \ln C_i}{dx} + zF\frac{d\psi}{dx} \right] \tag{3.17}$$

since $\tilde{\mu}_{i0}$ is not a function of x.

If we let V_i = the velocity (cm/s) of ion i across the membrane with respect to a fixed coordinate system, the molar flux of ion i, j_i, with respect to a fixed coordinate system is

$$j_i = C_i V_i = \left(\frac{g\,mol}{cm^3}\right)\left(\frac{cm}{s}\right) = \frac{g\,mol}{cm^2 s} \tag{3.18}$$

where

$$V_i = -\frac{D_i}{RT}\frac{d\tilde{\mu}_i}{dx} \qquad (3.19)$$

where D_i is the diffusion coefficient of ionic species, i, in the membrane. We also know that $d \ln C_i = dC_i/C_i$, so that we can write

$$V_i = -\frac{D_i}{RT}\left[RT\left(\frac{1}{C_i}\frac{dC_i}{dx}\right)+zF\frac{d\psi}{dx}\right] \qquad (3.20)$$

Then

$$j_i = V_iC_i = -\frac{D_iC_i}{RT}\left[RT\left(\frac{1}{C_i}\frac{dC_i}{dx}\right)+zF\frac{d\psi}{dx}\right]$$

and

$$j_i = -D_i\left[\frac{dC_i}{dx}+\frac{zF}{RT}C_i\frac{d\psi}{dx}\right] \qquad (3.21)$$

The current carried by this type of ion as it crosses the membrane is

$$I_i = j_izF \left(C/cm^2 s\right) \qquad (3.22)$$

If we now assume that the *electrical potential gradient* across the membrane is linear (called the *Goldman* assumption), we can write

$$\frac{d\psi}{dx} = \frac{\Delta\psi}{\Delta x} \qquad (3.23)$$

where $\Delta\psi = \psi_2 - \psi_1$ and $\Delta x = x_2 - x_1$ (Δx is the membrane thickness).
Substituting Equation 3.23 into Equation 3.22 and using Equation 3.21 gives the current, I, carried by each ion (dropping the subscript, i) as

$$I = -zFD\frac{dC}{dx}-\frac{(zF)^2 DC\Delta\psi}{RT\Delta x} \qquad (3.24)$$

For constant I and $\Delta\psi$, this is a first-order ordinary differential equation:

$$\frac{dC}{dx}+\left(\frac{zF\Delta\psi}{RT\Delta x}\right)C = -\frac{I}{zFD} \qquad (3.25)$$

This can be solved using standard methods (e.g., multiply through by the integrating factor $e^{(zF\Delta\psi/RT\Delta x)x}$ and integrate both sides with respect to x across the membrane). Solving for the current, I, gives

$$I = -\frac{(zF)^2 D\Delta\psi \left[C_2 - C_1 e^{-zF\Delta\psi/RT} \right]}{RT\Delta x \left[1 - e^{-zF\Delta\psi/RT} \right]} \qquad (3.26)$$

Here the subscripts 1 and 2 refer to conditions on either side of the membrane. When this derivation is applied to cell membranes, position 2 is taken as being inside the cell and position 1 is outside the cell.

Equation 3.26 represents the current carried by each diffusing ion. When we consider the current carried by all ions, at equilibrium, we must have electroneutrality. That is, at equilibrium,

$$\sum_i I_i = 0 \qquad (3.27)$$

We apply this relationship over all the positive and negative ions. Equation 3.25 when combined with condition (Equation 3.26) gives us the *general Nernst equation* (also called the Goldman–Hodgkin–Katz or constant field equation) for the ionic potential across a membrane. For many cells, the only important ions in establishing their resting membrane potential are Na^+ and K^+. Applying Equation 3.23:

$$I_{Na} + I_K = 0 \qquad (3.28)$$

Using Equation 3.27 in Equation 3.28 gives

$$-\frac{(zF)^2 D_{Na}\Delta\psi \left[C_{2,Na} - C_{1,Na} e^{-zF\Delta\psi/RT} \right]}{RT\Delta x \left[1.0 - e^{-zF\Delta\psi/RT} \right]} - \frac{(zF)^2 D_K\Delta\psi \left[C_{2,K} - C_{1,K} e^{-zF\Delta\psi/RT} \right]}{RT\Delta x \left[1.0 - e^{-zF\Delta\psi/RT} \right]} = 0$$

We now define permeability coefficients for Na and K (Chapter 2) as

$$P_{Na} = \frac{D_{Na}}{\Delta x} \quad \text{and} \quad P_K = \frac{D_K}{\Delta x}$$

We also let $z = +1$ for Na and K. We can now divide through and eliminate $(zF)^2 \Delta\psi/RT \left[1.0 - e^{-(zF\Delta\psi/RT)} \right]$ from both expressions. This gives

$$-P_{Na} \left[C_{2,Na} - C_{1,Na} e^{-F\Delta\psi/RT} \right] = P_K \left[C_{2,K} - C_{1,K} e^{-F\Delta\psi/RT} \right]$$

Clearing the parentheses, collecting like terms, and rearranging give

$$e^{-F\Delta\psi/RT} = \frac{P_{Na}C_{2,Na} + P_K C_{2,K}}{P_{Na}C_{1,Na} + P_K C_{1,K}} \tag{3.29}$$

Taking the natural logarithm of both sides and solving for $\Delta\psi$ give

$$\Delta\psi = -\frac{RT}{F}\ln\frac{[C_{2,K} + \alpha C_{2,Na}]}{[C_{1,K} + \alpha C_{1,Na}]} \tag{3.30}$$

where $\alpha = P_{Na}/P_K$.

In nerve and muscle cells, P_K is about 100 times larger than P_{Na} ($\alpha = 0.01$) and the resting membrane potential, $\Delta\psi$, is mainly attributed to the K diffusion potential. Since $\Delta\psi$ is primarily due to the K diffusion potential, it can be effected by the extracellular K concentration, $[K]_o$. The normal $[K]_o$ is about 4 mM and $\Delta\psi$ is in the range −60 to −80 mV. Increasing $[K]_o$ makes $\Delta\psi$ more positive. A change in $\Delta\psi$ in the positive direction is called *depolarization*. Similarly, reducing $[K]_o$ makes $\Delta\psi$ more negative and is called *hyperpolarization*. Figure 3.7 shows the dependance of resting membrane potential on $[K]_o$ assuming $[K]_i = 140$ mM for various $[Na]_o$ and $[Na]_i$. The Nernst potential for the various $[K]_o$ is shown for comparison. It is evident that at low $[K]_o$, the membrane potential deviates from what would be expected from the K gradient alone.

Not all cells have a membrane potential determined by the K gradient. In red blood cells, the membrane is relatively impermeable to K and to Na but is more permeable to Cl because of the presence of an anion (HCO_3^-/Cl^-)

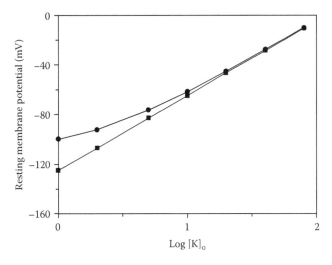

FIGURE 3.7
Membrane potential versus log $[K]_o$.

exchanger. In red cells, the membrane potential, $\Delta\psi$, is about $-12\,mV$, which reflects the Cl gradient across the cell ($[Cl]_o = 110\,mM$; $[Cl]_i = 70\,mM$).

3.5.5 Channels

Ions rapidly diffuse across cell membranes because of the presence of ion specific *channels*. Channels are membrane proteins or complexes of several membrane proteins that only allow particular ions to move through them. In general, channels consist of three functional units: a *selectivity filter* that determines which ions are conducted by the channel, a *pore* that provides an aqueous pathway for ion movement across the membrane, and a *gating mechanism* that determines when the channel is open or closed.

Channels are classified into two broad categories depending on their gating mechanism:

- *Voltage-gated channels*: These open and close in response to changes in membrane potential ($\Delta\psi$). Examples are Na, Ca, and K channels of nerve and muscle. These are important in propagation of the action potential and muscle contraction.

The molecular structures of many voltage-gated channels have similar features. The voltage-gated K channels consist of complexes of several repeating subunits (oligomers) with each subunit consisting of six α-helical transmembrane regions. The region between the fifth and sixth transmembrane regions inserts into the membrane as a hairpin structure called a p-loop (Figure 3.5). The association of p-loops in the complex is thought to form the channel pore. The fourth transmembrane segment contains positively charged Arg or Lys residues at every third position and is thought to be important in the gating mechanism.

Some K channels undergo spontaneous *deactivation* that is a time-dependent loss of activity after their initial opening. This may involve the 19-amino acid NH_2-terminal section that interacts with the pore after opening and blocks further activity.

Na channels consist of four separate regions each having the structural features of K channels. Na channels are activated by membrane depolarization and, after a delay, spontaneously *deactivate*. The deactivation mechanism is thought to be similar to the K channel deactivation mechanism.

- *Ligand-gated channels*: Here we have two types: *extracellular* ligand-gated channels and *intracellular* ligand-gated channels. The first type open or close in response to changes in the *extracellular* concentration of a *hormone* (a *first messenger*). An example is the *nicotinic acetylcholine receptor* found on the postsynaptic membrane of neuromuscular junctions. Channels of this type are activated by binding neurotransmitters such as acetylcholine (ACh), γ-amino isobutyric acid (GABA), and serotonin, among others. These channels play a crucial role in transmission of nerve impulses.

Intracellular receptors respond to changes in *intracellular* concentrations of an ion or peptide (a *second messenger*). An example of this kind is the *calcium ryanodine receptor* found on the ER membrane. This channel opens in response to ryanodine (an alkaloid extracted from tree bark), caffeine, and a transient increase in cytosolic Ca. *Activation of sytosolic Ca²⁺ release* initiates *contraction* in cardiac and skeletal muscle.

3.5.6 Signal Transduction

We now know that many different receptors convey instructions from hormones or other extracellular first messengers by stimulating one or more receptors to release a membrane-bound protein. This protein, in turn, acts on other intermediates called *effectors*. Often, the effector is an enzyme that converts an inactive precursor molecule to an active second messenger that can diffuse through the cytoplasm and carry the signal beyond the membrane in a process called *signal transduction*. The second messenger triggers a cascade of molecular reactions leading to a change in the cells' behavior. For example, the cell might secrete a hormone or release glucose or contract.

3.5.7 G-Proteins

A great many first messengers including *histamine* and NE rely on a class of molecules called *G-proteins* to direct the flow of signals from the receptor to the rest of the cell. G-proteins are named that way because they bind to *guanine* nucleotides.*

    a. More than 150 receptors have been identified that convey messages through G-proteins.

    b. At least 20 distinct forms of G-proteins have been identified. The different forms are identified by subscripts (G_i, G_s, etc.).

    c. A G-protein consists of a complex of three subunits, designated as the α, β, and γ subunits. The G-protein is bound to the intracellular (inner surface) side of the membrane in close proximity to (but not in contact with) the receptor. Only the α subunit is mobile. The β and γ subunits are fixed. The α subunit is the largest of the three, with the β subunit of intermediate size and the γ subunit the smallest.

    d. The nucleotide guanosine diphosphate (GDP) is bound to the α subunit.

    Figure 3.8 is a schematic diagram that illustrates the sequence of events that lead to signal transduction in a cardiac muscle cell.

---

* A guanine nucleotide, like all nucleotides, consists of an organic base (in this case, guanine), a sugar, and one or more phosphates.

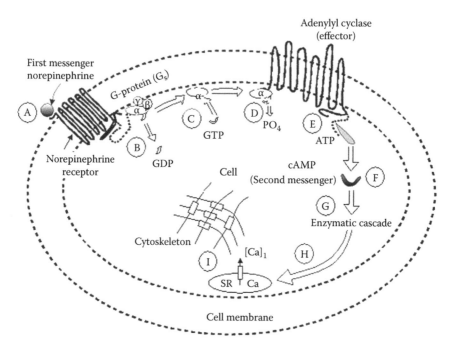

FIGURE 3.8

Ⓐ The first messenger (e.g., NE, histamine) binds to its receptor binding site on the extra-cellular surface of the membrane. Ⓑ This changes the 3-D configuration of the receptor and signals the G-protein to exchange GDP for *guanosine triphosphate* (GTP), which activates the G-protein. Ⓒ The G-protein dissociates freeing the α subunit, which then diffuses along the membrane. A GTP unit attaches to the vacant GDP site. Ⓓ The α subunit then interacts with an effector (e.g., adenylyl cyclase), activating the effector. After a few seconds, the α subunit converts GTP to GDP, thereby inactivating itself. The inactivated α subunit will then reassociate with a nearby β–γ complex. Ⓔ The effector adenylyl cyclase converts ATP to the second messenger, c-AMP. Ⓕ cAMP then initiates a cascade of enzymatic reactions. Ⓖ The details of these reactions have not, as yet, been worked out in cardiac muscle but may involve the second messengers IP_3 and diacylglyceride (DAG), as well as protein kinases. Ⓗ The net result of the enzymatic cascade is to open Ca channels in the SR, which releases Ca into the cytoplasm of the cell and increases the intracellular concentration of free Ca, $[Ca]_i$. Ⓘ Increased $[Ca]_i$ allows an increase in the rate of cycling of actin–myosin crossbridges in the cytoskeleton of the cell, causing the cell to contract with a greater force.

3.5.8 G-Protein-Dependent Effectors

There are several G-protein-dependent effectors that have been identified. In general, each of the effectors is activated by one or more different G-proteins. A sample of physiological effects that are mediated by G-proteins is given in Table 3.2.

Each of the effectors acts to release a second messenger or messengers that initiate a cascade of intracellular events leading to a modification of the behavior of the cell. One of the first G-protein–effector–second messenger combinations identified was the *transductin–phosphodiesterase–cyclic guanosine*

TABLE 3.2

Physiological Effects Mediated by G-Proteins

Stimulus	Affected Cell	G-Protein	Effector	Effect
Epinephrine, glucagon	Liver cells	G_s	Adenylyl cyclase	Breakdown of glycogen
Epinephrine, glucagon	Fat cells	G_s	Adenylyl cyclase	Breakdown of fat
Leutinizing hormone	Ovarian follicles	G_s	Adenylyl cyclase	Increased synthesis of estrogen and progesterone
Antidiuretic hormone	Kidney cells	G_s	Adenylyl cyclase	Conservation of water by kidney
ACh	Heart muscle cells	G_i	Potassium channel	Slows heart rate and decreases pumping force
Enkephalins, endorphins, opioids	Brain neurons	G_i/G_o	Calcium and potassium channels, adenylyl cyclase	Change electrical activity of neurons
Angiotensin	Smooth muscle cells in blood vessels	G_q	PLC	Muscle contraction, elevation of blood pressure
Odorants	Neuroepithelial cells in nose	G_{Olf}	Adenylyl cyclase	Detection of odorants
Light	Rod and cone cells in retina	G_t	c-GMP, phosphodiesterase	Detection of visual signals
Pheromone	Baker's yeast	GPA1	Unknown	Mating of cells

Source: Gether, U. *Endocr. Rev.* 21, 90, 2000.

monophosphate (c-GMP) system. Light acts on rhodopsin (a receptor for light), which is present on the surface of cells in the retina. The activated rhodopsin then causes activation of the G-protein, transductin (G_t). The α-subunit of G_t then diffuses across the membrane to the effector enzyme phosphodiesterase.* Phosphodiesterase converts c-GMP (a second messenger) to GMP (c-GMP keeps sodium channels open, while GMP closes Na channels). The closing of Na channels hyperpolarizes the inside of the retinal cells by preventing Na^+ from entering the cell and increasing the negative charge inside of the cell. This generates an electrical signal that is relayed to the brain.

The effector enzyme *phospholipase C (PLC)* is also a membrane-bound enzyme. When activated by a G-protein, PLC breaks down a membrane-bound phospholipid into two very important second messengers, *1,2-diacylglyceride (DG)* and IP_3. Increase in $[IP_3]$ triggers the release of Ca^{2+}

---

* *Note:* By convention, any biochemical whose name ends in "ase" is an enzyme. That is, a catalyst that acts to increase the velocity of a particular biochemical reaction without being consumed itself.

(another second messenger) from the ER and/or SR of skeletal and cardiac muscle cells. Increases in intracellular [Ca^{2+}] lead to cell contraction (Section 3.7).

Example 3.7 illustrates the actions of *adenylyl cyclase* and the activation of K channels by ACh on the behavior of heart muscle cells (*myocytes*).

Example 3.7 NE and Ach Receptors Act through G-Proteins in Heart Muscle Cells

When NE binds to its receptor on the surface of a myocyte, the receptor activates the G-protein, G_s. The α subunit diffuses along the membrane and activates the effector enzyme *adenylyl cyclase* (Figure 3.8). Adenylyl cyclase is a membrane-bound effector enzyme with 12 transmembrane segments. It converts ATP to the second messenger, c-AMP. c-AMP activates *protein kinases*, which leads to increased phosphorylation of intracellular proteins and through a cascade of other intracellular pathways to increased sytosolic [Ca^{2+}]. An increase in [Ca^{2+}] leads to an increase in the force of contraction of the cell. c-AMP, through other pathways, leads to an increase in the *rate* of formation and breaking (turnover rate) of *actin–myosin crossbridges* that increases myocardial *contractility* (Section 3.8).

Figure 3.9 illustrates how ACh, acting through another G-protein (G_i), modulates the action of NE. When ACh binds to its receptor on the myocyte, it activates the α subunit of G_i to open K channels, repolarizing cells and leading to a decreased force of contraction. ACh also inhibits adenylyl cyclase by an, as yet, unidentified mechanism (possibly subunit exchange).

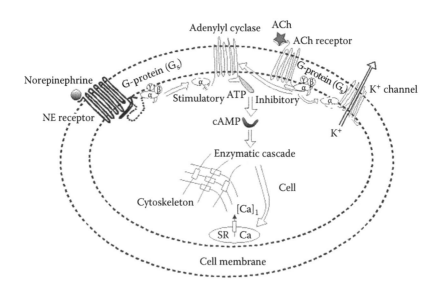

FIGURE 3.9
Schematic diagram showing the action of the inhibitory G-protein, G_i. Opening K^+ channels repolarizes the cell and reduces the force of contraction. The G_i protein also directly inhibits the stimulatory action of adenylyl cyclase by an, as yet, unidentified mechanism.

3.5.9 Calcium Sparks

Autoregulation of blood flow by arterioles involves a myogenic response to changes in intravascular pressure. Local increases in pressure lead to a graded depolarization of the membrane potential of the smooth muscle cells in the wall of the regulating arteriole. Over a limited pressure range, the smooth muscle cells will contract, decreasing the diameter of the arteriole, increasing local resistance to blood flow, and returning flows to near baseline levels in spite of the local pressure increase. Over the physiological range of membrane potentials (−60 to −30 mV), arterioles dilate when the membrane potential of the smooth muscle cells is made more negative (*hyperpolarized*). This happens because influx through voltage-gated Ca^{2+} channels decreases. Ca^{2+} entry through voltage-gated channels effects the spatially averaged $[Ca]_i$ in both cardiac and smooth muscle. Ca^{2+} is also sequestered and released through ligand-gated channels in the SR of muscle cells and the ER of myocytes (heart muscle cells). In cardiac muscle cells, release of Ca^{2+} through these channels (called *ryanodine-sensitive Ca^{2+} release channels*) directly contribute to the spatially averaged $[Ca]_i$ and to contraction (Example 3.6 and Figure 3.8). In arterial smooth muscle cells, local increases in intracellular calcium ion concentration, $[Ca]_i$, resulting from activation of the ryanodine-sensitive Ca^{2+} release channels in the SR cause arterial dilation. Ryanodine-sensitive local releases of Ca^{2+} (called *Ca^{2+} sparks*) occur just under the surface membrane in arterial smooth muscle cells. These local releases of Ca^{2+} have little effect on the spatially averaged $[Ca]_i$. However, they activate nearby K channels, hyperpolarizing the smooth muscle cell membrane and causing dilation (relaxation) of the resistance vessel [2].

3.6 Action Potentials

In nerve and muscle cells, when the membrane is depolarized above a certain value (called the *threshold value**), voltage-gated Na and K channels are opened (activated). However, the channels do not open immediately and there is a difference in the rate of opening of the Na and K channels. Na channels open faster than K channels. This difference in opening rate is an important feature of the action potential. It means that the initial response of the membrane to an increase in potential above the threshold value is a fast, very large increase in conductance† of Na ions (g_{Na}). This response is an *all-or-nothing* response.

---

* Action potentials require some depolarizing event to be initiated. In vivo, the initiation and propagation mechanisms involve specialized cells, receptors, and chemical neurotransmitters.
† Conductance is the reciprocal of resistance in an electrical circuit. It is a measure of the ability of an ion to carry an electrical current across the membrane.

What happens is that the threshold depolarization at first opens a small fraction of the Na channels. Na⁺ then rapidly diffuse down their concentration gradient through the open channels. This causes a further depolarization of the membrane and opens more Na channels in an accelerating, *positive feedback*, process. Within a short time, essentially all of the Na channels will be open (activated) and the membrane will be more permeable to Na than to K. Within this short period of time, the membrane potential becomes positive (called *overshoot*) and approaches the equilibrium potential for Na (about +50mV in nerve and muscle).

This process terminated by two mechanisms:

1. *Inactivation of the Na channels.* Na channels spontaneously inactivate when the membrane is depolarized (Figure 3.10). Na channels respond to depolarization by a rapid opening of their *activation gate* followed by a slower closing of their *inactivation gate*. This means that Na channels are only able to conduct ions for a short period of time after depolarization. After that they enter an *inactivated state* (Figure 3.10c). Inactivation of the Na channels leads to a reduction in the rate of Na⁺ entering the cell.

2. *Activation of the voltage-gated K channels.* K channels also open their activation gates in response to a depolarizing membrane potential.

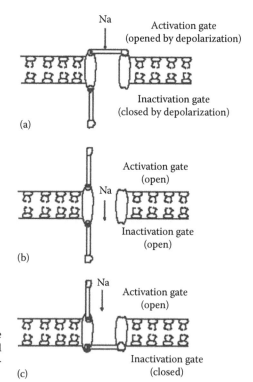

FIGURE 3.10
Voltage-gated sodium channel. (a) Before membrane depolarization, (b) initial phase of depolarization, and (c) spontaneous inactivation.

However, they do so more slowly than Na channels. When the Na channels open, K channels also open. The K ions diffuse through their channels down their concentration gradient making the membrane potential more negative. (Remember that the equilibrium potential for K$^+$ is −75 mV.)

The combination of opening K channels and inactivation of Na channels cause the membrane potential to become negative again. This is called *repolarization*.

3.6.1 The Hodgkin–Huxley (HH) Model

Membrane currents are studied by an electrophysiological technique called *voltage clamp*. In this technique, an electrode is inserted into a cell and is referenced to a similar electrode in the external solution. The membrane voltage across the cell is set at a given value by the experimenter and the amount of current that must be passed between the internal and external electrodes to maintain the desired voltage difference is measured. Measurements of Na and K currents in giant squid axons by Hodgkin and Huxley [3] in the 1940s and 1950s led to our present understanding of the role of Na and K currents underlying the action potential. These measurements led to a mathematical description of the behavior of Na and K currents during a nerve action potential called the *HH* model. The model results are in good agreement with experimental observations shown in Figure 3.12.

The HH model is an empirical kinetic description of the electrical responses of the voltage-gated Na and K channel activity that correctly models the major features of excitability such as the shape and conduction velocity of the action potential. Although it has several limitations, it is the starting point for much of the current theoretical work on voltage-gated channels.

The HH model has separate equations for sodium conductance (g_{Na}) and potassium conductance (g_K). In each case, there is an upper limit to the possible conductance, so that g_{Na} and g_K are expressed as maximum conductances $\bar{g}_{Na}$ and $\bar{g}_K$ multiplied by coefficients that represent the fraction of the maximum conductances actually expressed. The coefficients are numbers that vary between 0 and 1. All of the kinetics in the model appear as time variations of the multiplying coefficients. Conductance changes depend only on voltage and time, not on concentration of either Na or K or on the direction or magnitude of the current flow.

Experimentally, g_K follows an s-shaped (delayed) time course on depolarization and an exponential time course on repolarization. The HH model accounts for these properties by assuming the following model for the K channel:

$$I_K = n^4 \bar{g}_K \left(\Delta\psi - \Delta\psi_K \right)$$

where n is a multiplying coefficient, which varies between 0 and 1. One can show that if n rises exponentially from 0, n^4 follows an s-shaped curve mimicking experimental observation. Alternatively, when n falls exponentially to 0, n^4 also falls exponentially, mimicking experimental observation. The model further assumes that the K channel is voltage dependent and moves between its open and closed positions with first-order kinetics. This is modeled as

$$\text{(closed) ''}1-n\text{''} \underset{\beta_n}{\overset{\alpha_n}{\rightleftarrows}} n\,(\text{open})$$

The rate constants α_n and β_n are voltage dependant. If the initial values of the probability n are known, subsequent values can be calculated by solving the first-order differential equation:

$$\frac{dn}{dt} = \alpha_n(1-n) - \beta_n n$$

The HH model uses a similar scheme to describe I_{Na}. In this case, however, there are two opposing gating processes, activation and inactivation. The model uses two gating constants, called m and h. When the channel is in the open position, the sodium current is given by

$$I_{Na} = m^3 h \bar{g}_{Na}(\Delta\psi - \Delta\psi_{Na})$$

At rest, m is low and h is high. During depolarization, m rises rapidly and h falls slowly. Taking the cube of m causes a small delay in the rise and multiplying by the slowly falling h makes the coefficient m^3h eventually fall to a low value again. After the depolarization, m recovers quickly and h recovers slowly to their original values. The parameters m and h are assumed to undergo first-order reversible kinetics.

$$\text{(not activated) ''}1-m\text{''} \underset{\beta_m}{\overset{\alpha_m}{\rightleftarrows}} m\,(\text{activated})$$

$$\text{(activated) ''}1-h\text{''} \underset{\beta_h}{\overset{\alpha_h}{\rightleftarrows}} h\,(\text{not activated})$$

The transition rates are then determined by solving the following differential equations:

$$\frac{dm}{dt} = \alpha_m(1-m) - \beta_m m$$

$$\frac{dh}{dt} = \alpha_h(1-h) - \beta_h h$$

When the membrane potential is stepped to a new value and held there, the equations predict that m, h, and n relax exponentially to their new values.

The HH model treats activation and inactivation as entirely independent of each other. Both depend on membrane potential. Each can prevent a channel from being open, but their actions are independent of each other.

The complete HH model also contains a leakage current as

$$I_i = m^3 h \bar{g}_{Na}\left(\Delta\Psi - \Delta\Psi_{Na}\right) + n^4 \bar{g}_K\left(\Delta\Psi - \Delta\Psi_K\right) + \bar{g}_L\left(\Delta\Psi - \Delta\Psi_L\right)$$

where $\bar{g}_L$ is a fixed background leakage current.

3.6.1.1 The HH Model Can Predict Action Potentials

A schematic diagram of an action potential as computed from the sodium and potassium conductances by the HH model is shown in Figure 3.11. The Na and K currents are indicated by dashed lines.

The types of K channels responsible for the action potential in nerve and muscle do not undergo inactivation. Instead, as the membrane potential becomes more negative, the K channel activation gates close and g_K returns to its resting value. This does not occur right away, so that the K channels remain open for some time even after the membrane has repolarized. This leads to a period, called *hyperpolarization*, in which the membrane potential approaches the equilibrium potential for K^+.

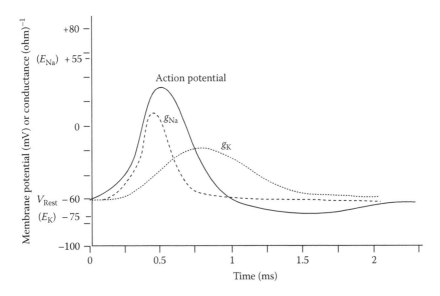

FIGURE 3.11
The action potential and the time course of the major currents that make up the ion flow during an action potential.

Surgeons use these properties during open heart surgery. The heart is arrested by infusion of a cold solution of saturated KCl (called a *cardioplegic solution*). This initially depolarizes the myocytes (see Figure 3.7) and then repolarizes the cells and arrests the heart in *diastole* (see Chapter 5). The high extracellular $[K^+]_o$ prevents further depolarization and propagation of the action potential.

3.6.2 Membrane Currents

A technique for measuring cell electrical properties called *patch clamping* has led to further insight into ion channel behavior by allowing recordings of single channel behavior. In this technique, an electrode is positioned to touch the surface of a cell and form a tight seal with the cell membrane. If the membrane contains a channel, individual channel openings and closings are observed under voltage clamp conditions. A typical recording of the behavior of a single Na channel during repetitive depolarizing pulses is shown schematically in Figure 3.12. Note that the individual channel openings occur more often during the initial part of the depolarizing pulse and less frequently during the later part. This reflects Na channel deactivation. The average conductance changes for many runs are similar to the current changes measured by Hodgkin and Huxley in giant squid axons.

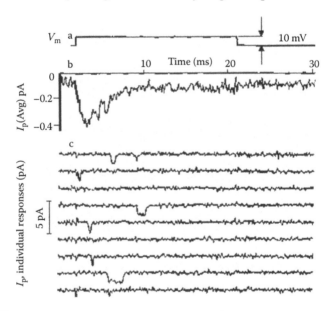

FIGURE 3.12
Typical patch clamp data. (a) Patch electrode recordings of currents recorded in rat muscle membranes in response to a 10 mV depolarization. (b) The average response to 300 individual recordings. The recordings in (c) show nine individual membrane patch currents. (Redrawn from Sigworth, F.J. and Neher, E. *Nature 287*, 447, 1980.)

3.6.3 Accommodation and Refractory Periods

The initiation of an action potential requires a rapid depolarization of the membrane potential. If depolarization occurs slowly, some Na channels have time to inactivate and the threshold for initiation of an action potential increases. At very slow depolarization rates, initiation of an action potential may be totally blocked. Holding a membrane in a depolarized state using a voltage clamp or increasing $[K]_o$ will inactivate Na channels and block normal Na conductance. Blocking the initiation of an action potential by increasing the threshold value is called *accommodation*.

There is a period of time after initiation of an action potential when it is impossible to initiate another action potential, no matter how intense the depolarizing stimulus. This is called the *absolute refractory period*. This period extends from the time of initiation of the action potential until about 0.5–1.0 ms after the peak. During the period following the peak of the action potential, the Na channels are inactivated and so, no increase in g_{Na} can occur following another stimulus. The *relative refractory period* refers to the period >1.0 ms after the peak. During this period, it is possible to elicit an increase in g_{Na}, with a depolarizing stimulus, since some of the Na channels have been reset. However, the threshold for a full blown action potential is higher than normal since not all of the Na channels have recovered from inactivation and it takes a stronger stimulus to initiate an all-or-nothing response.

3.7 How Do Muscles and Cells Contract?

Contraction of cells is an important process in cell motility and, in the case of cardiac and skeletal muscle, in providing force for blood flow and mechanical work. In intestinal cells, contraction provides the force to move the contents of the intestines through the bowels, allowing nutrients to be absorbed and providing for the elimination of solid waste.

3.7.1 Overview of Muscle Contraction

For our purposes muscle can be classified into three broad categories depending on anatomical structure and physiological function. These are *skeletal*, *cardiac*, and *smooth* muscles. The basic contractile mechanism is similar in all muscle types as well as in most nonmuscle cells such as *endothelial* cells.

In the *sliding filament-crossbridge model*, muscle contraction involves three steps:

a. Formation of *actin–myosin* crossbridges
b. Force development and crossbridge movement
c. Crossbridge cycling

The different anatomical arrangements and cell-to-cell communication mechanisms then give each of the different muscle types its unique mode of transmission of the mechanical force that is developed.

3.7.1.1 Anatomical Arrangements in Skeletal Muscle

In skeletal muscle, the organizational unit for mechanical force development is called the *sarcomere*. Sarcomere units are arranged in series in a skeletal muscle fiber so that a small shortening of each sarcomere unit adds to shorten the muscle fiber. On the other hand, crossbridge forces add in parallel, so that fiber force and muscle force increase with cross-sectional area (see Figure 3.13).

3.7.1.2 Control and Mechanics

Normal skeletal muscle activity is *neurogenic* (voluntary) and *multiunit*. A motor neuron (a nerve cell that terminates at a motor muscle) and the individual muscle fibers innervated by its branches form a functional group called a *motor unit*. Motor units contain fibers of the same type. The size of motor units varies, with smaller units (fewer muscle fibers per motor nerve) giving finer control. An action potential and signal transmission at the *neuromuscular* junction initiate excitation of the muscle fibers in the motor unit.

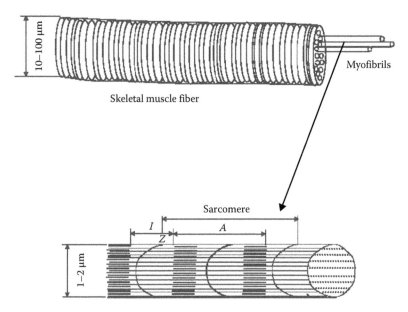

10–100 μm

Myofibrils

Skeletal muscle fiber

Sarcomere

I

Z

A

1–2 μm

FIGURE 3.13
Organization of skeletal muscle myofibrils.

3.7.1.3 Cellular Regulation of Contractile Activation

Depolarization of the muscle fiber surface membrane by a rapidly produced action potential produces a brief synchronized contraction called
a *twitch*. *Transverse-tubules* (T-tubules), which are transverse anatomical
structures, formed by the surface membrane conduct the action potential
inward.

Depolarization of the T-tubules initiates a series of *excitation–contraction*
coupling steps that activate crossbridge cycling.

1. Opening of Ca channels and rapid release of a large quantity of Ca,
   which is normally stored in the SR, adjacent to the T-tubules. This
   raises the intracellular free $[Ca^{2+}]$ from $<10^{-7}$ to $>10^{-6}$ M.

2. Ca^{2+} binds to troponin, releases the inhibiting effects of troponin on
   the thin filament, and permits contraction.

3. When excitation ends, a powerful active transport system (Ca–
   ATPase) in the SR reaccumulates Ca^{2+}. A Ca^{2+}-binding protein, *calse-
   questrin*, increases storage capacity for Ca^{2+} in the SR. Intracellular
   free $[Ca^{2+}]$ falls to resting levels, Ca^{2+} dissociates from troponin, and
   contraction is again inhibited during relaxation. Figure 3.14 illustrates the sequence of events.

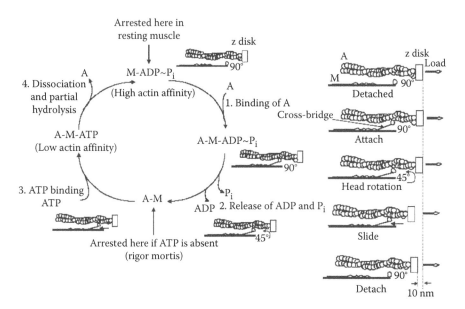

FIGURE 3.14
Crossbridge cycling and generation of force in skeletal and cardiac muscles.

3.7.1.4 Neural Control of Force

Two neural mechanisms control muscle force production:

Gradation by successive *recruitment* of motor units. Since motor units
are arranged in parallel, this increases the active cross-sectional area.
Fine control is achieved by tuning for motor unit size and fiber type.

Gradation by *frequency*. This increases the force per unit area. A single
twitch by a motor unit due to a single action potential is too brief
to stretch the elastic components of the muscle fibers and produce
maximum force. At high frequencies, the twitch force of each fiber
adds together to achieve a larger steady maximum force due to
more sustained intracellular Ca^{2+} activation. This is illustrated in
Figure 3.15.

Genetic defects in these control steps are present in many muscle diseases. Action potential changes due to abnormal Na or Cl channels cause

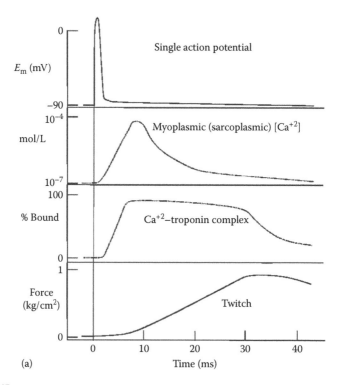

(a)

FIGURE 3.15

(a) Timing of events in the generation of a twitch force. Upper figure shows a single action
potential. Middle two figures show cytosolic free calcium concentration and formation of
Ca–troponin complex. Lower figure shows generation of twitch force. The electrical event triggers the chemical event, which precedes the mechanical event (the twitch force).

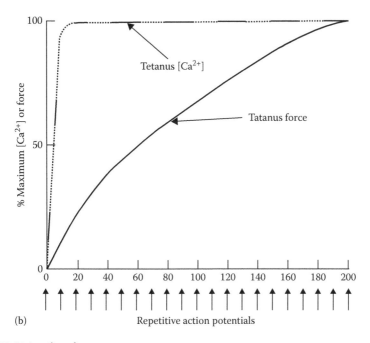

(b)   Repetitive action potentials

FIGURE 3.15 (continued)
(b) Generation of a sustained force due to frequency modulation of action potentials. Again, the electrical events trigger the chemical events that produce the mechanical event (sustained force in a skeletal muscle fiber).

several types of periodic *paralysis* or *myotonia* (hyperexcitability and stiffness). Increased surface membrane Ca permeability is related to deficient *dystrophin* and associated membrane proteins in *Duchenne muscular dystrophy*. Abnormal SR Ca release channels are associated with *halothane*-induced *malignant hyperthermia*. Halothane is a general anesthetic commonly used during surgery.

3.8 Mechanical Properties of Skeletal Muscle

When skeletal muscle is stretched from its resting (no-load) length (L_o) without being stimulated to contract, it exhibits passive *stress–strain* behavior similar to many *elastic* materials. The force (stress) required to stretch the muscle is proportional to the increase in length (strain) until a yield stress is reached. When skeletal muscle is stimulated to contract at constant length (*isometric contraction*), the maximum force (tension) that it develops occurs at approximately L_o. Active force decreases at both longer and shorter lengths. The decrease above L_o is attributed to increased overlap between myosin crossbridges and actin sites producing less than optimum force during contraction. The decrease below L_o is attributed to unfavorable activation and filament interaction. That is, most muscles operate in a lever system, inserting into skeletal elements across joints (e.g., biceps). Rotation around the joint as a pivot amplifies small muscle changes near L_o to larger movements (favorable force–length relation), although this action requires more force to lift a given load. The relationship between muscle length and isometric force for active, passive, and total force developed is illustrated in Figure 3.16a. The relationship between sarcomere length and force is illustrated in Figure 3.16b for skeletal muscle.

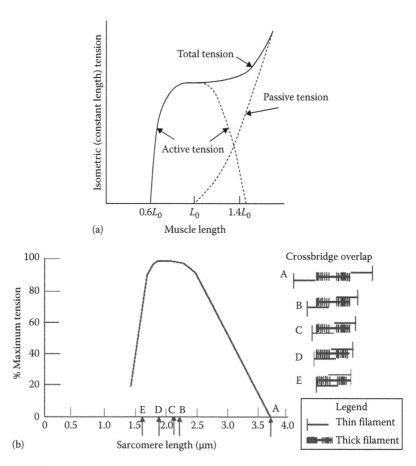

FIGURE 3.16
(a) When a muscle is stretched from some initial length, L_0, without being stimulated to contract, it follows the passive length–tension (stress) curve and acts like any distensible material (e.g., a rubber band). When the muscle is stretched and also stimulated to contract, it develops force (tension) along the active tension curve (see mechanism in Figure 3.16b). The total force developed is the sum of these two forces. (b) Force development in skeletal muscle. Within the physiological range, force development depends on the degree of overlap of the actin–myosin crossbridges. The most favorable overlap usually occurs at the resting length of the muscle.

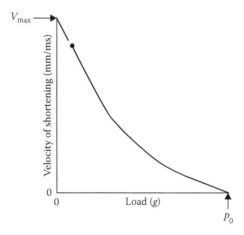

FIGURE 3.17
Relationship between load and velocity of shortening. At some load, P_0, the muscle cannot develop enough force (fiber stress) to lift the load and cannot shorten. At zero load, the velocity of shortening is maximal for a given muscle and *inotropic* state. V_{max} (the velocity of shortening at zero load) is a measure of the inherent ability of the muscle to develop force, called the *contractility* of the muscle.

3.8.1 Relationship between Force Developed and Velocity of Shortening

So far, we have been talking about isometric (constant length) contractions. When a muscle is given a load (weight) to lift and is stimulated to contract, it does two things: it develops just enough force (tension) to lift the weight (if it can) and then it shortens.

The relationship between the force developed by the muscle and its *velocity* of shortening is simplest to understand during *isotonic* (constant force) contractions. Figure 3.17 illustrates the relationship between velocity of shortening and load.

1. Velocity of shortening is maximum when the force (load) is zero.

2. Velocity decreases nonlinearly with increasing load, reaching zero at the maximum force that the muscle can develop (isometric contraction).

3. The molecular basis for the decreasing velocity with increased load is a decrease in *crossbridge cycling rates* when crossbridges must change conformation against an increasing counterforce.

3.9 Energetics of Skeletal Muscle Contraction

Muscle contraction transduces chemical energy (in the form of the high energy phosphate bonds of ATP) into mechanical work and heat. Mechanical work, defined as force times distance, only occurs when a muscle shortens

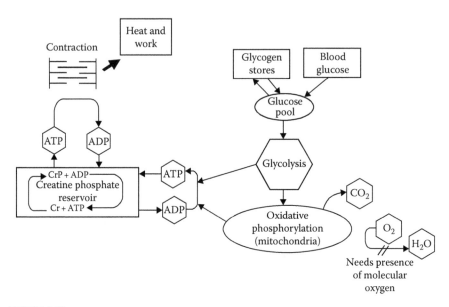

FIGURE 3.18
The energetics of muscle contraction.

against a load. The mechanical efficiency is quite high at moderate load and velocity. Isometric contractions (no shortening) do not involve mechanical work, but still serve important functions. Figure 3.18 is a schematic diagram showing the overall relationship between ATP production, mechanical work, heat, and ATP utilization in muscle contraction.

Muscle heat production includes

1. Resting heat, related to basal cell functions (e.g., active transport, biosynthesis of enzymes, and proteins)
2. Heat of activation (related mainly to excitation–contraction coupling steps)
3. Shortening heat (e.g., crossbridge cycling)
4. Recovery heat (metabolic reactions that renew the energy stores)

The chemical energy used in muscle contraction provided directly by ATP hydrolysis is ultimately supplied by metabolism. Different fiber types (fast and slow) utilize different pathways (glycolysis and oxidative phosphorylation) to different extents.

Skeletal muscle energetics are the major factor in body heat production and energy requirements, increasing many times with exercise and shivering and providing the major stimulus for regulation of cardiovascular and respiratory functions.

3.9.1 Cardiac Muscle

The cross-striation of cardiac muscle produces a sarcomere organizational structure, which is similar to that of skeletal muscle.

Cardiac muscle (and most smooth muscles) form sheets around hollow organs. Relatively short muscle fibers must function together to produce pressure on the contents of the hollow tube. This is in contrast to the lever action of the long skeletal muscle fibers. Therefore, fiber activity in cardiac muscle (and most smooth muscle) is unitary. Again, this is in contrast to the gradation and multiunit activity of skeletal muscle.

Unitary cardiac muscle activity depends on the organization of individual cardiac muscle cells (myocytes) into a functional whole unit with both mechanical and electrical continuity. The complex interdigitations of the fibers at intercalated disks form mechanically strong junctions. Specialized, high permeability junctions (gap junctions, nexus) allow electrical conduction between cells.

Cardiac (and smooth muscle) activity is *involuntary*. That is, not controlled directly by the central nervous system. The activity of cardiac muscle (and many smooth muscles) is *myogenic* (spontaneous) and only *modulated* by nerves. Activity in other smooth muscles is *neurogenic*, that is, similar to skeletal muscle.

Under normal physiological conditions, contractile activation of the heart is controlled by excitation of the plasmalemma by an action potential, excitation–contraction coupling, and Ca removal by the SR and plasmalemma. The normal pacemaker of the heart is the firing of the vagus nerve at the sinoatrial (SA) node and propagation of the action potential throughout the myocardium by a set of specialized conduction fibers (see Chapter 5).

3.9.1.1 Pacemaker Activity

Under conditions in which the electrical activity of the conduction pathways is abnormal (*arrhythmias, heart blocks*) some myocardial cells (*pacemaker cells*) undergo spontaneous (myogenic) depolarization and can act as the pacemaker for initiating contraction of the ventricle(s). Myogenic excitation is initiated by the pacemaker cells that spontaneously depolarize fastest and is then propagated to all myocardial cells. The spontaneous depolarization of pacemaker cells is a relatively slow process and under normal conduction conditions, the action potential initiated at the SA node is propagated through the entire left ventricle before any pacemaker cells can depolarize to their threshold potential.

Cardiac action potentials are of a much longer duration (typically lasting 300–400 ms) than action potentials generated at the neuromuscular junction.

Although the normal pacemaker is the firing of the vagus nerve at the SA node, the frequency of pacemaker activity can be altered by sympathetic firing rate and by circulating drugs and hormones (*chronotropic agents*).

3.9.1.2 Excitation–Contraction Coupling

Excitation–contraction coupling requires extracellular Ca^{2+} entry through voltage-dependent Ca channels (Section 3.7). This triggers Ca^{2+} release from the SR, which increases intracellular free $[Ca^{2+}]$, inhibits the cardiac troponin–tropomyosin complex, and allows actin–myosin interaction. When excitation ends, Ca^{2+} uptake by the SR and Ca transport out of the cell (mainly by a NCE) reduces intracellular free $[Ca^{2+}]$ and causes relaxation of the ventricle.

Unlike skeletal muscle, total force in cardiac muscle cannot be graded by recruitment (cardiac muscle cells act in concert) or by tetanic summation at high frequency (due to the long cardiac action potential). However, Ca^{2+} entry, Ca^{2+} uptake and release by the SR, Ca^{2+} binding to troponin, and Ca^{2+} transport out of the cell can all be regulated. These *inotropic* mechanisms allow regulation of the contractile state of the heart by hormones and drugs (*inotropic agents*).

3.9.1.3 Mechanical Properties

The mechanical properties of cardiac muscle are qualitatively similar to those of skeletal muscle. Isometric and nearly isotonic contraction each occur briefly during the cardiac cycle. The relationship between cardiac force and initial length operates over a narrower length range than skeletal muscle, partly due to a high passive tension. Increased force of contraction with increasing length (again, over a narrow range) is a functionally important regulatory mechanism (the *Frank–Starling* principle). Figure 3.19 shows the relationship between force of contraction and initial fiber length for cardiac muscle. Cardiac shortening velocity decreases with increasing force as in skeletal muscle. However, velocity of contraction is regulated inotropically.

3.9.2 Smooth Muscle

Smooth (unstriated) muscle does not have aligned sarcomeres with Z disks, but the basic contractile mechanism is similar in all muscle types. Actin and myosin filaments are organized into sarcomere-like structures by *dense bodies* (see Figure 3.20), which serve as functional analogs of Z disks. The small cells are linked mechanically as in cardiac muscle.

Regulation of contraction is myosin based and varies among the different smooth muscle types. In multiunit neurogenic muscle types (e.g., blood vessels), excitatory and inhibitory autonomic nerves control motor units. In unitary myogenic muscle types (e.g., gastrointestinal muscle, uterus), regulation is similar to cardiac muscle because of electrical continuity at gap junctions and spontaneous activity that is modulated by both excitatory and inhibitory signals from autonomic and intrinsic (*enteric*) nerves as well as hormones, drugs, and stretch. Activation of crossbridge interaction requires phosphorylation of a myosin *regulatory site*, controlled by a myosin kinase and a phosphatase (Figure 3.20). Here, the kinase activity depends on Ca^{2+}

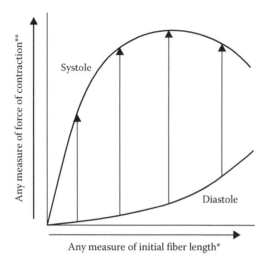

*End-diastolic volume, end-diastolic pressure
**Stroke volume, cardiac output

FIGURE 3.19

Relationship between force of contraction and stretch of the ventricle prior to contraction. When the ventricle is passively stretched (e.g., by an increase in venous return due to a transfusion or exercising skeletal muscle) over the physiological range, it responds with a larger force of contraction up to a certain point. This mechanism, called the *Frank–Starling* mechanism or *heterometric autoregulation*, is independent of neural control. It allows the heart to handle increases in venous return without invoking neural reflexes and is an important adaptive mechanism in the failing heart. The heart actually operates on one of a family of Frank–Starling curves, depending on its *inotropic* state (see Chapter 5).

binding to *calmodulin*, a troponin analog, and thus on free intracellular $[Ca^{2+}]$. With stimulus removal, SR and plasma membrane active transport systems and NCEs reduce free intracellular $[Ca^{2+}]$ to resting levels. Relaxation may be very slow.

3.9.3 Mechanical and Energetic Properties of Smooth Muscle

Mechanical properties mainly reflect differences in tissue organization in the different smooth muscle types as well as differences in myofilament organization and crossbridge cycling. Length changes passively after stretch (elastic behavior), but the active length–force and force–velocity relationships in isometric and isotonic contraction are similar to those in striated muscle. *Tone* (sustained basal force) results from sustained stimulation and slow relaxation. Force can be graded by frequency, with tetanic summation, since action potentials (if present) are brief relative to the twitch.

Energy utilization is much more economical than in striated muscle since crossbridge cycling is slower during force development and becomes extremely slow during force maintenance.

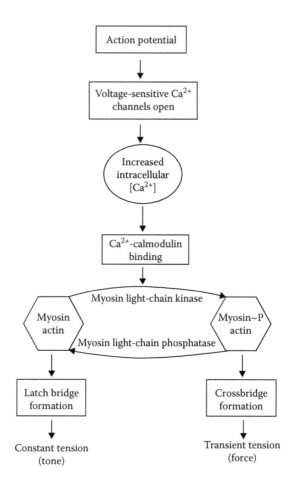

FIGURE 3.20
Actin–myosin crossbridge formation in smooth muscle, showing latch-bridge formation and formation of transient force. All larger blood vessels (arteries, arterioles, veins, venules) have smooth muscle cells, either embedded in their walls or nearby in the interstitial space. Latch-bridge formation allows these muscle cells to provide structural support for the blood vessel by providing a constant basal level of vasoconstriction, called *tone*. The myosin content is much lower and the actin content higher than in striated muscle. The ratio of thick to thin filaments is low. The thin filaments contain tropomyosin but no troponin.

3.10 Summary

In this chapter, we have provided some of the cellular and biochemical bases upon which physiological function rests. We have covered *free energy and electrochemical potential* as the driving force for transport of ions and molecules. We have discussed the major mechanisms by which ions and molecules get into and out of cells, *diffusion and carrier-mediated transport*.

We have considered *Gibbs–Donnan equilibrium* and the ionic basis for the action potential. In the final section, we covered the molecular basis for the mechanisms by which cells contract. We have discussed the electrical basis for the *initiation* of contraction and the ionic basis for *excitation–contraction coupling* and *relaxation*. We have outlined the contractile mechanisms for the three major muscle types, *skeletal*, *cardiac*, and *smooth*, and their similarities and differences. We have pointed out that they are similar in that contraction always involves mobilization of free intercellular $[Ca^{2+}]$ and is *myosin* based.

This chapter then sets the stage for our discussion of cardiovascular dynamics and mechanics in Chapter 5.

Problems

Osmotic Pressure

3.1 Calculate the osmolarity of 0.9% NaCl. Show how you got your answer.

3.2 Describe what will happen to cell volume when cells are placed in a large volume of
  a. 0.2% NaCl
  b. 0.9% NaCl
  c. 1.8% NaCl
  d. 0.3 M glucose. Assume that the glucose concentration inside the cell is zero at $t = 0$ and that the cell membrane is permeable to glucose.
  e. 0.6 M glucose
  f. 0.3 M glucose + 0.9% NaCl. Assume that NaCl acts as an impermeable ion because of the Na–K–ATPase pump.

3.3 Describe the events underlying the changes in red cell volume and Cl content when CO_2 is bubbled into a suspension of red blood cells in 0.9% NaCl. (*Note:* red blood cells are a rich source of the enzyme carbonic anhydrase.)
  a. Describe how the red blood cell volume changes as venous blood passes through the lungs.
  b. The "hematocrit" is the volume of red cells as a percent of total blood volume (total blood volume = plasma volume + red cell volume). The "normal" hematocrit is 45%. Which has the higher hematocrit, venous or arterial blood? Explain.

3.4 A solution of albumin, a positively charged protein, is placed in a dialysis bag and put into a very large volume of 10 mM NaCl, 1 mM K_2SO_4, and 1 mM $CaCl_2$. After dialysis overnight, a sample of the fluid inside the dialysis bag is withdrawn and assayed for Na; the [Na] in the dialysis bag was determined to be 5 mM.

a. What would be the concentrations of K, Ca, Cl, and SO_4 in the dialysis bag? (Assume that because of the large volume of the fluid outside the bag, the external concentrations are not changed.)
b. What is the membrane potential across the wall of the dialysis bag?

Hint: Since the charged protein is impermeable, the ion distribution must follow the Donnan ratio.

Equilibrium Potentials

3.5 Calculate how the membrane potential for eukaryotic cells changes when the cells are placed in media with the following Na and K concentrations. Assume that $P_{Na}/P_K = 0.01$ and that $[Na]_i = 10\,mM$ and $[K]_i = 120\,mM$ for all conditions.

a.

$[K]_o$ (mM)	$[Na]_o$ (mM)
1	149
2	148
5	145
10	140
20	130
40	110
80	70

b. Plot the membrane potentials obtained in part a versus log $[K]_o$.
c. How would the result in part b be different if $P_{Na} = 0$?

3.6 Calculate the equilibrium potentials for each of the ion concentrations given below.
a. $[K]_o = 5\,mM$; $[K]_i = 140\,mM$
b. $[Na]_o = 140\,mM$; $[Na]_i = 8\,mM$
c. $[Ca]_o = 2\,mM$; $[Ca]_i = 50\,nM$ ($1\,nM = 10^{-9}$)
d. $[Cl]_o = 120\,mM$; $[Cl]_i = 30\,mM$

Diffusion Potentials

3.7 Consider a membrane separating two solutions with the compositions given below:

A	B
$50\,mM$ K_2SO_4	$5\,mM$ K_2SO_4
$10\,mM$ NaCl	$100\,mM$ NaCl
$20\,mM$ $MgCl_2$	$5\,mM$ $MgCl_2$
$V = 1\,L$	$V = 1\,L$

What is the osmolarity difference across the membrane?

Calculate the approximate membrane potential (assuming that side B is the "internal" side) at 25°C assuming the following conditions:

a. The membrane is permeable only to K^+.
b. The membrane is permeable only to Mg^{2+}.
c. The membrane is permeable only to Cl^-.
d. The membrane is permeable only to SO_4^{2-}.
e. The membrane is permeable to Na^+ and K^+ but impermeable to other ions.

3.8 Under the various conditions of Problem 3.7, will the ion concentrations given remain stable or will they change with time? Explain your answer.

3.9 Will the membrane potential be positive or negative on side B under the conditions given below? (Assume that the membrane is impermeable to ions that are not specifically mentioned.) Explain your answer.

a. The membrane is five times more permeable to K^+ than to Na^+.
b. The membrane is five times more permeable to Na^+ than to K^+.
c. The membrane is impermeable to SO_4^{2-} but is equally permeable to all other ions.

3.10 The following data were recorded for the transmembrane flux of galactose as a function of galactose concentration:

$[C_i]$ (mol/L)	J_i (nmol/cm² min)
6.25×10^{-6}	15.0
7.50×10^{-5}	56.25
1.00×10^{-4}	60.0
1.00×10^{-3}	74.9
1.00×10^{-2}	75.0

a. Estimate J_{max} and K_m.
b. What would J_i be at $[C_i] = 2.5 \times 10^{-5}$ mol/L and at $[C_i] = 5.0 \times 10^{-5}$ mol/L?
c. Now assume that $[C_i] \ll K_m$, so that

$$J_i \cong K[C_i]$$

where $K = J_{max}/K_m$.
Further, suppose that for this case, we can approximate $J_i = -d[C_i]/dt$, so that

$$d[C_i]/dt = -K[C_i]$$

The half time, $t_{1/2}$, is the time required for the concentration of the substrate on the outside of the carrier to be reduced to half of its value.

Assuming that the initial substrate concentration is $[C_i]_0$ at time $t = 0$, determine the half time in terms of the parameter K.

3.11 The following data apply to an excitable cell in a bathing solution.

Ion	Intracellular Concentration (mM)	Extracellular Concentration (mM)	Relative Resting Membrane Conductance, g_i
Na⁺	12	120	0.05
K⁺	120	4	0.5
Cl⁻	4	120	0.45

When more than two ions are involved, the Goldman equation becomes very complex. For more than two ions, the resting membrane potential for a cell can be obtained by a weighted average of the equilibrium potentials for the three ions. The weighting factors are the relative conductance of each ion. The resting membrane potential is then calculated as

$$E_m = \frac{g_K}{\Sigma g}E_K + \frac{g_{Na}}{\Sigma g}E_{Na} + \frac{g_{Cl}}{\Sigma g}E_{C_i}$$

where $\Sigma g = g_K + g_{Na} + g_{Cl}$.

The E_i are the respective Nernst potentials for each ion.

a. What is the resting membrane potential (sign and magnitude)?

b. Is the inside positive or negative with respect to the outside?

3.12 Refer Figure P3.12 to answer the next question:

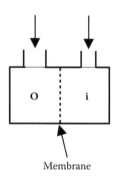

Membrane

FIGURE P3.12

If compartment **i** contains 100 mM sucrose, and the membrane is impermeable to that solute but highly permeable to water,

a. What concentration of NaCl would be needed in compartment **O** to prevent volume flow if the membrane was permeable to both Na^+ and Cl^-?
b. What concentration of NaCl would be needed in compartment **O**, if the membrane is impermeable to Na^+ but highly permeable to Cl^-?
c. If you add 20 mM $MgCl_2$ to compartment **O**, what would be the osmotic pressure difference between the two compartments if the membrane were impermeable to all solutes but highly permeable to water? Which way would water move?

References

1. Gether, U. Uncovering molecular mechanisms involved in activation of G protein-coupled receptors. *Endocr. Rev.* 21(1): 90–113, 2000.
2. Stern, M.D. and Cheng, H. Putting out the fire: What terminates calcium-induced calcium release in cardiac muscle? *Cell Calcium* 35(6): 591–601, 2004.
3. Hodgkin, A.L. and Huxley, A.F. A quantitative description of membrane current and its application to conduction and excitation in nerve. *J. Physiol.* 117: 500–544, 1952.
4. Sigworth, F.J. and Neher, E. Single-Na channel currents observed in cultured rat muscle cells. *Nature* 287: 447–449, 1980.

4

Principles and Biomedical
Applications of Hemodynamics

4.1 Introduction

Blood circulates throughout vessels and organs within a closed loop called the *circulatory system*. This system consists of two separate flow paths in series: the *systemic circulation* and the *pulmonary circulation*. Oxygenated blood leaving the pulmonary circulation flows to the left side of the heart. The oxygenated blood ejected by the left ventricle (the *cardiac output* [CO]) then travels via the *arterial* system through the systemic circulation to the head and all the major organs where oxygen is exchanged for carbon dioxide in the *microcirculation*.* Deoxygenated blood then returns to the right side of the heart via the *venous vessels* (the *venous return*) to complete the circuit. The pulmonary circulation supplies blood to lung tissue and exchanges carbon dioxide for oxygen. A series of four valves† (two on each side of the heart) ensures one-way flow through the circulatory system. Flow (CO) through the systemic circulation averages 4–6 L/min and is supplied by the *left ventricle*. Flow through the pulmonary circulation is supplied by the *right ventricle*. The left and right ventricles contract and pump blood in parallel so that the CO and venous return are equal. The pulmonary circulation is a relatively low-pressure system. The normal maximum pressure in the *pulmonary artery* is 25–28 mm Hg. In contrast, the normal maximum pressure in the *aorta* is 100–120 mm Hg. The pressure which the left ventricle must pump against to open the aortic valve and eject the stroke volume is 4–5 times greater than the pressure that the right ventricle must pump against to open the pulmonic valve. As a consequence, the right ventricular wall is less muscular and considerably thinner than the left ventricular wall. Ventricular walls and the walls of arterial blood vessels are *distensible*. That is, their equilibrium

---

* The microcirculation consists of all blood vessels less than 200 μm in diameter. This includes arterioles, true capillaries, and venules. Most blood–tissue mass transfer and regulation of flow take place at the microcirculatory level.
† The valves on the right side of the heart are the *atrioventricular* (AV) and the *pulmonic*. The valves on the left side are the atrioventricular and the *aortic*.

diameter at any time depends on their elastic (material) properties and a balance between the internal and external pressure forces acting on the wall. The difference between the internal and external pressure forces is called the *transmural* pressure. Vessels stretch when they are subject to an increase in transmural pressure.

In disease states such as *aortic insufficiency*, an incompetent aortic valve increases the working volume of the left ventricle (volume overload). This *dilation* leads to thickening of the ventricle wall (*hypertrophy*), increased stiffness, and an increase in both peak-systolic and end-diastolic pressures. The heart enlarges in such a way as to maintain a ratio of wall thickness to ventricular radius (h/R), which is similar to that in a normal heart. In uncontrolled *hypertension* (pressure overload), the left ventricular wall also thickens (*hypertrophy*). However, in this case the ratio h/R increases leading to *decreased* chamber volume, increased stiffness, increased peak-systolic and end-diastolic pressures, and decreased ventricular filling. In either case, the end result is *heart failure* (a decreased ability to maintain CO).

In the normal aging process as well as in disease states such as *atherosclerosis, remodeling* of the walls of arteries leads to increased stiffness with a resultant increase in flow resistance and arterial blood pressure. Atherosclerosis also causes a buildup of *plaque* in the walls of certain arteries (especially the coronary arteries). This decreases the cross-sectional area that is available for flow (known as *stenosis*), causes a local increase in flow resistance, and leads to lower than normal flow or absence of flow (*ischemia*). If this occurs in one or more of the coronary arteries, it is called *coronary artery disease*. When coronary artery flow is low enough so that heart failure is immanent, the patient is a candidate for either *balloon angioplasty with placement of a stent* or a *coronary artery bypass* operation.

Heart valves (particularly the aortic valve) may also undergo *calcification* and stenosis. This leads to an increase in chamber pressure and poor ejection. When the stenosis becomes large enough, the patient is a candidate for a *valve replacement*.

In health (e.g., exercise) as well as disease, the relationships among pressure, flow, and resistance (impedance) in arterial vessels are dynamic processes. These processes depend on factors such as the *pulsatile* nature of the ejection, *blood viscosity, mechanical properties* of the vessel wall, *tapering* and *bifurcation* of the vessels, the *network* properties of the circulatory vessels (series and parallel connections), and the reflex and local regulatory mechanisms that act to maintain a stable arterial blood pressure and blood volume.

In this chapter, we consider the relationships among pressure, flow, and fluid resistance both in large vessels such as arteries where we can use continuum mathematics and in the microcirculation where continuum mathematics may not apply.

4.2 Electrical Analogs of Steady Flow: Ohm's Law

Hemodynamics is a branch of fluid dynamics in which blood is the only fluid being considered. The flow of blood in the large vessels is, by nature, *pulsatile*. That is, the pressure, velocity, and the volumetric flow rate in the large arteries and the ventricles vary with time over a single cardiac cycle. However, under normal resting conditions, these variations are approximately *periodic* so that the cycle repeats itself, at least over short time periods. For certain purposes, it has proven useful to consider the time-average values of pressure, velocity, and flow as *steady-state* parameters.

In the steady state, we can write Ohm's law for the rate of transport of a quantity between any two points as

$$\text{Rate} = \frac{\text{driving force}}{\text{equivalent resistance}}$$

Consider the resistive DC circuit shown in Figure 4.1. In this circuit, the flow is that of electrons (or current, I), the driving force is the potential difference across the circuit (ΔE), and the equivalent resistance (R_{eq}) is that of the series–parallel arrangement of the resistors. For the circuit in Figure 4.1, we can write Ohm's law as

$$I = \frac{\Delta E}{R_{eq}}$$

By analogy, we can write Ohm's law for the steady laminar volumetric flow of blood between any two points in the circulation as

$$Q\left(\text{L/min}\right) = \frac{\Delta P \, \text{mm Hg}}{R_{eq}\left[\dfrac{\text{mm Hg}}{\text{L/min}}\right]} \tag{4.1}$$

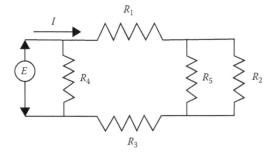

FIGURE 4.1
Series–parallel electrical circuit. Resistors R_1, R_2, and R_3 form a series connection, while R_4 and R_5 are parallel pathways for the flow of current.

The volumetric flow (Q) between the two points equals the driving force for flow (the pressure difference between the two points, ΔP) divided by the equivalent hemodynamic resistance of the vessels and organs in the flow path (R_{eq}), where $\Delta P = P_1 - P_2$, and the units of equivalent resistance are specified so that the units on either side of Equation 4.1 are equal.

What are some of the factors that determine equivalent hemodynamic resistance? They are analogous to those factors that determine equivalent electrical resistance.

a. The physical properties of the fluid, in particular the fluid viscosity

b. The length of the path between Points 1 and 2

c. The diameter(s) of the vessel or vessels in the path

d. The particular series–parallel arrangement of the vessels in the network

e. The distensible nature of the vessel walls

We recognize that Equation 4.1 is only an approximation. Flow and pressure are both time-varying (pulsatile), and the vessels have resistive, capacitive, and (perhaps) inductive elements.

A more realistic approach would be to treat the flow between any two points as an analogous AC circuit and write Ohm's law in terms of the *impedance* of the network:

$$Z(\omega) = \frac{\Delta P(\omega)}{Q(\omega)} \tag{4.2}$$

where ω is some characteristic frequency of the pressure wave.

Since impedance is a frequency concept, it can be defined in several ways. We will return to Equation 4.2 in a later section where we consider pulsatile flow in a tube.

Example 4.1 Total Peripheral Resistance

The equivalent resistance of all of the vessels in the series–parallel network between the left side of the heart and the right side of the heart is called the *total peripheral resistance* (R_{TP}). For a standard person, the flow between those two points is the CO. The mean (time-averaged) pressure in the aorta is 100 mm Hg. The pressure at the right atrium is, by convention, 0 mm Hg (atmospheric pressure). Calculate R_{TP}.

Solution

Using Equation 4.1,

$$R_{TP} = \frac{\Delta P}{Q} = \frac{100 - 0}{5} = 20 \left(\frac{mm\,Hg}{L/min} \right)$$

Total peripheral resistance increases with age and in certain disease states where the arteries stiffen. R_{TP} decreases in exercise where local regulatory processes act to dilate the arterioles (the major resistance vessels) in skeletal muscle.

4.3 Newton's Law of Viscosity

Blood is a complex, multiphase fluid. It contains an aqueous phase (plasma) with dissolved proteins (albumin, the globulin, and fibrinogen) and electrolytes as well as suspended particles (red blood cells, leukocytes, and platelets). The mechanical properties of whole blood depend on the volume fraction of red cells (the *hematocrit*) as well as the amount and distribution of the protein content of the plasma. In large vessels, the pulsatile nature of the flow keeps the cells suspended and we approximate blood as a homogeneous fluid. In the microcirculation, especially the smaller vessels, the red cells and plasma clearly form separate phases.

In the next sections, we develop the equations of motion for a homogeneous viscous fluid and apply them to flow in the large vessels.

Figure 4.2a shows a pair of large parallel plates of length, *L*, and width, *W*, separated by a small distance, *d*. The top plate is fixed and the bottom

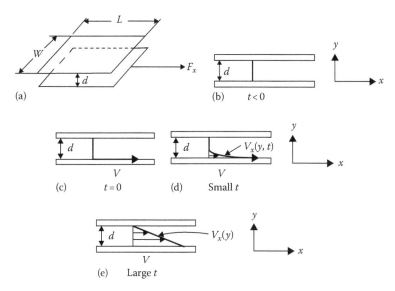

FIGURE 4.2
Parallel plate laminar flow. (a) Dimensions of parallel plates. $L,W \gg d$. F_x = Force on plates in the x-direction (b) Velocity in the y direction at $t < 0$ (c) At $t = 0$, the force, F_x pulls the bottom plate in the x-direction with a constant velocity, *V*. The top plate remains fixed. (d) X-Velocity profile as a function of y and t, $V_x(y, t)$ at small values of time (unsteady state). (e) Steady state X-velocity profile (only a function of y) at long times.

plate can be pulled in the x-direction. The space between the plates is filled with a viscous fluid. At $t = 0$, the velocity of the fluid is 0 everywhere in the space between the plates (Figure 4.2b). At $t = 0$, we pull the lower plate in the x-direction with a constant force, F_x (Figure 4.2c), which causes the lower plate to move at a velocity, V. We assume that the fluid nearest to the wall of the plate clings to the wall and moves with the same velocity as the wall. This is called the "no-slip at the wall" boundary condition. For small values of time, a transient condition occurs and the velocity profile appears as in Figure 4.2d. At large values of time, a steady-state velocity profile ensues as in Figure 4.2e. It has been experimentally determined that, at steady state, the applied force per unit area, F_x/A, is proportional to the velocity, V, and inversely proportional to the distance, d. We can write this as

$$\frac{F}{A} \propto \frac{V}{d} \tag{4.3}$$

where $A = LW$ (the surface area parallel to the flowing fluid).

A force per unit area is called *stress*. In this case, the force is applied parallel to the surface of the plates so that the stress is a *shear* stress and is denoted by the symbol τ_{yx}. In this notation, the second subscript refers to the orientation of the force (the x-direction) and the first subscript refers to the direction perpendicular (normal) to the surface on which the shear force acts. In this case, the shear force acts on a surface that is oriented in the x–z-direction. The normal to that surface is the y-direction. We also note that the top plate is moving with a velocity of zero, so that the difference in velocity between the top and bottom plates is (top − bottom) $0 - V$. If we choose the origin of the y-axis to be at the bottom plate, the difference in height between the top and bottom plates is $d - 0$. Relationship (4.3) can now be written as

$$\tau_{yx} \propto -\left(\frac{0-V}{d-0}\right) \tag{4.4}$$

The negative sign has been inserted so that relationships (4.3) and (4.4) are equal. In order to make an equation out of relationship (4.4), we need to multiply the right-hand side by a proportionality factor. We will call this proportionality factor, μ. Relationship (4.4) now becomes

$$\tau_{yx} = -\mu\left(\frac{0-V}{d-0}\right) \tag{4.5}$$

We now take the limit as $d \to 0$ and recognize that the terms in brackets on the right-hand side of Equation 4.5 satisfy the definition of the derivative of the x-component of velocity with respect to y:

$$\tau_{yx} = -\mu\left(\frac{dv_x}{dy}\right) \tag{4.6}$$

Equation 4.6 is a one-dimensional (1-D) representation of a more general constitutive equation known as *Newton's law of viscosity*. Fluids that obey the three-dimensional (3-D) generalization of this equation over their entire flow regimes are called *Newtonian fluids*. Other constitutive equations are available for fluids, which *do not* obey Newton's law of viscosity (*non-Newtonian* fluids). We shall discuss several of these (a *power-law* fluid, a *Casson* fluid, and a *Bingham* fluid) in the next section.

Equation 4.6 can be interpreted on several levels. On one level, we can use this equation as a definition of viscosity. That is, the viscosity of a Newtonian fluid is a phenomenological coefficient which when multiplied by the negative of its observed velocity gradient equals the applied shear stress. On another level, we can view Equation 4.6 as an example of a linear rate law (see Equation 2.22). In that case, viscosity is interpreted as a material parameter, that is, a physical property of the fluid. It is this interpretation of viscosity that is most useful.

Units of viscosity
Equation 4.6 requires that the units on the right-hand side must be equal to the units on the left-hand side. The units on the right-hand side are stress (force/area) while the units on the left-hand side are velocity gradient (velocity/length). The units of viscosity must be such that the units on both sides are equal. In the CGS system, force/area can be represented as dynes/cm^2. Velocity gradient has units of cm/s cm. From Equation 4.6, **viscosity** (μ) must have **units** of

$$\text{Dynes/cm}^2 = (\text{dynes s/cm}^2)(\text{cm/s cm}) \tag{4.6a}$$

Since a dyne is a g cm/s^2 force, the units of viscosity are often written as *g/cm s*. One g/cm s is called a *poise* in honor of Jean L. M. Poiseuille (1799–1869), a French physician who was one of the pioneers in the quantitative study of blood flow in tubes. However, the poise yields a rather large number for ordinary fluids, so viscosities are usually reported in *centipoise*. One centipoise (cps) = 10^{-2} poise. To provide some comparison, the viscosity of water at 25°C and 1 atm is about 1 cps. The viscosity of blood plasma at 37°C is about 1.2–1.5 cps and the viscosity of whole blood at 45% (normal) hematocrit and 37°C is 3.5–4.5 cps.

4.4 Laminar Flow and the Viscosity of Blood

Rheology is the study of the mechanical properties of fluids. Fluids, including blood, are classified according to their flow, or rheological, behavior when a continuous force is applied to the fluid.

In the experiment described above, the fluid between the plates moves in *laminar flow*. That is, the fluid moves in thin sheet-like layers (called *lamellae*) that slide over one another in an orderly fashion. A laminar layer of fluid moving at a higher velocity exerts a shear force on its nearest neighbor moving at a lower velocity and, in turn, is pulled forward by a shear force exerted by its nearest neighbor moving at an even higher velocity. Since each such shear force generates an equal and *opposite* force, fluid layers moving at a higher velocity experience a reactive force from adjacent fluid layers moving at a lower velocity. This reactive force acts to retard the forward movement of the higher velocity layer. The magnitude of the retarding force depends on the *intermolecular attractive* force of the fluid molecules for each other. When we measure the force required to achieve a given velocity gradient for a particular fluid in a fixed geometry, the ratio of these two measurements is a measure of the intermolecular attractive forces in the fluid. We define this as the *viscosity*. Thus, viscosity plays the same role in fluids that the coefficient of sliding friction plays in solids. Since the viscosity of a fluid depends on intermolecular attractive forces, it is a physical property of the fluid.

If a fluid obeys Equation 4.6, we can plot the applied shear stress on the *y*-axis and the negative of the measured velocity gradient (or strain rate) on the *x*-axis and obtain a straight line whose slope is the (Newtonian) viscosity.

Different fluids exhibit different types of rheological behavior. Figure 4.3 shows some possible responses. Fluid A represents a Newtonian fluid. Fluids B and C both represent different non-Newtonian fluid behavior. In Fluid B, the viscosity (defined as the slope of the curve at any point) decreases with increasing shear stress. This occurs in fluids containing dissolved polymers, which can form microstructures at low shear rates (e.g., plasma proteins and fibrinogen) that are then broken down at the higher shear rates. Curve C is a non-Newtonian fluid with a *yield stress*, τ_0. A yield stress is often exhibited by

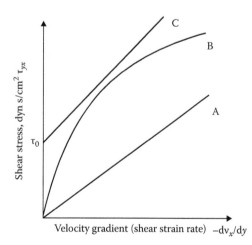

FIGURE 4.3
Rheological behavior of fluids.

heterogeneous fluids containing a particulate phase, which aggregates into some mechanical structure at low shear rates. At some finite shear stress, τ_0, the aggregate is broken down and the fluid starts to flow. Normal whole blood is a heterogeneous fluid and red cells are capable of aggregation at low shear rates. If the fluid approximates Newtonian behavior at shear stresses greater than τ_0, it is called a *Bingham plastic*.

Although whole blood exhibits a yield stress at low shear rates, the value is quite small (~0.05 dynes/cm²). This means that a pressure difference of less than 0.02 mm Hg would be sufficient to maintain flow in a capillary of 50 µm in diameter and 0.5 cm in length. However, in a capillary bed equivalent to 500 such vessels in series, if the flow were suddenly stopped by local vaso-constriction of the terminal arterial supplying that bed, approximately 10 mm pressure drop would be required to restart the flow. It is extremely difficult to measure the yield stress of blood, since it is affected by factors such as the concentration of anticoagulant, the concentration of several large proteins (fibrinogen, albumin), the hematocrit, and the wide variation in properties from specimen to specimen.

Because of the complex nature of blood, various constitutive relationships (rheological models) have been used to describe its stress–strain behavior over a wide range of red cell concentrations and shear rates. Each of these models have advantages and disadvantages when used over different ranges of shear stress and red cell concentration. The most popular models are described in the following.

4.4.1 The Casson Model

The Casson equation is the most popular model for representation of the non-Newtonian behavior of blood over a wide range of red cell concentrations and shear rates of 1–100,000 s⁻¹. It is particularly useful in the low shear-rate region for modeling the yield stress:

$$\tau^{1/2} = \eta^{1/2}\dot{\gamma}^{1/2} + \tau_y^{1/2} \tag{4.7}$$

where
  τ is the shear stress
  τ_y is the yield stress
  $\dot{\gamma}$ is the shear rate
  η is the Casson or "apparent" viscosity

Figure 4.4a shows the changes in viscosity as red cells are added to plasma. The viscosity of whole blood increases in a nonlinear way with increasing *hematocrit (Hct)*. Therefore, whole blood, over a large range of hematocrits, behaves as a non-Newtonian fluid. Figure 4.4b shows the presence of a yield stress at low shear rates.

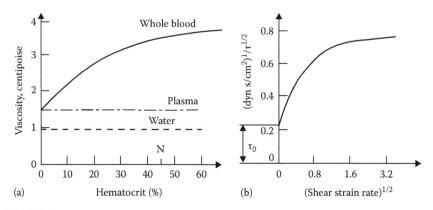

FIGURE 4.4
(a) Blood behaves as a non-Newtonian fluid over a wide range of hematocrits. (b) At low shear rates, whole blood shows a small, but finite, yield stress due to red cell aggregation.

The normal range of Hct is 35%–45%, with a mean value of 40%. Figure 4.4 shows that blood can be approximated as a Newtonian fluid over the normal hematocrit range. This is not true for severely *anemic* patients (Hct < 30) or for patients with blood diseases that appreciably increase Hct, such as *polycythemia*.

4.4.2 The Power-Law Fluid Model

The power-law model is an empirical relationship between shear stress and strain rate, which simulates the rheological behavior of blood over a wide range of hematocrits. It is not applicable at very low shear rates.

$$\tau_{rz} = -m \left| \frac{dv_z}{dr} \right|^{n-1} \frac{dv_z}{dr} \tag{4.8}$$

The power-law equation reduces to Newton's law of viscosity with $n = 1$ and $m = \mu$. The deviation of n from unity is thus a measure of the deviation of the fluid from Newtonian behavior. For values of $n < 1$, viscosity decreases with increasing shear rate and this behavior is termed *pseudoplastic*. Blood behaves as a pseudoplastic fluid over a large range of shear rates.

4.4.3 The Bingham Plastic Model

$$\tau_{rz} = -\mu_0 \frac{dv_z}{dr} \pm \tau_0 \quad \text{if } |\tau_{rz}| > \tau_0 \tag{4.9a}$$

$$\frac{dv_z}{dr} = 0 \quad \text{if } |\tau_{rz}| < \tau_0 \tag{4.9b}$$

The positive sign is used when τ_{rz} is positive and the negative sign is used when τ_{rz} is negative. A Bingham plastic remains rigid when the shear stress is less than the yield stress, τ_0, but behaves somewhat like a Newtonian fluid when the shear stress is greater than τ_0.

4.5 Flow in a Tube

In the laminar flow experiment described in Section 4.3, it is clear that the x-component of velocity is changing with position between the bottom and top plates (in the y-direction). Since blood is an *incompressible fluid* (ρ does not change with pressure), the x-component of the mass flux (i.e., the product of fluid density, ρ, and x-component of velocity, v_x), is also changing in the y-direction. This also means that the *momentum flux* of the fluid layers is changing in the y-direction. Remember that momentum flux is the product of mass flux (ρV_x) and velocity (V_x). From the description of the velocity distribution between the plates, we can say that the fluid layers *lose their forward momentum* (x-direction) as we move from the bottom plate to the top plate (in the y-direction). In this thought "experiment," the fluid layers at the higher velocities have greater x-momentum than the fluid layers at the lower velocities. In other words, momentum is being transported from the faster-moving fluid to the slower-moving fluid and is being dissipated by the forces' necessity to overcome the intermolecular attractive forces between adjacent fluid layers (the viscosity). This molecular mechanism is one of the two mechanisms by which momentum is transported in a fluid. The other mechanism involves changes in ρ with pressure in a *compressible fluid*. For liquids at ordinary pressures, this second mechanism can be neglected.

In a footnote to Table 2.3, we showed that shear stress has the same dimensional units as momentum flux. In fact, shear stresses and normal stresses in a flowing fluid are direct measures of momentum flux.

We can now proceed to analyze the steady laminar flow of a viscous fluid in a rigid tube using a momentum balance.

4.5.1 The Hagan–Poiseuille Equation (Poiseuille's Law)

Figure 4.5 is a schematic representation of steady laminar flow of a fluid of constant density in a rigid tube. We use cylindrical coordinates since they are the most convenient to describe the position in a cylindrical tube. To avoid consideration of entrance and exit effects, we specify that the tube length, L, is large compared with its radius, R. We position the tube in the horizontal plane so as to neglect gravitational forces.*

---

* If the tube is oriented in the vertical direction, we add a gravitational force, ρg_z to the momentum balance.

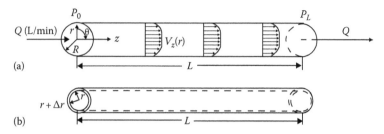

FIGURE 4.5
(a) Momentum balance in steady, laminar flow in a tube. (b) A cylindrical shell of fluid at position r and thickness Δr, in the middle of the following fluid, is the control volume for the momentum balance.

4.5.1.1 Visualize the Flow

We first attempt to visualize the flow pattern. Since $P_0 > P_L$, the bulk flow is in the z-direction, the fluid being forced through the tube by the pressure difference $(P_0 - P_L)$. We also recognize that the fluid at the walls moves with the velocity of the walls (the "no-slip at the wall" condition). Since the tube is rigid and stationary, the fluid at $r = \pm R$ is moving at zero velocity in the z-direction, while the fluid in the center of the tube $(r = 0)$ is moving with the highest velocity in the z-direction. That is, $V_z = f(r)$. We also recognize that with an incompressible fluid such as blood, the rigid walls do not allow a velocity component in the r-direction. Since the tube is not rotating, we can also eliminate any consideration of a velocity component in the θ direction.

We consider an imaginary cylindrical shell of thickness Δr and length, L, somewhere in the middle of the flowing fluid in Figure 4.5 as our control volume. A steady state *is defined* as any time period during which all accumulations (mass, energy, etc.) within the control volume are zero. At steady state, the general balance equation (see Chapter 2) for the z-component of momentum can be written as

$$\begin{matrix} \text{rate of momentum} \\ \text{into shell} \end{matrix} - \begin{matrix} \text{rate of momentum} \\ \text{out of shell} \end{matrix} + \begin{matrix} \text{sum of external forces} \\ \text{acting on shell} \end{matrix} = 0$$

The only external force we will consider is the pressure force acting on the shell surfaces at $z = 0$ and L.

We now list the contributions to the momentum balance in the z-direction:
Rate of momentum* into shell across cylindrical surface at position r:

$$\left(2\pi r L \tau_{rz}\right)\big|_r \tag{4.10a}$$

---

* The notation $(2\pi r L \tau_{rz})|_x$ is read as the quantity enclosed in parentheses $(2\pi r L \tau_{rz})$ evaluated at position x.

Rate of momentum out of shell cylindrical surface at position $r + \Delta r$:

$$\left(2\pi r L \tau_{rz}\right)\big|_{r+\Delta r} \tag{4.10b}$$

Rate of momentum into shell across annular surface* at $z = 0$:

$$\left(2\pi r \Delta r V_z\right)\left(\rho V_z\right)\big|_{z=0} \tag{4.10c}$$

Rate of momentum out across annular surface at $z = L$:

$$\left(2\pi r \Delta r V_z\right)\left(\rho V_z\right)\big|_{z=L} \tag{4.10d}$$

Pressure force acting on the annular surface at $z = 0$:

$$\left(2\pi r \Delta r\right)P_0 \tag{4.10e}$$

Pressure force acting on the annular surface at $z = L$:

$$-\left(2\pi r \Delta r\right)P_L \tag{4.10f}$$

By convention, we take "in" and "out" to be in the positive direction of the axes.

We now add up all the contributions to the z-component of the momentum balance:

$$\left(2\pi r L \tau_{rz}\right)\big|_{r} - \left(2\pi r L \tau_{rz}\right)\big|_{r+\Delta r} + \left(2\pi r \Delta r V_z\right)\left(\rho V_z\right)\big|_{z=0}$$

$$-\left(2\pi r \Delta r V_z\right)\left(\rho V_z\right)\big|_{z=L} + \left(2\pi r \Delta r\right)P_0 - \left(2\pi r \Delta r\right)P_L \tag{4.10g}$$

We observe that the product $(2\pi r \Delta r)(\rho V_z)$ is the mass flow (g/s) of any differential laminar layer (at any r-position) at any z-position. At steady state, mass flow must remain constant within the control volume. For an incompressible fluid, ρ does not change with pressure (e.g., z-position). For the product ρV_z to remain constant at any r-position, V_z also cannot change with z-position. The third and fourth terms in Equation 4.10g then add up to zero. We now divide

---

* If we look down at our control volume shell, the fluid is flowing through an annular region composed of two concentric circles. The annular area between these circles is: $A = \pi(r + \Delta r)^2 - \pi r^2 = \pi(r^2 + 2r\Delta r + \Delta r^2) - \pi r^2 = \pi(2r\Delta r + \Delta r^2)$. If we now assume that $r \gg \Delta r$, we can neglect Δr^2, and $A \approx 2\pi r \Delta r$.

Equation 4.10g by $2\pi L\Delta r$ and take the limit as Δr (the thickness of the control volume) goes to zero. This gives

$$\lim_{\Delta r \to 0}\left(\frac{(r\tau_{rz})|_{r+\Delta r} - (r\tau_{rz})|_r}{\Delta r}\right) = \left(\frac{P_0 - P_L}{L}\right)r \tag{4.11}$$

We recognize that the left-hand side of this equation is just the definition of the first derivative of $(r\tau_{rz})$ with respect to r. Equation 4.11 then becomes

$$\frac{d}{dr}(r\tau_{rz}) = \left(\frac{P_0 - P_L}{L}\right)r \tag{4.12}$$

We can now integrate Equation 4.12 with respect to r and divide both sides by r, to give

$$\tau_{rz} = \left(\frac{P_0 - P_L}{2L}\right)r + \frac{C_1}{r} \tag{4.13}$$

where C_1 is a constant of integration. To evaluate C_1, we argue on physical grounds that the momentum flux, τ_{rz}, cannot be infinite at $r = 0$. This means that C_1 must be zero. Then Equation 4.13 becomes

$$\tau_{rz} = \left(\frac{P_0 - P_L}{2L}\right)r \tag{4.14}$$

Equation 4.14 tells us that the momentum flux, τ_{rz}, is a linear function of radial position in a tube. The maximum momentum flux occurs at the walls (i.e., at $r = \pm R$), and the momentum flux is zero at the tube center (i.e., at $r = 0$). This is illustrated in Figure 4.6.

The important point here is that the maximum shear stress occurs at the wall and is given by the relationship

$$\tau_{rz,max} = \left(\frac{P_0 - P_L}{2L}\right)R \tag{4.14a}$$

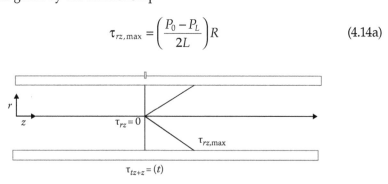

FIGURE 4.6
Stress distribution in a tube showing linear relationship between τ_{rz} and radial position.

At this point, the derivation applies to any fluid in steady, laminar flow. If we now assume that we are approximating blood by a Newtonian fluid, we can substitute Newton's law of viscosity (e.g., Equation 4.6) for τ_{rz}. If we are approximating blood rheological properties using one of the non-Newtonian constitutive equations, we would substitute the appropriate relationship between shear stress (momentum flux) and velocity gradient for τ_{rz}. For a Newtonian fluid, the constitutive equation is

$$\tau_{rz} = -\mu \frac{dV_z}{dr} \tag{4.15}$$

Using 4.15 in 4.14 gives

$$\frac{dV_z}{dr} = -\left(\frac{P_0 - P_L}{2\mu L} \right) r \tag{4.16}$$

We can now separate the variables and integrate with respect to r:

$$V_z = -\left(\frac{P_0 - P_L}{4\mu L} \right) r^2 + C_2 \tag{4.17}$$

where C_2 is a constant of integration. We evaluate this constant using the "no-slip at the wall" boundary condition. That is, at $r = R$, $V_z = 0$. This gives

$$C_2 = \frac{(P_0 - P_L)R^2}{4\mu L} \tag{4.18}$$

Using Equation 4.18, the velocity distribution becomes

$$V_z = \left(\frac{P_0 - P_L}{4\mu L} \right)(R^2 - r^2) = \frac{(P_0 - P_L)R^2}{4\mu L}\left[1 - \left(\frac{r}{R} \right)^2 \right] \tag{4.19}$$

Equation 4.19 is the equation of a parabola. This tells us that when there is steady, incompressible, laminar flow in a rigid tube the fluid has a parabolic velocity profile.

The maximum velocity occurs at the center of the tube (at $r = 0$) and is given by

$$V_{z,max} = \frac{(P_0 - P_L)R^2}{4\mu L} \tag{4.20}$$

Figure 4.7 is a schematic representation of the velocity distribution.

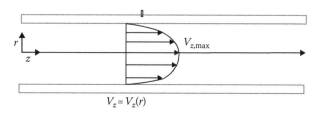

FIGURE 4.7
Velocity profile for laminar flow in a tube. The coordinate axes are chosen so that the center of the tube is $r = 0$ and the left position is $z = 0$. Since $V_z \neq V_z(z)$, by continuity, the same parabolic velocity profile exists at all z positions along the tube.

The integrated average velocity, $\langle V_z \rangle$, is calculated by integrating all the velocities over the cross section and dividing by the cross-sectional area.

$$\langle V_z \rangle = \frac{\int_0^{2\pi} \int_0^R V_z r \, dr \, d\theta}{\int_0^{2\pi} \int_0^R r \, dr \, d\theta} \tag{4.21}$$

where V_z is given by Equation 4.19, and

$$\int_0^{2\pi} \int_0^R r \, dr \, d\theta = \left(\int_0^R r \, dr \right) [\theta]_0^{2\pi} = \left(\int_0^R r \, dr \right)(2\pi) = \left(\frac{r^2}{2} \right)_0^R (2\pi) = \pi R^2$$

$$\int_0^{2\pi} \int_0^R V_z r \, dr \, d\theta = \left(\int_0^R V_z r \, dr \right)[\theta]_0^{2\pi} = 2\pi \int_0^R V_z r \, dr$$

Using Equation 4.19 for V_z, we can write

$$2\pi \int_0^R V_z r \, dr = \frac{2\pi(P_0 - P_L)}{4\mu L} \int_0^R (R^2 - r^2) r \, dr$$

$$= \frac{2\pi(P_0 - P_L)}{4\mu L} \left[R^2 \int_0^R r \, dr - \int_0^R r^3 \, dr \right]$$

Evaluating the two definite integrals gives

$$R^2 \int_0^R r\, dr - \int_0^R r^3\, dr = R^2 \left[\frac{r^2}{2} \right]_0^R - \left[\frac{r^4}{4} \right]_0^R$$

$$= \frac{R^4}{2} - \frac{R^4}{4} = \frac{R^4}{4}$$

Substituting these results back into Equation 4.21 gives

$$\langle V_z \rangle = \frac{\left[\dfrac{2\pi(P_0 - P_L)}{4\mu L} \right]}{\pi R^2} \left[\frac{R^4}{4} \right] = \frac{(P_0 - P_L)R^2}{8\mu L} \tag{4.22}$$

Notice that $\langle V_z \rangle$ is just half of $V_{z,\max}$.

The volumetric rate of flow, Q (L/min), is the product of the average velocity and the cross-sectional area available for flow. For a tube, the cross-sectional area is just πR^2, so that

$$Q = \frac{\pi R^4 (P_0 - P_L)}{8\mu L} \tag{4.23}$$

This result is called the Hagan–Poiseuille equation or Poiseuille's law, after the two scientists credited with its independent derivation at around the same time (1839–1841).

This result is very important, physiologically, for the following reasons:

1. If we compare Equation 4.23 with Ohm's law (Equation 4.1), we can identify the specific factors that contribute to the resistance to flow in a rigid tube. Comparing 4.1 with 4.23 yields

$$R_{eq} = \frac{8\mu L}{\pi R^4} \tag{4.24}$$

   This equation tells us that the resistance to laminar flow of a single rigid vessel is directly proportional to the vessel length and the viscosity of the fluid and inversely proportional to the tube radius raised to the fourth power. Additionally, Figure 4.4a shows us that the viscosity of blood is a nonlinear function of the hematocrit (volume% of red cells). The fourth power dependence means that small changes in vessel radius have large effects on resistance to flow. In fact, vasoconstriction and vasodilation of arterioles are the major mechanisms by which the CO is redistributed to tissues and organs in response to changing metabolic demands (as in exercise).

2. Even though Equation 4.23 was derived for laminar flow in rigid tubes, it has been demonstrated experimentally that the fourth power inverse relationship between vessel resistance and vessel radius holds in vivo as well, that is, in distensible vessels. In vivo, the factor $(8/\pi)$ no longer applies.

4.6 General Forms of the Equations of Motion

While a shell balance is useful in analyzing flow problems with simple geometries, it gets rather complex and messy when dealing with complex shapes, 3-D flow and non-steady-state problems. A more useful approach is to derive the general equation of continuity (non-steady-state mass balance) and equations of motion (non-steady-state momentum balance) for three dimensions. We can then use these as starting points for the analysis of any complex flow problem. The equations are derived in a rectangular coordinate system. However, for many problems, cylindrical coordinates (flow in large vessels) or spherical coordinates (left ventricular stress distributions) are more appropriate. In Appendix B, we summarize many of the important relationships in all three coordinate systems. This makes the process of setting up and solving flow problems a relatively straight forward procedure of eliminating those terms in the equations that do not apply to the problem of interest and solving the resulting set of simplified equations using appropriate boundary and initial conditions.

In this section, we use vector and tensor notation, primarily to shorten very lengthy mathematical results and to focus on the interpretation of the equations rather than the mathematics. However, the reader will find that a detailed knowledge of vector and tensor mathematics is not required to utilize the equations in Appendix B, to solve problems.

In the last part of this section we transform the equations of change into dimensionless form. This allows us to focus on a small number of dimensionless parameters that characterize fluid systems. In particular, we see that the *Reynolds* number, a dimensionless parameter that allows us to compare flow fields in different size vessels, appears naturally in the nondimensional form of the equations of motion. The value of the Reynolds number is used with an empirical rule to distinguish between laminar flow (a nice, orderly flow field) and turbulent flow (a more chaotic flow field). This is important, since the equations of motion we derived only apply to laminar flow. There is *no generally accepted theory of liquid turbulence*. Hence, all problems in turbulent flow must be solved using empirical correlations. These correlations are derived from experimental data taken on a system that is dimensionally similar to the one you are studying.

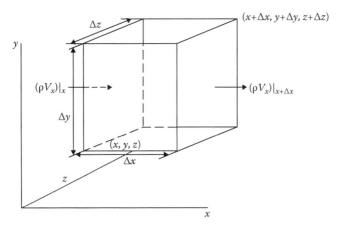

FIGURE 4.8
Differential control volume (Δx, Δy, Δz) in the middle of a 3-D flowing fluid.

4.6.1 The Equation of Continuity

To derive the equation of continuity, we write a mass balance over a differential volume element (Δx, Δy, Δz) somewhere in the middle of a 3-D flowing fluid (see Figure 4.8):

$$
\left\{
\begin{array}{c}
\text{rate of mass} \\
\text{accumulation in} \\
\text{volume element}
\end{array}
\right\}
=
\left\{
\begin{array}{c}
\text{rate of mass} \\
\text{into volume} \\
\text{element}
\end{array}
\right\}
-
\left\{
\begin{array}{c}
\text{rate of mass} \\
\text{out of volume} \\
\text{element}
\end{array}
\right\}
$$

The rate of mass across the face $\Delta y \Delta z$ at position x is $(\rho V_x)|_x \Delta y \Delta z$. The rate of mass leaving the face $\Delta y \Delta z$ at position $x + \Delta x$ is $(\rho V_x)|_{x+\Delta x} \Delta y \Delta z$. Similar expressions can be written for the other two pairs of faces ($\Delta x \Delta z$ and $\Delta x \Delta y$). The rate of accumulation of mass within the volume element is $\Delta x \Delta y \Delta z (\partial \rho / \partial t)$. Substituting the mass rates in and out for each of the faces and the accumulation rate into the differential balance equation (above), dividing the entire equation by the volume, $\Delta x \Delta y \Delta z$, and taking the limit as Δx, Δy, and Δz approach zero, gives

$$
\frac{\partial \rho}{\partial t} = -\left[\frac{\partial}{\partial x} \rho V_x + \frac{\partial}{\partial y} \rho V_y + \frac{\partial}{\partial z} \rho V_z \right]
\tag{4.25}
$$

This is the equation of continuity (conservation of mass). It describes the rate of change of density of a flowing fluid at a fixed point in space that results from changes in the mass velocity vector, $\rho \mathbf{V}$. In vector notation, Equation 4.25 can be written as

$$
\frac{\partial \rho}{\partial t} = -(\nabla \cdot \rho \mathbf{V})
\tag{4.25a}
$$

Here ∇ is a vector quantity called the "divergence" and has units of reciprocal length. **V** is the velocity vector. The components of ρ**V** (the mass flux) in Cartesian (rectangular) coordinates are given in Equation 4.25. The components of $\nabla\rho$**V** in Cartesian, cylindrical, and spherical coordinates are given in Appendix B. Physically, the divergence of the mass flux is just the net rate of mass *outflow* per unit volume. Equation 4.25a states that the rate of increase of the density of a fluid within a small volume element fixed in space equals the net rate of mass *inflow* (hence the negative sign) to the volume element per unit volume.

The general form of the continuity equation (4.25a), applies in those circumstances where the fluid or material is compressible, as in gas flow in the respiratory system or where there is remodeling of tissue due to a sustained increase in stress as in the left ventricle or aorta during volume overload or pressure overload hypertension. Otherwise, we use a special form of the continuity equation which applies to fluids whose density does not change with pressure (*incompressible fluids*), at least over the pressure ranges encountered physiologically. For constant density fluids (blood, cerebrospinal fluid, and sinovial fluid),

$$\frac{\partial \rho}{\partial t} = 0,$$

and the continuity equation becomes

$$(\nabla \cdot \mathbf{V}) = 0. \tag{4.26}$$

Equation 4.26 is not as restrictive as first appears. In order for Equation 4.26 to apply, it is only necessary that ρ be constant for a fluid element as it moves along a streamline.

4.6.2 The Equations of Motion

To derive the equations of motion, we again use a differential volume element $\Delta x\Delta y\Delta z$, somewhere in the middle of a 3-D flowing fluid. The fluid can move through all six faces of the volume element in any arbitrary direction. The non-steady-state momentum balance can be written as

$$\begin{Bmatrix} \text{rate of} \\ \text{momentum} \\ \text{accumulation} \\ \text{in volume element} \end{Bmatrix} = \begin{Bmatrix} \text{rate of} \\ \text{momentum} \\ \text{into volume} \\ \text{element} \end{Bmatrix} - \begin{Bmatrix} \text{rate of} \\ \text{momentum} \\ \text{out of volume} \\ \text{element} \end{Bmatrix} + \begin{Bmatrix} \text{sum of forces} \\ \text{acting on} \\ \text{system} \end{Bmatrix}.$$

We recognize that since we have 3-D flow, this is a *vector* equation with components in the x-, y-, and z-directions.

The momentum balance is similar to the one that was used in a previous section to derive the Hagan–Poiseuille equation, except now we put in an accumulation term to allow for *unsteady-state* behavior. Again, we have to account for the momentum transfer due to two mechanisms:

a. At the molecular level, by viscous transport (velocity gradient) among adjacent laminar layers, τ_{ij}.
b. At the macroscopic level, by convective transport (bulk flow of fluid), since momentum flux is mass flux times velocity, $(\rho \mathbf{V}) \cdot \mathbf{V}$.

We begin by looking at the contributions to the momentum balance equation in the x-direction (see Figure 4.9). The y- and z-components are done in the same way.

The rate at which the x-component of momentum enters the plane $\Delta y \Delta z$ at position x by convection is

$$\rho V_x V_x |_x \, \Delta y \Delta z.$$

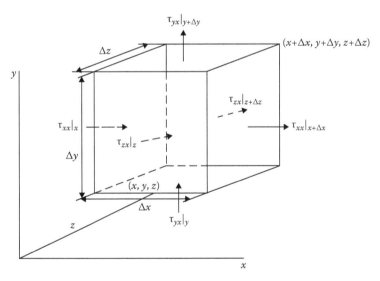

FIGURE 4.9
Control volume ($\Delta x \Delta y \Delta z$) in the middle of a 3-D flowing fluid. The arrows point in the direction that the x-component of momentum is transported through the differential surfaces. There are three components of the stress *tensor** in each of the x-, y-, and z-directions.

---

* Physical quantities can be placed in three distinct mathematical categories: scalars such as temperature, volume, and time that have only magnitude and not direction; vectors such as velocity, momentum, acceleration, and force, that have both magnitude and three possible directions; and second-order tensors, such as the shear stress and momentum flux tensors that have nine components, three components of magnitude in each of the three possible directions.

The rate at which it leaves the plane at position $x + \Delta x$ is

$$\rho V_x V_x \big|_{x+\Delta x} \Delta y \Delta z$$

The rate at which the x-component of momentum enters the plane $\Delta x \Delta z$ at position y by convection is

$$\rho V_y V_x \big|_y \Delta x \Delta z$$

The rate at which it leaves the plane at position $y + \Delta y$ is

$$\rho V_y V_x \big|_{y+\Delta y} \Delta x \Delta z$$

The rate at which the x-component of momentum enters the plane $\Delta x \Delta y$ at position z by convection is

$$\rho V_z V_x \big|_z \Delta x \Delta y$$

The rate at which it leaves the plane at position $z + \Delta z$ is

$$\rho V_z V_x \big|_{z+\Delta z} \Delta x \Delta y$$

In addition, the rate at which the x-component of momentum enters the plane $\Delta y \Delta z$ at position x by molecular transport is

$$\tau_{xx} \big|_x \Delta y \Delta z$$

The rate at which it leaves the plane at position $x + \Delta x$ is

$$\tau_{xx} \big|_{x+\Delta x} \Delta y \Delta z$$

The rate at which the x-component of momentum enters the plane $\Delta x \Delta z$ at position y by molecular transport is

$$\tau_{yx} \big|_y \Delta x \Delta z$$

The rate at which it leaves the plane at position $y + \Delta y$ is

$$\tau_{yx} \big|_{y+\Delta y} \Delta x \Delta z$$

The rate at which the x-component of momentum enters the plane $\Delta x \Delta y$ at position z by molecular transport is

$$\tau_{zx} \big|_z \Delta x \Delta y$$

The rate at which it leaves the plane at position $z + \Delta z$ is

$$\tau_{zx}\big|_{z+\Delta z} \Delta x \Delta y$$

These momentum fluxes represent stresses. τ_{xx} represents the *normal (perpendicular)* stress on the x-plane. τ_{yx} represents the *shear (or tangential) stress* resulting from the viscous force in the x-direction on the y-plane. Similarly, τ_{zx} is also a shear (or tangential) stress resulting from the viscous force in the x-direction on the z-plane.

The major forces acting on a flowing fluid are *pressure forces, which act at surfaces and gravity, which acts at the center of mass.* The x-components of the pressure force are

at position x on the plane $\Delta y \Delta z$:

$$p\big|_x \Delta y \Delta z$$

at position $x + \Delta x$ on the same plane:

$$p\big|_{x+\Delta x} \Delta y \Delta z$$

The x-component of the gravitational force that acts on the control volume, $\Delta x \Delta y \Delta z$:

$$g_x \Delta x \Delta y \Delta z$$

where g_x is the x-component of the gravitational force per unit mass, g. Pressure in a flowing fluid is defined by an equation of state $p = p(\rho, T)$, where ρ is the fluid density, and T is the absolute temperature (°K).

The rate of accumulation of x-momentum in the control volume is

$$\Delta x \Delta y \Delta z \frac{\partial (\rho V_x)}{\partial t}$$

We can now substitute these expressions into the momentum balance equation, divide the resulting expression by $\Delta x \Delta y \Delta z$, and then take the limit as the volume $\Delta x \Delta y \Delta z$ approaches zero to give us the differential equation describing the x-component of the equation of motion in rectangular coordinates as

$$\frac{\partial}{\partial t}(\rho V_z) = -\left(\frac{\partial}{\partial x} \rho V_x V_x + \frac{\partial}{\partial y} \rho V_y V_x + \frac{\partial}{\partial z} \rho V_z V_x \right)$$

$$-\left(\frac{\partial}{\partial x} \tau_{xx} + \frac{\partial}{\partial y} \tau_{yx} + \frac{\partial}{\partial z} \tau_{zx} \right) - \frac{\partial p}{\partial x} + \rho g_x \qquad (4.27)$$

The y- and z-components can be obtained by similar momentum balances in those directions over the differential control volume. These are listed in Appendix B.

The terms ρV_x, ρV_y, and ρV_z are the three components of the mass velocity vector $\rho\mathbf{V}$. The terms $\partial/\partial x$, $\partial/\partial y$, and $\partial/\partial z$ are the three components of a vector ∇p, called the gradient of pressure. The terms $\rho V_x V_x$, $\rho V_x V_y$, $\rho V_x V_z$, $\rho V_y V_z$, etc., are the nine components of the convective momentum flux $\rho\mathbf{VV}$, which come about by taking the dot product of $\rho\mathbf{V}$ and $\mathbf{V}$. The terms τ_{xx}, τ_{yx}, τ_{zx}, τ_{yz}, etc., are the nine components of the stress tensor, τ.

In vector notation, the equations of motion can be written as

$$\frac{\partial}{\partial t}(\rho\mathbf{V}) \quad = \qquad - \quad [\nabla\cdot\rho\mathbf{VV}] \qquad -[\nabla\cdot\tau]$$

rate of increase of momentum per unit volume	rate of momentum gain by convection per unit volume	rate of momentum gain by viscous forces per unit volume

$$-\nabla p \qquad\qquad + \quad \rho g. \qquad\qquad (4.28)$$

	pressure force on fluid element per unit volume	gravitational force on fluid element per unit volume

Equation 4.28 describes the rate of momentum gain per unit volume for a control volume *fixed in space* somewhere in the middle of the flowing fluid. The three components of the equations of motion are given in *Appendix B* for Cartesian (x, y, z), cylindrical (r, θ, z), and spherical (r, θ, ϕ) coordinate systems.

With the help of the continuity equation, Equation 4.28 can be rearranged to reflect the changes in momentum experienced by a control volume *moving with the velocity of the fluid*. In that case, the equations of motion become identical to Newton's second law for the flowing fluid (mass × acceleration = sum of the forces). The equations state that a small control volume moving with the fluid is accelerated because of the forces acting on it. The forces are those due to pressure (surface forces), gravity (body forces), and viscous drag (normal and shear stresses).

It should be kept in mind that Equations 4.25a and 4.28 *are perfectly general; they apply to any continuous medium*. We will encounter a different form of the equations of motion when we consider the motion of the wall of a distensible vessel in response to a pulsatile pressure gradient. In that case, we are interested in displacements (changes in vessel diameter) rather than velocity.

In order to use Equation 4.28 to solve velocity distributions, we must supply constitutive relationships between the various stresses and their velocity gradients and fluid properties. A set of constitutive relationships for the nine components of the stress tensor that applies to complex, 3-D flow of Newtonian fluids is

$$\tau_{xx} = -2\mu \frac{\partial V_x}{\partial x} + \frac{2}{3}\mu(\nabla \cdot \mathbf{V}) \qquad (4.29a)$$

$$\tau_{yy} = -2\mu \frac{\partial V_y}{\partial y} + \frac{2}{3}\mu(\nabla \cdot \mathbf{V}) \qquad (4.29b)$$

$$\tau_{zz} = -2\mu \frac{\partial V_z}{\partial z} + \frac{2}{3}\mu(\nabla \cdot \mathbf{V}) \qquad (4.29c)$$

$$\tau_{xy} = \tau_{yx} = -\mu\left(\frac{\partial V_x}{\partial y} + \frac{\partial V_y}{\partial x}\right) \qquad (4.29d)$$

$$\tau_{yz} = \tau_{zy} = -\mu\left(\frac{\partial V_y}{\partial z} + \frac{\partial V_z}{\partial y}\right) \qquad (4.29e)$$

$$\tau_{zx} = \tau_{xz} = -\mu\left(\frac{\partial V_z}{\partial x} + \frac{\partial V_x}{\partial z}\right) \qquad (4.29f)$$

Substituting Equations 4.29a through 4.29f in Equation 4.28 gives the general equations of motion for a Newtonian fluid with varying density and viscosity.

The equations of motion and continuity in Appendix B, Equations 4.29a through 4.29f, an equation of state $p = p(\rho)$ for compressible fluids, the density dependence of viscosity $\mu = \mu(\rho)$, and the boundary and initial conditions, *completely determine the pressure, density, and velocity components in a flowing isothermal fluid.*

The complete set of equations are almost never used to set up flow problems. Instead, simplified versions of the equations are used as starting points for the analysis of particular flow problems. The two most common assumptions used are those of constant ρ (incompressibility) and constant μ (Newtonian fluid). For gas flow, $\mu = 0$ (inviscid fluid), is often assumed.

4.6.3 Constant ρ and Constant μ: The Navier–Stokes Equations

For constant ρ, the equation of continuity becomes

$$[\nabla \cdot \mathbf{V}] = 0$$

If we also assume a Newtonian fluid of constant μ the equations of motion simplify to

$$\rho \frac{D\mathbf{V}}{Dt} = -\nabla p + \mu \nabla^2 \mathbf{V} + \rho g \tag{4.30}$$

Here, $D\mathbf{V}/Dt$ is the substantial derivative (the derivative following the motion of the fluid) and, in Cartesian (rectangular) coordinates it is defined as

x-component

$$\frac{\partial V_x}{\partial t} + V_x \frac{\partial V_x}{\partial x} + V_y \frac{\partial V_x}{\partial y} + V_z \frac{\partial V_x}{\partial z}$$

y-component

$$\frac{\partial V_y}{\partial t} + V_x \frac{\partial V_y}{\partial x} + V_y \frac{\partial V_y}{\partial y} + V_z \frac{\partial V_y}{\partial z} \tag{4.31}$$

z-component

$$\frac{\partial V_z}{\partial t} + V_x \frac{\partial V_z}{\partial x} + V_y \frac{\partial V_z}{\partial y} + V_z \frac{\partial V_z}{\partial z}$$

The components of the Navier–Stokes equations of motion for a Newtonian fluid of constant ρ and μ are given in Appendix B.

4.6.4 The Euler Equations: For μ = 0 [or (∇ · τ) = 0]

For inviscid flow (very low viscosity gases), the equations of motion simplify to

$$\rho \frac{D\mathbf{V}}{Dt} = -\nabla p + \rho g \tag{4.32}$$

Use of the general equations of motion and continuity to set up flow problems

1. To set up constant-density, constant-viscosity flow problems
   a. Start with visualization of the flow problem. Determine which set of coordinates apply (e.g., *for flow in a tube use cylindrical coordinates*). Decide which flow directions will be nonzero.

b. Write down the general form of the nonzero component(s) of the equations of motion for constants ρ and μ from Appendix B.

c. Eliminate those terms that are zero (or nearly zero). Here you may have to use some judgment and perhaps make additional assumptions about the behavior of the flow field based on the physical nature of the problem you are solving.

d. Write down the equation of continuity in the coordinate system you are using from Appendix B. Eliminate the terms that are zero.

e. If possible, use the results from Step d to further simplify the equation(s) of motion obtained in Step c.

f. Integrate the resulting equation(s) using appropriate initial and boundary conditions. This will then yield the pressure and velocity distributions.

2. To set up problems with variable density and viscosity

a. Start with visualization of the flow problem. Determine which set of coordinates apply. Decide which flow directions will be nonzero.

b. Write down the general form of the nonzero component(s) of the equations of motion from Appendix B, in terms of the components of the stress tensor.

c. Pick a fluid model (Newtonian, power law). Look up* a set of constitutive relationships between shear stress and velocity gradient for this fluid model in the coordinate system that you are using. Simplify the constitutive relationships by eliminating zero velocity or zero velocity gradient terms.

d. Write down the equation of continuity in the coordinate system you are using from Appendix B. Eliminate the terms that are zero.

e. If possible, use the results from Step d to further simplify the equation(s) obtained in Steps b and c.

f. If the fluid is compressible† (a gas), obtain an equation of state for the fluid in the form: $\rho = \rho(p)$.

g. Integrate the resulting equation(s) using appropriate initial and boundary conditions. This will then yield the pressure, velocity, and density (for a compressible fluid) distributions.

---

* One-dimensional constitutive relationships for a Casson, a power-law, and a Bingham fluid are given in Equations 4.7 through 4.9. The nine components of the stress tensor for a Newtonian fluid are given in Equations 4.29 in rectangular coordinates. Complete sets of components of the stress tensor for each of these fluids in several coordinate systems may be found in textbooks of Rheology.

† Most liquids at ordinary pressures (including blood) are incompressible, so you can eliminate this step if you are considering blood flow.

4.6.4.1 A Word about Boundary Conditions

When we to carry out the integrations in steps 1.f and 2.g above, several integration constants appear. These are evaluated using "boundary conditions." Boundary conditions are mathematical statements that describe the physical situations that occur at the boundaries of the system of interest. For flow in tubes, these occur at the inlet, outlet, at the walls, and in the center of the tube. In general, you need as many different boundary conditions for each differential equation as the number of times you must integrate the equation to arrive at the velocity distribution for that directional component. For steady laminar flow in a rigid tube, we only have one directional component (the z-component) and $V_z = V_z(r)$. We have to integrate the z-component of velocity with respect to the r-direction twice. Therefore, we require two boundary conditions: one at the walls ($r = R$) and one at the tube center ($r = 0$).

The most commonly used boundary condition for flow in tubes is called the "no-slip at the wall" boundary condition. This states that at a *solid–fluid* interface the fluid velocity equals the velocity with which the solid surface is moving.

Example 4.2 Steady Laminar Flow in a Tube

We now use the general equations of motion to set up the equation that was previously derived for steady laminar flow in a tube using a shell balance. As before, we first visualize the flow field and decide on cylindrical coordinates. We also decide that the z-component of velocity is the only nonzero component and V_z is only a function of the radial distance, r. For a Newtonian fluid of constant ρ and μ, we write down the z-component of velocity from Appendix B as

$$\rho\left(\frac{\partial V_z}{\partial t} + V_r\frac{\partial V_z}{\partial r} + \frac{V_\theta}{r}\frac{\partial V_z}{\partial \theta} + V_z\frac{\partial V_z}{\partial z}\right)$$

$$= -\frac{\partial p}{\partial z} + \mu\left[\frac{1}{r}\frac{\partial}{\partial r}\left(r\frac{\partial V_z}{\partial r}\right) + \frac{1}{r^2}\frac{\partial^2 V_z}{\partial \theta^2} + \frac{\partial^2 V_z}{\partial z^2}\right] + \rho g_z$$

The first term is zero, since we are at steady state. The second and third terms are zero, since the flow visualization concludes that both V_r and V_θ are zero. The second term in brackets on the right-hand side is zero, as is the gravity term since we are dealing with a horizontal tube.

We now consider the equation of continuity in cylindrical coordinates from Appendix B.

$$\frac{\partial \rho}{\partial t} + \frac{1}{r}\frac{\partial}{\partial r}\left(\rho r V_r\right) + \frac{1}{r}\frac{\partial}{\partial \theta}\left(\rho V_\theta\right) + \frac{\partial}{\partial z}\left(\rho V_z\right) = 0$$

The first three terms are zero by steady state because both V_r and V_θ are zero.

Now, since we are dealing with an incompressible fluid, ρ is constant and can be taken outside the differential sign. Then, we can divide both sides by ρ. Furthermore, since there is only one integration variable, z, we can change the partial derivative to an ordinary derivative. Thus, the continuity equation becomes

$$\frac{dV_z}{dz} = 0$$

We now return to the reduced equation of motion to try further simplification.

$$\rho\left(V_z \frac{\partial V_z}{\partial z}\right) = -\frac{\partial p}{\partial z} + \mu\left[\frac{1}{r}\frac{\partial}{\partial r}\left(r\frac{\partial V_z}{\partial r}\right) + \frac{\partial^2 V_z}{\partial z^2}\right]$$

The first term is zero by the results of the continuity equation. The last term is zero since we can write

$$\frac{\partial^2 V_z}{\partial z^2} = \frac{\partial}{\partial z}\left(\frac{\partial V_z}{\partial z}\right)$$

and then use the results of continuity.

We then say that the pressure drop across the tube is linear, so we can approximate

$$\frac{\partial p}{\partial z} \approx \frac{P_L - P_0}{L}$$

We are then left with only one integration variable, r, and we can convert the partial derivatives to ordinary derivatives to give

$$0 = \frac{P_0 - P_L}{L} + \mu\frac{1}{r}\frac{d}{dr}\left(r\frac{dV_z}{dr}\right)$$

Integrating twice with respect to r and using the boundary conditions $V_z = 0$ at $r = R$ (no-slip at the wall) and $V_z =$ finite at $r = 0$ (τ_{rz} is finite at $r = 0$) give

$$V_z = \frac{(P_0 - P_L)R^2}{4\mu L}\left[1 - \left(\frac{r}{r}\right)^2\right]$$

This is the same result as in Equation 4.19.

4.7 Unsteady Laminar Flow in a Tube

While steady-state analysis of flow in a tube is useful to gain some insight into the shape of the velocity profiles and the parameters that are important, the pressures generated by the ventricles that drive flow in the systemic circulation is *pulsatile*. That is, both pressure and flow change with time over a single cardiac cycle. Figures 5.7, 5.9, and 5.10 illustrate the pulsatile nature of the pressure, pressure gradient, and velocity. In addition, local and reflex mechanisms adjust peripheral resistance, heart rate, and force of contraction so that pressure levels and CO (flow) also change in response to changing metabolic needs. The method outlined in the previous section can readily be used to set up unsteady-state flow problems. The bad news is that *unsteady-state flow problems generate partial differential equation models*, as compared with steady-flow problems that generate ordinary differential equation models. The solution of a partial differential equation model will depend on its boundary and initial conditions. For most practical cases, no analytical solution of the model will exist and one must resort to numerical methods such as *finite differences* and *finite element*. Since numerical methods only generate particular solutions for a single set of parameter values, many cases must be simulated in order to gain insight into the general behavior of the flow field. Several commercially available *computational fluid dynamics (CFD)* software packages (e.g., ADINA®, COMSOL® Multiphysics®, and FLUENT®) will solve the equations of motion for particular cases using a finite element mesh for the geometry of the vessel. These also have extensive sets of constitutive equations that will allow the simulation of vessels with different mechanical properties.

For pulsatile laminar flow in a rigid tube, the velocity field will still only have a z-component. However, now $V_z = V_z(r, t)$. Using the z-component of the equation of motion for a Newtonian fluid of constant ρ and μ in cylindrical coordinates along with the equation of continuity in cylindrical coordinates from Appendix B, and simplifying as before give

$$\rho \frac{\partial V_z}{\partial t} = -\frac{\partial p}{\partial z} + \mu \frac{1}{r} \frac{\partial}{\partial r} \left(r \frac{\partial V_z}{\partial r} \right) \tag{4.33}$$

Equation 4.33 is the starting point for many analyses of pulsatile flow.

When the tube wall is not rigid, additional equations for the motion of the wall and, perhaps, the r-component of the equation of motion for the fluid must be considered.

4.8 Pulsatile Flow in a Thin-Walled Tube

Womersley [1,2] studied pulsatile flow in a thin-walled tube. For pulsatile flow in a thin-walled tube, we represent the pressure gradient per unit length as a simple harmonic motion of amplitude $\bar{P}$, frequency ω, and phase lag ϕ. We assume that either

$$-\frac{dP}{dz} = \bar{P}\cos(\omega t - \phi) \tag{4.34}$$

Or

$$-\frac{dP}{dz} = \bar{P}\sin(\omega t - \phi) \tag{4.35}$$

This is convenient since in the linear case, we can use a *Fourier** analysis to reconstruct any measured pressure gradient as a linear sum of sines and cosines with different frequencies. We can include both possibilities for the pressure gradient by writing

$$-\frac{dP}{dz} = \bar{P}\left[\cos(\omega t - \phi) - j\sin(\omega t - \phi)\right] \tag{4.36}$$

where $j = \sqrt{-1}$.

If the real and imaginary terms retain their identity throughout the analysis, then the real part of the particular integral will be the integral for $\bar{P}\cos(\omega t - \phi)$ and the imaginary part will be the particular integral for $-j\bar{P}\sin(\omega t - \phi)$.

We can simplify the algebra by expanding both terms on the right-hand side of Equation 4.36 and substituting the trigonometric identities.

$$\bar{P}\cos(\omega t - \phi) = \bar{P}[\cos\omega t \cos\phi + \sin\omega t \sin\phi]$$

and

$$-j\bar{P}\sin(\omega t - \phi) = -j\bar{P}[\sin\omega t \cos\phi - \cos\omega t \sin\phi]$$

---

* *Fourier analysis* (see Chapter 6) allows us to express any periodic function as the sum of a series of sines and cosines (or any set of orthogonal functions) of different frequencies. The terminology that is used to describe these frequencies is borrowed from the way we describe *musical notes*. The fundamental frequency is the steady state or "DC" component. We add to that sines and cosines of increasing frequency until we recreate a given periodic function as closely as we wish (see Appendix C).

We then multiply through, collect terms and factor to arrive at the result

$$-\frac{dP}{dz} = \bar{P}[\cos(\omega t - \phi) - j\sin(\omega t - \phi)] = \bar{P}[\cos\phi - j\sin\phi][\cos(\omega t) - j\sin(\omega t)]$$

We can now use the Euler relationship, $e^{j\omega t} = \cos(\omega t) + j\sin(\omega t)$ and define a complex amplitude as $A^* = \bar{P}[\cos\phi - j\sin\phi]$ to put the pressure gradient in the complex exponential form

$$-\frac{dP}{dz} = \frac{P_0 - P_L}{L} = A^* e^{j\omega t} \tag{4.37}$$

where $\bar{P}$ is the modulus of A^* and ϕ its phase angle. $\text{Re}(A^* e^{j\omega t}) = \bar{P}\cos\phi$ $\cos(\omega t - \phi)$ and $\text{Im}(A^* e^{j\omega t}) = -\bar{P}\sin\phi\sin(\omega t - \phi)$.

Here, Re(.) represents the real part of the quantity enclosed in parentheses and Im(.) represents the imaginary part of the quantity enclosed in parentheses.

The pressure gradient is then periodic in time with a frequency

$$f = \frac{\omega}{2\pi} \tag{4.38}$$

The complex exponential form of the pressure gradient in Equations 4.33 and 4.37 is the starting point for Womersley's analysis of pulsatile flow in a thin-walled tube. Substituting the value for $-dP/dz$ in (4.33) gives

$$\rho\frac{\partial V_z}{\partial t} = -A^* e^{j\omega t} + \mu\frac{1}{r}\frac{\partial}{\partial r}\left(r\frac{\partial V_z}{\partial r}\right) \tag{4.39}$$

Using the chain rule to expand

$$\frac{1}{r}\frac{\partial}{\partial r}\left(r\frac{\partial V_z}{\partial r}\right)$$

and rearranging gives

$$\mu\left[\frac{\partial^2 V_z}{\partial r^2} + \frac{1}{r}\frac{\partial V_z}{\partial r}\right] - \rho\frac{\partial V_z}{\partial t} = -A^* e^{j\omega t} \tag{4.40}$$

If we now make the transformation

$$V_z = u e^{j\omega t} \tag{4.41}$$

where u is a function of only r (separation of variables), Equation 4.40 becomes

$$\frac{d^2 u}{dr^2} + \frac{1}{r}\frac{du}{dr} - \frac{j\omega}{\nu}u = -\frac{A}{u} \qquad (4.42)$$

Since time does not explicitly appear in (4.42), the partial derivatives become ordinary derivatives. Also, since $j^2 = -1$, we can write this equation in the form

$$\frac{d^2 u}{dr^2} + \frac{1}{r}\frac{du}{dr} + \frac{j^3\omega}{\nu}u = -\frac{A^*}{\mu} \qquad (4.43)$$

where $\nu = \mu/\rho$, is called the "kinematic viscosity." In this form, the differential equation for u turns out to be a form of Bessel's equation. The solution to this equation is available in many mathematical texts. Using the boundary conditions

$$\text{at } r = 0, \quad V_z = \text{finite} \qquad (4.44)$$

and

$$\text{at } r = R, \quad V_z = 0 \text{ (no-slip at the wall)} \qquad (4.45)$$

the solution for u is

$$u = \frac{A^*}{\rho\omega j}\left[1 - \frac{J_0\left(r\sqrt{\frac{\omega}{\nu}}j^{\frac{3}{2}}\right)}{J_1\left(R\sqrt{\frac{\omega}{\nu}}j^{\frac{3}{2}}\right)}\right] \qquad (4.46)$$

where
J_0 is a Bessel function of the first kind of order zero and complex argument
J_1 is a Bessel function of the first kind of order one and complex argument

These are tabulated functions whose values and properties are elaborated in several places [3].
   We now introduce the dimensionless parameters

$$\alpha = R\sqrt{\frac{\omega}{\nu}} \quad \text{and} \quad y = \frac{r}{R}$$

The velocity distribution, V_z, becomes

$$V_z = \frac{A^*}{\rho\omega j}\left[1 - \frac{J_0\left(\alpha y j^{\frac{3}{2}}\right)}{J_1\left(\alpha j^{\frac{3}{2}}\right)}\right]e^{j\omega t} \tag{4.47}$$

Equation 4.47 is still in complex form. If we were now to take the pressure gradient as the real part of $A^* e^{j\omega t}$, the corresponding flow would be the real part of (4.47).

The velocity distribution can then be integrated over the cross-sectional area of the tube to give the following result for the volumetric flow, Q (mL/s):

$$Q = \frac{A^*\pi R^2}{j\omega\rho}\left[1 - \frac{2J_1\alpha j^{\frac{3}{2}}}{\alpha j^{\frac{3}{2}}J_0\alpha j^{\frac{3}{2}}}\right]e^{j\omega t} \tag{4.48}$$

The separation of Equations 4.47 and 4.48 into real and imaginary parts leads to an inconvenient form for computation. Since the Bessel functions in Equations 4.47 and 4.48 are complex variables, it is more convenient to express them in terms of a modulus (amplitude) and phase.

$$J_0\left(\alpha y j^{\frac{2}{3}}\right) = M_0(y)e^{j\theta_0(y)}$$

$$J_0\left(\alpha j^{\frac{2}{3}}\right) = M_0 e^{j\theta_0}$$

$$J_1\left(\alpha j^{\frac{2}{3}}\right) = M_1 e^{j\theta_1}$$

Then if the real part of $A^* e^{j\omega t} = \bar{P}\cos(\omega t - \phi)$, we can write Equation 4.48 in the form

$$Q = \frac{\pi R^4\bar{P}}{\mu\alpha^2}\left[\sin(\omega t - \phi) - \frac{2M_1}{\alpha M_0}\sin(\omega t - \phi + \delta)\right] \tag{4.49}$$

where

$$\delta = 135° - \theta_1 + \theta_0 \tag{4.50}$$

Tables of M_0, M_1, and δ are given in references such as [3] and [4].

Steady, laminar flow in a rigid tube occurs when $\omega \to 0$. In that case, Equations 4.48 and 4.49 reduce to the Hagan–Poiseuille relations for velocity and flow given by Equations 4.19 and 4.23.

4.8.1 Calculation of Flow from Measured Pressure Gradient

The calculation of flow from pressure gradients measured in vivo using Equations 4.49 and 4.50 can be quite lengthy. Figure 4.7 shows the flow velocity in the femoral artery of a dog (upper curve) and the pressure gradient between the aorta and the femoral artery computed from simultaneous measurements of pressure at the two sites.

The first step in calculation of the corresponding flow rate is to represent this pressure as a Fourier series. If T is the time for one cardiac cycle

$$\theta = \frac{2\pi t}{T} \tag{4.51}$$

so that θ, measured in degrees, runs from $0°$ to $360°$ during one cardiac cycle. Womersley used 24 values of θ located $15°$ apart. A brief outline of the method for computation of the Fourier coefficients is given in Appendix C. Usually, the fundamental and a small number of harmonics are required to fit a well-behaved function (see Appendix C). Womersley used six harmonics to fit similar data. These are then converted to modulus and phase form. The flow is then the sum of the six terms of the form

$$Q = \frac{\pi R^4 \overline{P}}{\mu \alpha^2} \left[\sin(\omega\theta - \phi) - \frac{2M_1}{\alpha M_0} \sin(\omega\theta - \phi + \delta) \right] \tag{4.52}$$

The simple analysis given above contains two very important limitations. First, the artery is regarded as a rigid tube, neglecting the change in radius which would occur in a distensible vessel due to a pulsatile pressure gradient. Second, the pressure gradient is assumed to be only a function of time, whereas it is generated by a pulsatile wave of finite velocity. In reality, there will be a radial component of velocity as well as an axial component. Therefore, a second partial differential equation for the radial component of the fluid velocity must be added as well as radial and axial components of the equation of motion for the vessel wall.

4.9 Computational Fluid Dynamics

It is clear from the above analysis that analytical solutions to realistic flow problems will not be possible, except under very restrictive conditions. The analysis of flow problems that have clinical relevance will require numerical methods. The most promising of these is CFD with *fluid–structure interaction* (FSI) capability. This approach, along with *finite element* mesh programs that are compatible with MRI, CAT scans, and other medical imaging modalities makes it possible to compute *patient-specific* velocity profiles and wall stresses for complex vessel shapes. In addition, large deformations of vessel walls can be simulated by incorporating Fung's [5–7] strain energy function for the nonlinear anisotropic properties of the vessel wall.

We have used ADINA CFD, FSI software, version 8.5 (ADINA R&D, Watertown, MA) to simulate pulsatile flow in a linearly elastic vessel (approximating a large artery, neglecting gravity).

The blood properties used in the simulation:

Density = 1050 kg/m^3
Viscosity = 1.05 g/s cm
Velocity = 5–10 m/s
Pressure (max) = 120 mm Hg
Pressure (min) = 80 mm Hg

Vessel properties used in the simulation:

Young's modulus = 2.8 MPa
Poisson's ratio = 0.37
Density = 1370 kg/m^3
Internal diameter = 2.5 cm
Wall thickness = 0.15 cm
Length = 30 cm

An input pressure waveform was created in MATLAB® and imported through the text file subroutine in ADINA. The pressure wave form is illustrated in Figure 4.10.

The mesh model created in ADINA is illustrated in Figure 4.11.

Wall motion in response to the input pressure waveform (Figure 4.10) is illustrated in Figure 4.12:

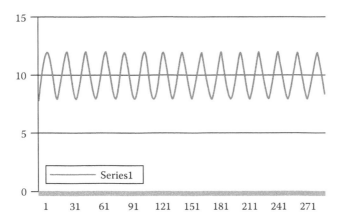

FIGURE 4.10
Pressure waveform used as an input. X-axis is in s, and y-axis is in mm Hg $\times 10^{-1}$.

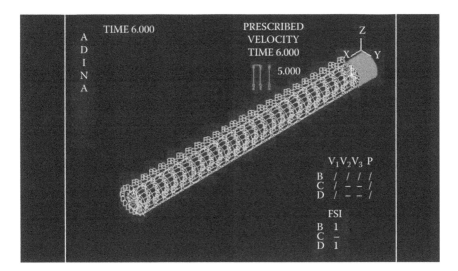

FIGURE 4.11
Mesh model for pulsatile flow in an elastic tube.

4.9.1 Wall Motion and Fluid–Structure Interactions

A force balance on the vessel wall is required to obtain the equations of motion for the wall. For example, the radial component of wall displacement, u_r (neglecting inertial terms) is given by

$$\rho_W \frac{\partial u_r}{\partial t^2} = \frac{\partial T_{rr}}{\partial r} + \frac{\partial T_{rz}}{\partial z} + \frac{T_{rr} - T_{\theta\theta}}{r}$$

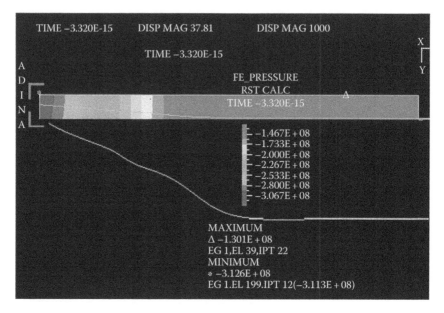

FIGURE 4.12
Wall motion in response to pressure wave in Figure 4.10.

The equation of incompressibility is given by

$$\frac{\partial u_z}{\partial z} + \frac{\partial u_r}{\partial r} + \frac{u_z}{r} = 0$$

where u_x, u_y, and u_z are the x-, y-, and z-components of the wall velocity. The definitions of the stress components in each direction, T_{rr}, T_{rz}, and $T_{\theta\theta}$, depend on whether the wall properties are linearly elastic and exhibit small strains or nonlinear anisotropic and large strains are allowed. In the latter case, a strain energy function is used and the anisotropic material properties must be available [5–7].

Solving problems that involve pulsatile flow in distensible vessels requires addressing some additional complex issues:

- If large deformations are involved, start with the equations of motion for the wall in vector–tensor notation.
- The vessels may be constrained at the ends or along their length. In these cases, one or more of the wall velocity equations may be neglected.
- The vessel may be totally or partially imbedded in active muscle. Surface coronary arteries are an example. The external pressure on the vessel wall may be different at the $\pm R$ positions.

- The aorta has both ascending and descending parts that have different directional vectors for the gravitational force. Also, flow out of the ascending aorta to the carotid and coronary arteries has to be considered, if the entire aorta is being modeled.
- The common carotid artery has a bifurcation at the end with flow into both the external and internal carotid arteries. In general, these have different radii.
- The vessel may have an implanted stent, which changes the wall's mechanical properties (and the boundary conditions) for a portion of the flow path.

All of these considerations will lead to different boundary conditions at the fluid–structure interface. Two common boundary conditions at the fluid–structure interface are

- No-slip at the wall.
- The components of the velocity vector have to be equal for the fluid and the wall.

In addition, the inlet pressure wave must reflect the amplitude and frequency of the inlet pressure in vivo at the vessel of interest.

4.9.2 Some Recent Results

Torii et al. [8] have used a CFD model of the right coronary artery and MRI-based geometry to compute patient-specific wall shear stress (WSS) and oscillatory shear stress (OSS) using CFX11® (ANSYS, Canonsburg, PA).

Spilker and Taylor [9,10] used three-element windkessel models to extend a CFD model for 3-D flow in the abdominal aorta using a finite element mesh generated by MeshSim® (Simmetrix, Inc., Clifton Park, NY). A Newton–Raphson iterative procedure at each step was used to fit model predictions to experimental values for patient-specific geometries.

Krittian et al. [11] used a patient-specific simulation model for 3-D blood-flow modeling in the human heart. The authors used FLUENT 6.3.26 for the CFD computations, Abaqus® 6.7-1 to generate the mesh, and MpCCI® 3.0.5 software for an explicit coupling of the two software systems. The KaHMo model does the implicit coupling internally.

Leach et al. [12] used a patient-specific biomechanical model to assess the rupture potential in vulnerable plaque in the internal carotid artery. The fine mesh was computed by Hypermesh (Altair Engineering, Troy, MI) and the FSI computations by ADINA (ADINA R & D, Watertown, MA) software.

It is clear from these illustrations that in vivo flow modeling has come closer to clinical usefulness through more realistic patient-specific computational models. It is also clear that each model requires a great deal of work to implement as well as close collaboration between engineers and clinicians.

4.10 Dimensionless Forms of the Equations of Change

The dimensionless forms of the equations of motion and continuity are important for two reasons. First, writing the equations in dimensionless form minimizes the number of independent parameters that have to be measured in an experiment. Second, if the dimensionless parameters are identical in two different systems, then both systems are described by identical dimensionless differential equations. If, in addition, the dimensionless boundary and initial conditions are the same, the dimensionless velocity and pressure distributions will be the same in each. This allows comparison of velocity and pressure distributions in vessels of different size that are geometrically similar.

For many flow systems, we can identify a characteristic length, D, and a characteristic velocity, $\langle V \rangle$. For flow in a circular vessel, D can be taken as the vessel's luminal diameter and $\langle V \rangle$ as the average velocity. The choice is arbitrary, but they must be carefully specified. We can then identify the following dimensionless quantities:

$$\mathbf{V}^* = \frac{\mathbf{V}}{\langle V \rangle}; \quad p^* = \frac{p - p_0}{\rho \langle V \rangle^2}; \quad t^* = \frac{t \langle V \rangle}{D}$$

$$x^* = \frac{x}{D}; \quad y^* = \frac{y}{D}; \quad z^* = \frac{z}{D}$$

The equations of continuity and motion for a Newtonian fluid of constant density and viscosity in vector notation are given by Equations 4.26 and 4.30, respectively. We can now define the dimensionless differential gradients as

$$\nabla^* = D\nabla = \left(\delta_x \frac{\partial}{\partial x^*} + \delta_y \frac{\partial}{\partial y^*} + \delta_z \frac{\partial}{\partial z^*} \right)$$

$$\nabla^{*2} = D^2 \nabla^2 = \left(\frac{\partial^2}{\partial x^{*2}} + \frac{\partial^2}{\partial y^{*2}} + \frac{\partial^2}{\partial z^{*2}} \right)$$

$$\frac{D}{Dt^*} = \left(\frac{D}{\langle V \rangle} \right) \frac{D}{Dt}$$

Here δ_x, δ_y, and δ_z are unit vectors in the x-, y-, and z-directions, respectively.

Substituting these definitions back into Equations 4.26 and 4.30 by noting that $\mathbf{V} = \langle V \rangle \mathbf{V}^*$, $p - p_0 = p^*(\rho \langle V \rangle^2)$, etc., and rearranging the result gives

$$(\nabla^* \cdot \mathbf{V}^*) = 0 \tag{4.53}$$

$$\frac{D\mathbf{V}^*}{Dt^*} = -\nabla^* p + \left(\frac{\mu}{D\rho \langle V \rangle} \right) \nabla^{*2} \mathbf{V}^* + \left(\frac{gD}{\langle V \rangle^2} \right) \frac{g}{g} \tag{4.54}$$

where g is the acceleration due to gravity, and in the cgs system its value is $980.665 \, cm/s^2$.

In these dimensionless forms of the equations of motion, two dimensionless groups appear as the coefficients of the $\Delta^{*2}V^*$ and $(g/g)^*$ terms. These dimensionless groups occur very often in fluid mechanics and they have been given names in honor of two pioneers in the field.

$$Re = \frac{\rho D \langle V \rangle}{\mu} = \text{Reynolds number} \qquad (4.55)$$

$$Fr = \frac{\langle V \rangle^2}{gD} = \text{Froude number} \qquad (4.56)$$

It has been established experimentally that *laminar* flow occurs in flow systems where the value of the Reynolds number < 2100. Systems with $Re > 2100$ may exhibit unstable laminar flow or, usually, *turbulent*[†] flow. The velocity profiles are very different in laminar and turbulent flow.

Example 4.3 Reynolds Number Calculations
Compare the Reynolds number in a capillary with that in the aorta.
The average velocity is given by

$$\langle V \rangle (\text{cm/s}) = \frac{Q(\text{mL/s})}{A(\text{cm}^2)} \qquad (4.57)$$

where
Q is the CO in mL/s
A is the total cross-sectional area available for flow in cm^2

a. We will use the value 1.78 cm for the inside diameter of the aorta, since this corresponds to a cross-sectional area of $2.5 \, cm^2$ (assuming a circular cross section). Assume a CO of 6.0 L/min. This is 6000 mL/60 s or 100 mL/s. The density of blood is about $1.04 \, g/cm^3$ and its viscosity is about 2.5 centipoise at a normal hematocrit of 45%. Remember 1 centipoise $= 1 \times 10^{-2} \, g/cm \, s$.

$$\langle V \rangle = \frac{100(\text{mL/s})}{2.5(\text{cm}^2)} = 40 \, (\text{cm/s})$$

---

[*] The term (g/g) is just a unit vector in the direction of gravity.
[†] Whereas laminar flow is an orderly process characterized by discrete layers of fluid slipping past each other with no mixing of the layers, turbulent flow is characterized by chaotic motion of the fluid and a large degree of mixing between adjacent fluid layers.

The Reynolds number is then:

$$Re = \frac{\rho D \langle V \rangle}{\mu} = \frac{\left(1.04\,\text{g/cm}^3\right)\left(1.78\,\text{cm}\right)\left(40\,\text{cm/s}\right)}{2.5 \times 10^{-2}\,\text{g/cm s}} = 2962$$

Since $Re > 2100$, we conclude that flow in the ascending aorta is just into the lower range of turbulent flow. If we had used a CO of 5.0 L/min, the Reynolds number would have been 2468, close to the upper limit for laminar flow.

a. A typical capillary has a diameter of 8×10^{-10} cm and a total cross-sectional area of $2500\,\text{cm}^2$. A CO of 100 mL/s would have an average velocity of 0.04 cm/s. If we assume that the density and viscosity of blood in the microvessels are the same as in the large vessels,* we can then calculate a Reynolds number for a typical capillary as

$$Re = \frac{\left(1.04\,\text{g/cm}^3\right)\left(8 \times 10^{-10}\,\text{cm}\right)\left(0.04\,\text{cm/s}\right)}{2.5 \times 10^{-2}\,\text{g/cm s}} = 1.33 \times 10^{-9}$$

Clearly, this Reynolds number indicates that capillary flow occurs in the very low flow regime.

4.11 Turbulence

There is no general theory of turbulence. Therefore, all insights into the behavior of turbulent flow fields depend on experimental data and empirical correlations. To gain some insight into the effects of turbulence on the velocity distribution for flow in a tube, we imagine that we have a fully developed turbulent flow field, and introduce a time-smoothed velocity, $\overline{V}_z$ and the average of the time-smoothed velocity, $\langle \overline{V}_z \rangle$. It has been determined experimentally that

$$\frac{\overline{V}_z}{\overline{V}_{z,\text{max}}} \approx \left(1 - \frac{r}{R}\right)^{\frac{1}{7}} \tag{4.58}$$

$$\left\langle \frac{\overline{V}_z}{\overline{V}_{z,\text{max}}} \right\rangle \approx \frac{4}{5} \tag{4.59}$$

---

* This is not necessarily a good assumption as blood does not behave as a continuum in vessels whose diameter approaches the size of a red cell. In particular, capillary hematocrit may not be the same as that of large vessel hematocrit due to red cell exclusion and stearic hindrance effects in the smaller capillaries.

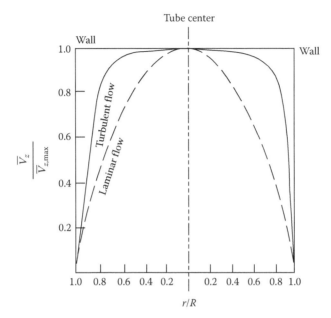

FIGURE 4.13

Qualitative comparison between laminar and turbulent flow velocity profiles in a tube. The turbulent velocity profiles are much flatter. This is due to chaotic nature of turbulent flow and the mixing of adjacent fluid layers. Similar flat velocity profiles are observed in laminar flow of two phase fluids such as blood, particularly in places like branches, bends, and in the venous system at the confluence of two collecting veins. The red cells tend to travel preferentially with the faster velocity plasma stream and stay near the centerline of the vessel in those areas.

Compare these results with those for laminar flow in a tube, Equations 4.19, 4.20, and 4.22. Figure 4.13 shows a qualitative comparison between laminar and time-smoothed velocity profiles for flow in a tube.

As indicated by the Reynolds number calculations above, there are very few places in the circulation that exhibit turbulent flow under normal resting conditions. Rather, when turbulent flow does occur, it is usually an indication of some pathological state. This observation has been used clinically to detect and diagnose some common hemodynamic problems associated with heart valve abnormalities as well as the ordinary measurement of systolic and diastolic blood pressures using a blood pressure cuff (*sphygmomanometer*). Some of the pathological conditions that lead to turbulent flow and heart sounds (called *murmurs*) are discussed below. Murmurs can be detected with the aid of a simple stethoscope or displayed as an electronic signal, using a microphone. The particular sounds have been characterized for the commonly occurring murmurs and aid in suggesting an initial diagnosis of the underlying pathology.

4.11.1 Commonly Occurring Heart Sounds

1. *Normal heart sounds*: Two heart sounds can be detected in most normal individuals, when a stethoscope (or microphone) is placed on the left side of the sternum (breastbone) at the third or fourth intercostal space (over the apex of the heart). The first heart sound, called S_1, occurs at the beginning of systole because of the abrupt closure of the atrioventricular (AV) valves. This produces a vibration of the cardiac structures and an abrupt stoppage of the blood in the ventricular chambers. This is similar to the "water hammer" that is produced in a water pipe when the valve is abruptly closed. The second heart sound, called S_2, arises from the closure of the aortic and pulmonic valves at the beginning of the period of isovolumic relaxation. These two sounds are described by many people as the "lub"–"dub" sounds of the heart as heard through a stethoscope. A third heart sound, S_3, is not normally heard through a stethoscope in adults. S_3 may be heard in children and occurs shortly after S_2, during the period of rapid passive ventricular filling. When S_3 is heard in

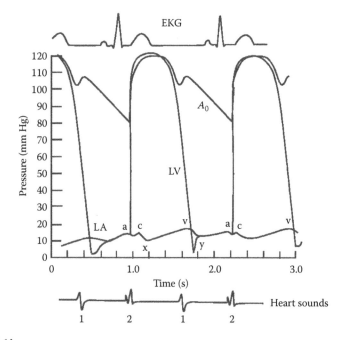

FIGURE 4.14
Normal pressures in the left atrium, left ventricle, and aorta. The electrocardiogram (EKG) is presented for timing reference. The normal heart sounds are indicated as 1 and 2 at the bottom. The labels a, c, and v on the atrial pressure wave are a = increase in pressure due to atrial contraction; c = increase in pressure due to ventricular contraction; and v = maximum atrial pressure associated with closing of aortic valve.

adults, it is usually associated with left ventricular failure. Figure 4.14 shows the normal pressures in the left ventricle, left atrium and aorta, as well as the timing of the two normal heart sounds.

2. *Abnormal heart sounds*

   a. *Mitral stenosis.* The mitral valve lies between the left atrium and the left ventricle. A *stenosis* is a narrowing of the opening between the atrium and the ventricle, decreasing the area available for flow during ventricular filling (diastole). During diastole, the pressure in the left atrium is elevated and there is a significant pressure gradient* between the atrium and the ventricle in order to keep the blood flowing across the obstructed valve opening. Abnormal heart sounds are present. There is a diastolic opening snap (OS) that occurs just after S_2 and corresponds to the opening of the mitral valve. This is followed by a murmur which decreases in intensity (decrescendo). There is an accentuation of the murmur just before S_1, which is due to the increased pressure gradient when the left atrium contracts (Figure 4.15).

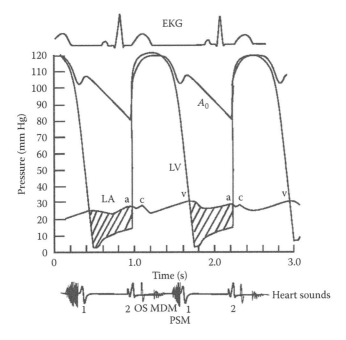

FIGURE 4.15
Pressure changes in mitral stenosis. 1 and 2 show the positions of the normal heart sounds. OS, opening snap; MDM, mid-diastolic murmur; PSM, presystolic murmur; A_0, aortic pressure.

---

* The normal pressure gradient across the mitral valve is very small (<1 mm Hg).

b. *Mitral regurgitation. Regurgitation* is the leakage or backflow of blood during systole because the valve leaflets do not form a tight enough seal. Normal closure of the mitral valve during systole requires the coordinated action of the valve leaflets, *chordae tendineae*, and the *papillary muscles.* Degeneration of any of these components as well as structural abnormalities in the valve opening will lead to a defective valve seal between the atrium and the ventricle during systole. *Acute* (short term) mitral regurgitation results in a high left atrial pressure and pulmonary edema. In *chronic* (long term) regurgitation, the heart compensates (*remodels*) by increasing the left atrial volume, thus reducing the left atrial pressure and the symptoms of pulmonary congestion. The remodeling also includes enlargement of the left ventricle and *hypertrophy* (thickening of the ventricular wall) due to the volume overload (since some of the output is going back into the atrium due to the incompetent valve, the forward CO increases to maintain systemic flow). The heart sound produced is a murmur that begins at the first heart sound and continues at the same level of intensity right through to the second heart sound (Figure 4.16).

c. *Aortic stenosis.* The aortic valve lies between the left ventricle and the aorta. In aortic stenosis, blood flow across the aortic valve

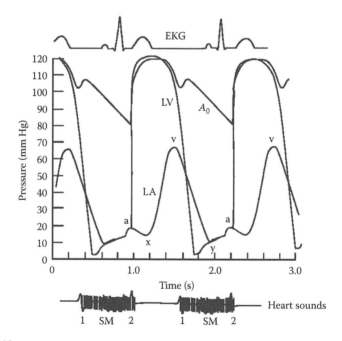

FIGURE 4.16
Pressure changes in chronic mitral regurgitation. SM = systolic murmur.

is restricted during systole. When the area available for flow is decreased by more than 50%, significant elevation of left ventricular pressure is necessary to drive blood into the aorta. Over time, the left ventricle remodels itself and increases its wall thickness (concentric hypertrophy) to compensate for the pressure overload and reduce ventricular wall stresses. However, the mechanical properties of both the ventricle and the atrium change. Both structures become thicker (hypertrophy) and stiffer due to the increased pressure. As the disease progresses, the increased atrial pressure is reflected back into the pulmonary venous system. This directly affects pulmonary capillary pressure, increases pulmonary blood–tissue filtration of fluid, and leads to pulmonary congestion and pulmonary edema (these are the symptoms of congestive heart failure). The large pressure gradient produces a systolic murmur that starts at S_1 and increases in intensity until about the middle of ejection, then decreases in intensity (crescendo–decrescendo). The murmur does not extend beyond S_2, and S_2 is diminished in intensity (Figure 4.17).

d. *Aortic regurgitation* (also called aortic insufficiency). This results from diseases of the aortic leaflets or dilatation of the aortic root. In aortic regurgitation, there is backflow of blood through the incompetent valve leaflets from the aorta to the left ventricle

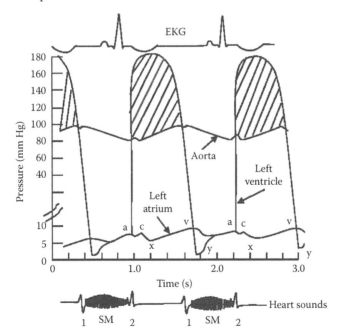

FIGURE 4.17
Aortic stenosis pressure changes. SM, systolic murmur.

during diastole when the pressure is greater in the aorta than in the ventricle. The heart must then pump the regurgitated volume plus the normal volume of blood returning from the left atrium. In acute aortic regurgitation, the left ventricle is of normal size, but stiffer due to the volume overload. The diastolic pressure in the ventricle rises because of the increased volume and stiffness. This increased pressure is reflected back into the left atrium and pulmonary circulation, resulting in pulmonary congestion or edema. In chronic aortic regurgitation, both the atrium and ventricle remodel themselves to accommodate the volume overload (there is mostly dilation, with only a modest amount of ventricular hypertrophy). The increased volumes allow the aortic (and therefore the systemic arterial) diastolic pressure to drop substantially. The combination of high left ventricular stroke volume (and therefore high systolic arterial pressure) and low diastolic pressure, increases the pulse pressure (the pulse pressure is the difference between the systolic and diastolic pressures), which is characteristic of aortic regurgitation. The lowered diastolic pressure relieves the symptoms of pulmonary congestion and edema. The abnormal heart sound consists of a diastolic murmur starting at S_2 and decreasing in intensity (decrescendo), which corresponds with the regurgitant flow (Figure 4.18).

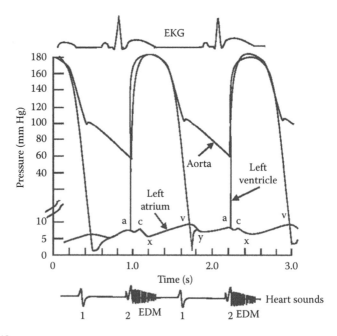

FIGURE 4.18
Pressure changes in chronic aortic regurgitation. EDM, end-diastolic murmur.

e. *Diseases of the right heart valves.* The tricuspid valve is the atrioventricular valve located between the right atrium and the right ventricle. Tricuspid stenosis is usually caused by rheumatic fever and produces a murmur similar to that of mitral stenosis, but heard closer to the sternum. The murmur intensifies on inspiration because of the increase in venous return during inspiration. Neck veins are also distended. Tricuspid regurgitation is usually functional and develops because of right ventricular enlargement due to either pressure or volume overload. Tricuspid regurgitation produces a systolic murmur that is best heard at the lower left sternal border. It is often soft but increases in intensity during inspiration.

The *pulmonic* valve is located between the right ventricle and the pulmonary artery. The cause of pulmonic *stenosis* is almost always congenital deformity of the valve. Pulmonic *regurgitation* often develops with severe pulmonary hypertension because of dilation of the valve ring by the enlarged pulmonary artery. There is a high pitched murmur along the left sternal border that decreases in intensity (decrescendo). It is often difficult to distinguish between pulmonic and aortic regurgitation on the basis of heart sounds. However, they are easily differentiated by *Doppler echocardiography*.

Gravitational effects become important in pulmonary blood flow. The pulmonary circulation is a low resistance circuit. Therefore, the right ventricle needs to generate only about 20% of the pressure of the left ventricle in order to eject the stroke volume into the pulmonary artery. Anatomically, the pulmonary artery lies at the base of the lung. Under normal, resting conditions, the systolic or diastolic pressures in the pulmonary artery are 25/8 mm Hg, with an average pressure of 15 mm Hg. Compare these with values of 120/80 mm Hg (systolic/diastolic) and 100 mm Hg average for the systemic circulation. In the normal, upright adult, the apex of the lungs (the highest point) is about 30 cm above the base (the lowest point). Since 1 mm Hg = 1.36 cm H_2O, a column of blood 30 cm high would produce a hydrostatic pressure head due to gravity ($\rho g \Delta h$) of about 22 mm Hg, that opposes the pressure in the pulmonary artery (note that blood is slightly denser than water, about 1.04–1.08 g/mL at 37°C). This pressure head is unevenly distributed, about 8 mm Hg being below the heart and about 15 mm Hg being above the heart. The consequences of this is that the arterial pressures in the uppermost portion of a standing adult are 15 mm Hg below those in the pulmonary artery and the pressures in the lowermost portion of the lungs are 8 mm Hg above those in the pulmonary artery. The consequences of this gravitationally caused pressure gradient to blood flow in the lung is that under normal, resting conditions there is no flow in the uppermost (approximately one-third) portion of the lung during any part of the cardiac cycle (Zone I) since the pulmonary artery pressure in this region is less than the alveolar pressure and the arterial vessels are closed. In approximately the middle third of the lung (Zone II), there is only blood flow during the systolic phase of the cardiac cycle when the arterial pressure is greater than the alveolar pressure and the arterial vessels are open. In approximately the lower third of the lung (Zone III) there is continuous blood flow since the arterial pressure here is always greater than the alveolar pressure. The good news is that almost two-thirds of the lung is readily available as a reserve for blood–tissue gas exchange when required by exercise or other factors that increases metabolic demand, simply by raising pulmonary artery blood pressure and perfusing more of the normally closed arterial vessels in the upper parts of the lung.

Shear stress and endothelial cells. Every blood vessel is lined by a single, continuous layer of *endothelial* cells, called the *endothelium*. True capillaries are vessels that are formed from a single layer of endothelial cells. The blood–tissue interface is thus separated by a distance approximating the thickness of an endothelial cell (about 1–3 μm). This close proximity facilitates diffusional transport of substances between the blood and tissues. Larger vessels, such as arteries and arterioles, have an endothelium as part of a multilayered composite structure. The endothelium is the layer of cells, which is always at the blood–tissue interface. The endothelium performs many functions. Among others, the endothelium provides a non-thrombogenic (non-clotting) surface for blood flow. Mechanical stresses, and in particular shear stress, affect the chemical response of endothelial cells. Equation 4.14a and Figure 4.6 show that the maximum shear stress exerted in a flowing fluid occurs at the tube wall, precisely at the endothelium. Endothelial cells respond relatively rapidly to shear stress changes. Second messengers, such as c-GMP and intracellular Ca initiate a cascade of intracellular events that alter gene expression, protein synthesis, cell turnover, and cell motility. The endothelial cells align themselves in the direction of the flow field. Low time-averaged shear stresses (oscillating around zero), particularly in spatial regions around bifurcations and in high curvature vessels, have been implicated in *atherogenesis* (the initiation of atherosclerotic plaques).

Neutrophils (and other *leukocytes*) are the first line of defense of our immune system. They are rapidly mobilized and attracted to sites of tissue injury and inflammation. Leukocytes are subject to large shear forces generated by the flow of blood through the circulatory system. In order for the rapidly circulating leukocytes to reach the site of injury, they first bind to the endothelial surface near the site, then roll across the surface to the site, attracted by mediators (*chemoattractants*) that are released at the injury site, and then migrate through the endothelium to the site of injury or inflammation. The binding and initial rolling of the leukocytes are mediated by a class of surface molecules called *selectins*, which are expressed on the surface of the leukocytes and have complementary recognition sites expressed on endothelial cells. The strengthening of the adhesion and the migration of the leukocytes are mediated by another class of adhesion molecules called *integrins*. Integrins are involved in the mechanism whereby the leukocytes can migrate between the endothelial cells to the site of injury without leaving a hole in the intact endothelial layer. Chemoattractant molecules are responsible for the transition from rolling to adhesion and migration.

4.12 Chapter Overview

In this chapter, we applied the principle of conservation of momentum and used a shell balance to derive the continuity equation and the equations of motion for steady laminar flow in a tube. We then extended this to derive the continuity equation for a compressible fluid and the general equations of motion for unsteady laminar flow in several regular geometries (rectangular, cylindrical, and spherical). We briefly discussed the use of CFD software with FSI capability when realistic hemodynamic problems are addressed. We also discussed the role of turbulent flow and pressure variations in the atrium, ventricles, and aorta in heart valve dynamics for the cases of valve stenosis and regurgitation.

Problems

4.1 Consider the parallel flow circuit illustrated in Figure P4.1. The following information is given:

$Q_A = 1300 \, \text{mL/min}$ $\bar{P}_1 = 100 \, \text{mm Hg}$.
$Q_B = 1100 \, \text{mL/min}$ $\bar{P}_2 = 4 \, \text{mm Hg}$.
$Q_C = 1350 \, \text{mL/min}$.

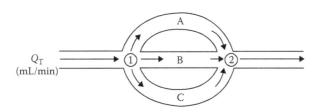

FIGURE P4.1
Parallel flow circuit.

a. What is the equivalent resistance of this circuit?

b. If the radius of Channel A decreases by 20%, and the pressure at Point 2 does not change. What must the pressure at Point 1 be in order for the total flow through the circuit to remain constant? Assume that Ohm's law and Poiseuille's law hold for this circuit.

c. If we require that ∇P remains constant, and the radius of Channel A increases by 20% and the radius of Channel C increases by 25%, what would the new flows be in Channels A, B, and C? Fill in the following table.

Channel	Flow after Dilation/ Contraction	Resistance before Dilation/ Contraction	Resistance after Dilation/ Contraction	Fraction of Total Flow before Dilation/ Contraction	Fraction of Total Flow after Dilation/ Contraction
A					
B					
C					

4.2 In the region of a plaque, velocity increases (kinetic energy) at the expense of pressure (potential energy). Figure P4.2 is a schematic diagram of a plaque in the aorta:

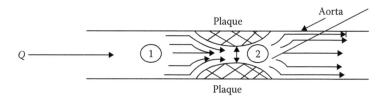

FIGURE P4.2
Schematic diagram for pressure drop and flow through a plaque in the aorta.

where

$$D_1 = 1.0\,\text{cm} \quad D_2 = 0.5\,\text{cm}$$

$$\bar{P}_1 = ? \quad \bar{P}_2 = 100\,\text{mm Hg}$$

$$\bar{V}_1 = 63.66\,\text{cm/s} \quad \bar{V}_2 = ?$$

D is the diameter of vessel, cm
$\bar{P}$ is the average pressure at indicated points, mm Hg
$\bar{V}$ is the average velocity at indicated points, cm/s
Q is the average flow rate, mL/s
a. Use Bernoulli's equation and the continuity equation to estimate the ventricular pressure (Position 1) and the velocity at the outlet (Position 2) across the plaque.

Given

Bernoulli's equation:

$$\Delta\left(\frac{1}{2}\rho\bar{V}^2\right) + g\Delta h + \Delta\bar{P} = 0$$

where

ρ is density of blood $= 1.056\,\text{g}/\text{cm}^3$

g is acceleration due to gravity $= 980\,\text{cm}/\text{s}^2$

h is height above reference pane (right atrium) $= 0$

Continuity equation:

$$\Delta(\bar{V}A) = 0$$

where A is the cross-sectional area at indicated points cm² and conversion factor $= 7.5 \times 10^{-4}$ mm Hg/(g/cm s²).

4.3 Flow in a nonlinear elastic tube:

a. Starting with the equations of motion and continuity, derive an equation for flow in a tube if the pressure drop, dP/dz, is constant. List all assumptions.

b. Assume that the cross-sectional area of the vessel is a nonlinear function of the transmural pressure as follows:

$$P = k\left[\frac{(A - A_0)^2}{A_0} + 1\right]$$

where

A is cross-sectional area of the tube at any time

A_0 is the reference area of the tube (constant)

k is the empirical constant

P is the transmural pressure across the tube wall

Starting with the results in Part a, derive an expression for flow (Q) in terms of the area, A and dA/dz. You can assume that the tube remains circular [i.e., $A(r) = \pi r^2$].

c. Integrate the equation from $z = 0$ to L (e.g., A_L), where L is the vessel length.

d. Use the relationship above to eliminate A_0 and A_L and derive an expression for Q in terms of an inlet pressure P_0 (at $z = 0$) and an outlet pressure P_L (at $z = L$).

4.4 Starting with the general equations of motion and continuity, derive expressions for the steady, laminar velocity profile and volumetric flow rate for a Bingham fluid in a rigid tube.

The constitutive equation for a Bingham fluid is given by

$$\tau_{rz} = -\mu_B \frac{dV_z}{dr} + \tau_0 \quad \text{for } |\tau_{rz}| > \tau_0 \quad \text{and} \quad \frac{dV_z}{dr} = 0 \quad \text{for } |\tau_{rz}| < \tau_0$$

where
  μ_B is the viscosity of the fluid
  τ_0 is a (constant) yield stress

NOTE: This means that there is a core of fluid in the center where $|\tau_{rz}| < \tau_0$ and V_z is *not* a function of r (V_z = constant). See Figure P4.4: Assume that the radius at that point is $r_0 = kR$, some fraction, k, of the tube radius, R.

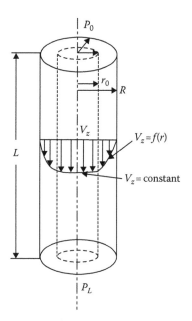

FIGURE P4.4
Velocity profile of a Bingham fluid in a right tube.

4.5 A stainless steel stent is to be inserted into a patient after balloon angioplasty of their left anterior descending (LAD) coronary artery. The internal diameter of the stent is 2.0 mm and its length is 10 mm. Use the equations of motion and continuity to set up the flow problem for calculation of the following hemodynamic parameters:
  a. The pressure drop through the stent
  b. The shear stress at the wall
  c. The average flow through the LAD coronary artery
  d. What would the Hagan–Poiseuille equation predict for Parts a, b, and c?
  e. Assuming that blood behaves as a Newtonian fluid whose viscosity is 2.5 times that of water and whose density is 1.08 g/cm³, calculate the Reynolds number for the flow in this stent (you will have to estimate the average velocity through the LAD). Does this correspond to laminar or turbulent flow?
  f. Explain the differences between laminar, turbulent, and pulsatile flow.

NOTE: *This problem cannot be solved analytically if pulsatile flow is assumed.*

References

1. Womersley, J.R. Method for the calculation of velocity, rate of flow and viscous drag in arteries when the pressure gradient is known. *J Physiol*. 127: 553–563, 1957.
2. Womersley, J.R. Oscilatory flow in arteries: The constrained elastic tube as a model of arterial flow and pulse transmission. *Phil Mag*. 46: 199–208, 1955.
3. Abbromowitz, M. and Stegen, I.A. *Handbook of Mathematical Functions, with Formulas, Graphs and Mathematical Tables*. National Bureau of Standards, U.S.A., 1960.
4. Jahnke, E. and Emde, F. *Funktionentafeln*, 3rd ed. Teubner Publishers, Leipzig, 1936, pp. 262, 266.
5. Fung, Y.C. Inversion of a class of nonlinear stress–strain relationships of biological soft tissue. *J. Biomech. Eng*. 101: 23–27, 1979.
6. Fung, Y.C. *Biomechanics. Material Properties of Living Tissues*. Springer Verlag, New York 1993.
7. Sim, W. and Sachs, M.S. Finite element implementation of a generalized Fung-elastic constitutive model for planar tissues. *Biomech. Model. Mechanobiol*. 4: 190–199, 2003.
8. Torii, R., Keegan, J., Wood, N.B., Dowsey, A.W., Hughes, A.D., Yang, G.-Z., Firmin, D.N., Thom, S.A.M., and Xu, X.Y. MR image-based geometry and hemodynamic investigation of the right coronary artery with dynamic vessel motion. *Ann. Biomed. Eng*. 38: 2606–2620, 2010.
9. Spilker, R.L. and Taylor, C.A. Tuning multidomain simulation to match physiological measurements. *Ann. Biomed. Eng*. 38: 2635–2648, 2010.
10. Figueroa, C.A., Vignon-Clementel, I.E., Jansen, K.E., Hughes, T.J.R., and Taylor, C.A. A coupled momentum method for modeling blood flow in 3-D deformable arteries. *Comput. Methods Appl. Mech. Eng*. 195: 5685–5706, 2006.
11. Krittian, S., Schenkel, T., Janoske, U., and Oertel, H. Partitioned fluid-solid coupling for cardiovascular blood flow: Validation study of pressure-driven fluid-domain deformation. *Ann. Biomed. Eng*. 38: 2676–2689, 2010.
12. Leach, J.R., Vitaliy, L.R., Soares, B., Wintermark, M., Mofrad, M.R.K., and Saloner, D. Carotid atheroma rupture observed in vivo and FSI predicted stress distribution on pre-rupture imaging. *Ann. Biomed. Eng*. 38: 2748–2765, 2010.

Further Readings

Adams, D., Barakch, J., Laskey, R. and Breeman, C.V. Ion channels and regulation of intracellular calcium in vascular endothelial cells. *FASEBJ3*: 2389–2400, 1989.

Asakura, T. and Karino, T. Flow patterns and spatial distribution of atherosclerotic lesions in human coronary arteries. *Circ. Res*. 66: 1045–1066, 1990.

Bayliss, W. On the local reaction of the arterial wall to changes in internal pressure. *J. Physiol*. 28: 220–231, 1992.

Berridge, M.J. Inositol triphosphate and diacylglycerol: two interacting second messengers. *Annu. Rev. Biochcm*. 56: 159–163, 1987.

Ferris, C. and Snyder, S. Inositol 1,4,5, triphosphate-activated calcium channels. *Annu. Rev. Physiol.* 54: 469–488, 1992.

Kirber, M., Walsh, J. and Singer, J. Stretch-activated ion channels in smooth muscle, a mechanism for initiation of stretch-induced contraction. *Pflugers Arch.* 412: 339–346, 1988.

Langille, B.L. and Adamson, S.L. Relationship between blood flow direction and endothelial cell orientation: arterial branch sites in rabbits and mice. *Circ. Res.* 58: 481–488, 1981.

Letsou, G., Rosales, O., Maitz, S., Vog, A., and Sumpio, B. Stimulation of adenylate cyclase activity in cultured endothelial cells subjected to stretch. *J. Cardiovasc. Surg.* 31: 634–639, 1990.

Lilly, L.S. Heart failure. In: Lilly, L.S. (Ed.) *Pathophysiology of Heart Disease.* Williams and Wilkins, Baltimore, 1998, p. 201.

Mo, M., Eskin, S., and Schilling, W. Flow-induced changes in Ca^{2+} signaling of vascular endothelial cells: effect of shear stress and ATP. *Am. J. Physiol.* 260: H1698–H1707, 1991.

Moore, J.E., Ku, D.N., Zarins, C.K., and Glagov, S. Pulsatile flow visualization in the abdominal aorta under differing physiological conditions: Implications for increased susceptibility to atherosclerosis. *J. Biomed. Eng.* 114: 391–397, 1992.

Morris, C. Mechanosensitive ion channels. *J. Membr. Biol.* 113: 93–107, 1990.

Nollert, M.U., Hall, E.R., Eskin, S.G., and McIntire, L.V. Effect of flow on arachidonic acid metabolism in human endothelial cells. *Biochimi. Biophysi. Acta.* 1005: 72–78, 1989.

Nollert, M.U., Panaro, N.J., and McIntire, L.V. Regulation of genetic expression in shear stress-stimulated endothelial cells. *Ann. NY Acad. Sci.* 665: 94–104, 1992.

Okano, M. and Yoshida, Y. Endothelial cell morphometry of atherosclerotic lesions and flow profiles at aortic bifurcations in cholesterol fed rabbits. *J. Biomed. Eng.* 114: 301–308, 1992.

Rosales, O., Sumpio, B. Changes in cyclic strain increase inositol triphosphate and diacylglycerole in endothelial cells. *Am. J. Physiol.* 262: C956–C962, 1992.

Rubayni, G.M., Romero, J.C., and van Houtte, P.M. Flow-induced release of endothelium-derived relaxing factor. *Am. J. Physiol.* 250: Hl145–H1149, 1986.

Schwartz, L.B., O'Donohue, M.K., and Purut, C.M. Myointimal thickening in experimental vein grafts is dependent on wall tension. *J. Vasc. Surg.* 15: 176–186, 1992.

Sumpio, B., Banes, A., Levin, L. Mechanical stress stimulates aortic endothelial cells to proliferate. *J. Vasc. Surg.* 6: 252–256, 1987.

Sumpio, B. and Widmann, M. Enhanced production of an endothelium derived contracting factor by endothelial cells subject to pulsatile stretch. *Surgery* 108: 277–282, 1990.

White, G.E. and Fujiwara, K. Expression and intracellular distribution of stress fibers in aortic endothelium. *J. Cell. Biol.* 103: 63–70, 1986.

5

The Cardiovascular System

5.1 Introduction

The cardiovascular system consists of the *heart* and the *systemic circulation*. While many important organ systems such as the lungs, kidneys, and brain are not part of the cardiovascular system, much of the work in *biomedical engineering* centers on the cardiovascular system. Therefore, we have chosen to use this system to develop and illustrate some basic principles of *biomechanics and hemodynamics* in both large vessels and *myocardium*. These principles can readily be applied to other organ systems, vessels, and cellular processes by considering their individual anatomical arrangement and functional properties.

The cardiovascular system is too complex to be described in simple terms. For example, to describe the heart simply as a mechanical pump is a gross oversimplification as anyone who has tried to develop an artificial heart will testify. Yet, one of the primary functions of the ventricles is to circulate blood. There are two types of descriptions of the cardiovascular system, which are most *prevalent* in the literature depending on the focus of the discussion: *anatomical* and *functional*. Anatomical descriptions are useful for orientation and to justify the geometry to be used in the analysis. Functional descriptions are usually concerned only with those functional properties of the organ or tissue, which are to be analyzed or modeled (e.g., blood–tissue transport of *macromolecules* in the coronary *microcirculation*). Within these general guidelines, both anatomical and functional descriptions can become quite detailed.

Figure 5.1 is a schematic diagram of the circulatory system (which includes the cardiovascular system). The following observations will help us.

- The vessels form a series–parallel flow network. Flow through the network is a function of the impedance (resistive and *capacitive* properties) of the individual vessels (vessel radius and length) as well as network properties (the series and parallel connections of the vessels). Network properties can vary with time and the metabolic requirements of the tissue or organ that is being supplied by the exchange vessels.

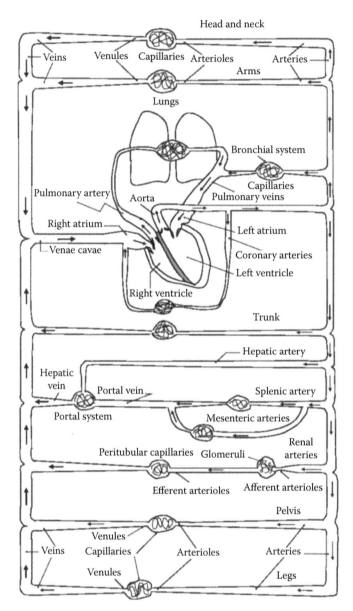

FIGURE 5.1
Vessel and organ arrangement in the cardiovascular system.

- Not all capillary beds are open (flowing) all the time due to *vasomotion* of the *terminal arterioles*.

- Vessels are tapered and form branches creating a tree-like structure.

- Daughter vessels *increase* in number and *decrease* in diameter as blood moves away from the left side of the heart until the level of the capillaries.

TABLE 5.1

Typical Dimensions of Circulatory System Vessels in Humans

Vessel	Approximate Number	Diameter (cm)	Total Cross-Sectional Area (cm²)	Avg. Velocity (cm/s)	Wall Thickness (cm)
Ascending aorta	1	2.0–3.2	2.5	63–27[a]	0.2
Large arteries	1900	0.2–0.6	20	20–50[a]	0.1
Arterioles	40×10^6	3×10^{-9}	40	1.55[b]	2×10^{-8}
Capillaries	12×10^9	8×10^{-10}	2500	0.05–0.1[b]	1×10^{-9}
Venules	80×10^6	2×10^{-9}	250	1.5–3.0[b]	2×10^{-9}
Veins	1900	0.5	80	1520[b]	0.05
Venae cavae	2	2–3	8	11–16[b]	0.15

[a] Peak average velocity.
[b] Time average velocity.

- Postcapillary vessels *increase* in diameter and *decrease* in number as the blood returns to the right side of the heart.
- The branching pattern is such that the *total* cross-sectional area available for flow increases from the aorta through the capillaries and then decreases from the *venules through the venae cavae*. Table 5.1 lists some typical dimensions and numbers of vessels in the circulatory system.

5.1.1 Vessels Are Identified by Their Anatomical Structure and Size

Arteries: These vessels are thick walled. The structure of the wall consists of three concentric cylindrical layers. The innermost layer, *the tunica intima*, consists of a single, continuous layer of endothelial cells resting on a basement membrane below which are collagen and elastin fibers and smooth muscle cells. The middle layer, *the tunica media,* also contains elastin and collagen fibers and smooth muscle cells. It is separated from the inner layer by the *internal elastic lamina*. The outer layer, *the tunica adventitia,* consists of loose connective tissue containing elastin and collagen. The walls of large arteries are nourished by small blood vessels, *the vasa vasorum*, which form capillary networks within the adventitia or outer media. The large arteries have relatively small numbers of smooth muscle cells in their walls and are not significantly affected by *sympathetic neuromuscular regulation*. They do, however, include *mechanoreceptors* (the aortic arch and carotid sinus baroreceptors) that are part of the *negative feedback* reflex arc that regulates systemic arterial pressure through the *vasomotor control center* in the *medulla*. The large arteries range from approximately 400 to 100 μm* in diameter. The primary function of the large and intermediate-sized arteries (200–400 μm) is to act as

---

* $1\,\mu m = 10^{-9}\,cm$.

conduit vessels for the distribution of blood to the organs and tissues of the body. The resistive and capacitive elements of the aorta and the large arteries convert pulsatile flow from the left ventricle into a continuous, steady flow in the smaller arteries and arterioles in much the same way that the springs and shock absorbers of an automobile suspension system dampens the bumps in a rough road. Recent measurements indicate that smaller arteries (100–150 μm) may also participate in blood–tissue exchange of highly diffusible gasses such as oxygen.

Arterioles: These vessels have thinner walls than the true arteries. The intima layer of the smaller arteries consists of a single, continuous layer of endothelial cells, which is separated from the media only by the internal elastic lamina.* The media contains relatively less elastin and collagen and more smooth muscle cells than do the large arteries. The smooth muscle cells of the arterioles are spirally arranged around the long axis of the vessel with relatively little connective tissue between them. The layers of smooth muscle cells are in direct contact with the endothelial cells in the intima. These smooth muscle cells are innervated by sympathetic neuromuscular junctions on the adventitial side. Arterioles range in size from approximately 150 to <15 μm in diameter. These are the major resistance vessels in the circulatory system and their major functions are the regulation of blood pressure and the redistribution of the cardiac output (CO) in response to changing metabolic requirements of organs and tissues. They may also participate in gas exchange. The smallest arterioles called *terminal arterioles* directly feed a single capillary bed.

Capillaries: These vessels are characterized by the fact that their wall consists of a *single, continuous layer of endothelial cells* enclosed by a basement membrane. *There are no smooth muscle cells in capillary walls.* Endothelial cells are very thin ranging in thickness from about 0.1 to 0.3 μm. The cells are separated from each other by *interendothelial junctions*. These junctions vary widely among different tissues in a way that appears to be closely related to their functional needs. Interendothelial junctions are believed to be one of the major pathways for blood–tissue exchange of water and small solutes. Capillary sizes range from approximately 10 to <5 μm in diameter. The major function of the capillaries is blood–tissue exchange of small solutes (e.g., Na and glucose) and water.

---

* Arterioles are often also characterized as first order, second order, etc. This refers to branching position with respect to the largest arteriole in the field of interest. The largest arteriole is designated as a first-order arteriole. The first branch of the first-order arteriole is designated as a second-order arteriole, and so on. Terminal arterioles are those arterioles that directly feed a capillary bed. The smaller arterioles (<40 μm in diameter) exhibit *vasomotion*. This is a rhythmic fluctuation in flow through the arteriole with a period of 1–10 s brought about by the periodic motion of the arteriole wall. Vasomotion is an *active response* brought about by the contraction and relaxation of the smooth muscle in the arteriole wall in response to neurogenic and local regulatory processes.

Venules: Blood flow from the capillaries collects in the venules. The venules also have a wall that consists of a single layer of endothelial cells enclosed by a basement membrane. However, the cytoplasm of venule endothelial cells is very thin and fenestrated. That is, there are holes in the cytoplasm. Some of these fenestrate are closed by diaphragms (lids) and some are not. The basement membrane is discontinuous. The smaller venules (15–40 μm in diameter) are surrounded by a minimal adventitia and occasional contractile cells called *pericytes*. Because of their relatively loose anatomical structure, the major function of venules appears to be blood–tissue transport of large molecules (e.g., albumin) and fluid. The larger venules have smooth muscle and along with the veins there are principally collecting and storage vessels for return of blood to the right side of the heart.

Veins: The veins are characterized as distensible, thin-walled (relative to arteries) vessels. There are innervated smooth muscle cells in the walls of veins. However, the density of the smooth muscle cells is much less than that found in arterioles. The veins also have *one-way valves* that help prevent blood from moving away from the heart. Since the flow velocity and pressure are low in the veins, the valves help prevent pooling of the blood in the lower extremities under the influence of gravity, that is, when going from a prone to a standing position. The major function of the venous system is to act as a *blood reservoir*.

Under normal resting conditions, 64%–67% of the entire blood volume (total blood volume is approximately 5 L for a 75 kg man) is in the venous* system. Sympathetic neural stimulation of veins decreases the volume of blood stored so that during exercise, when the CO increases, the increased circulating volume is recruited from the venous system.

Table 5.2 shows the distribution of the blood volume in the cardiovascular system under normal resting conditions.

TABLE 5.2

Resting Blood Volumes

Organ	% of Total
Large arteries	8
Small arteries	5
Arterioles	2
Capillaries	5
Small veins, venules, and venous sinuses	25
Large veins and venous reservoirs	39
Heart	7
Pulmonary vessels	9

* At rest, the arterial system holds only about 10%–15% of the total blood volume.

5.2 Flow, Pressure, and Volume Relationships in the Cardiovascular System

The left ventricle and arterial circulation can be thought of as a pair of coupled mechanical systems whose output is regulated by several reflex and local feedback and anticipatory control systems. The overall mechanical behavior of the coupled system is determined by the mechanical properties of the individual units and the set points of the regulators. The equilibrium point for mechanical equilibrium is determined by the interactions of the two systems.

A model for the equilibrium that exists in pressure and flow between the left ventricle and the arterial circulation can be derived from several different perspectives. If we are interested in the long-term behavior, we may wish to use a time-average approach to the relationship between pressure and flow. This approach is less complex mathematically and yet can provide clinically useful information when following the course of a particular treatment modality in patients with heart failure or hypertension. On the other hand, a more mathematically complex analysis suitable for physiological and pharmacological studies of the heart and circulation would need to include instantaneous pressure and flow. In this chapter, we will consider the time-average approach as well as a *compartmental approach* to cardiovascular dynamics that allows us to gain some insight into the dynamics of the cardiovascular system. We discuss instantaneous pressure-flow models in Chapter 4, *Hemodynamics*.

The events of the cardiac cycle start with an electrical signal (excitation) and then proceed through *excitation–contraction coupling* (which involves chemical and mechanical events) to a contraction of the ventricles (pressure generation) and ejection of blood (flow) into the pulmonary and systemic circulations. The electrical events of the cardiac cycle can be measured on the body surface (the skin) by using surface electrodes. A system of surface electrodes originally proposed by Einthoven has evolved into a common clinical diagnostic tool called the *electrocardiogram*, which is abbreviated as ECG or EKG (after the German spelling). Because electrical events always precede mechanical events in the cardiac cycle, distortions of a part or parts of the electrical signal have been used as diagnostic indicators of both electrical and mechanical dysfunctions of the heart muscle.

5.2.1 Electrocardiogram

The electrical events that precede the mechanical contraction and relaxation of the atrial and ventricular muscles are generated by the movement of Na^+, K^+, and Ca^{2+} ions into and out of specialized conduction cells in the myocardium and in cardiac muscle cells, generating an electrical spike or *action potential*.

In terms of their electrical activity, we can classify cardiac muscle cells into two categories.

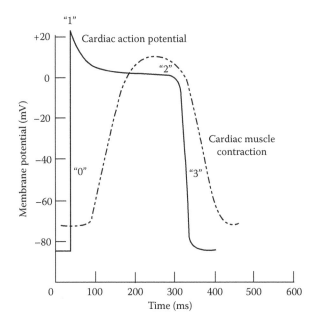

FIGURE 5.2

Relationship between membrane "action" potential and cardiac muscle contraction in a non-pacemaker cardiac muscle cell.

Nonpacemaker cells: Figure 5.2 is a schematic of the transmembrane potential that occurs in a typical nonpacemaker cardiac muscle cell. Note that the resting membrane potential is −85 mV (with respect to the outside of the cell). It means that when positively charged ions such as Na^+ move into this cell, its membrane potential increases toward zero. This process is called *depolarization*. An action potential is initiated when the neurotransmitters *acetylcholine and/or norepinephrine* act on their respective receptors on the surface of cardiac muscle cells to open a *voltage-gated* fast sodium channel. This allows sodium ions to rapidly enter the cell and cause rapid depolarization, which is indicated by the "0" phase in Figure 5.2. At some point near zero potential (indicated by "1"), the fast sodium channel closes. At this point, there is an increase in slow calcium/sodium conductance into the cell, stabilizing the membrane potential from "1" to "2." This is called the plateau phase of the action potential. During this phase, the relatively high calcium conductance allows Ca^{2+} to be transported down its electrochemical gradient from outside the cell to inside the cell increasing the intracellular concentration of free Ca^{2+}. The intracellular $[Ca^{2+}]$ then acts as a second messenger, initiating a cascade of intracellular events that eventually leads to contraction of the cytoskeleton of the cardiac muscle cell. A subsequent inactivation of the inward Ca/Na current and an increase in inward potassium conductance restore the membrane potential to its resting negative value (Phase "3"), where the cell waits for the next cycle to repeat.

Pacemaker cells: The *sinoatrial (SA) node* is a small group of specialized muscle cells that lie in the wall of the right atrium. Within this node lie *pacemaker cells* that are capable of spontaneous depolarization and under normal physiological conditions generate action potentials that determine the heart rate. They differ from nonpacemaker cells in that their resting potential slowly increases (probably due to an influx of Na from a slow, or "leakage," sodium channel) until a threshold value is reached. At this point, the fast sodium channel opens, causing a more rapid depolarization. Pacemaker cells also lack the slow calcium/sodium plateau phase, which is consistent with their role as generators of electrical signals rather than actively contracting units. Pacemaker cells are also found at the *atrioventricular* (AV) node and within the ventricular walls. Under normal physiological conditions, it is probable that only one pacemaker cell in the SA node, the most rapidly depolarizing one, goes through this spontaneous process all the way to completion. Once a single pacemaker cell fires, the action potential is propagated both from cell to cell contact and through specialized sets of cells that provide conduction pathways through the heart muscle. Ventricular muscle cells have long-duration action potentials (300–400 ms). Their mechanical activity begins 50–100 ms after the initiation of the electrical activity.

Under normal physiological conditions, the action potential is initiated by the firing of the *vagus* nerve at its junction with the SA node. The action potential is then spread radially through the atrial muscle cells by three mechanisms (see Figure 5.3):

1. Cell-to-cell contact.
2. *Bachmans bundle (BB)*. This is a specialized set of conduction fibers that help propagate the action potential from the right atrium to the left atrium (LA).

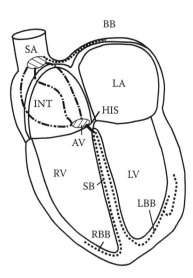

FIGURE 5.3
Action potential conduction pathways. See text for abbreviations.

3. *The internodal tracts (INT)*. While the main function of this conduction pathway is to conduct the excitation voltage from the SA node to the AV node, atrial depolarization is assisted by radial spread of depolarization from the INTs.

Shortly after the muscle cells are depolarized, they contract, expelling the contents of the atria into the ventricles. While this is happening, the action potential is being conducted from the SA node in the atria to the *AV node* in the ventricles through the INTs.

The conduction properties of the AV node are such that it delays impulse conduction by about 100 ms allowing the atria to contract. From the AV node, the impulses are conducted down the *Bundle of HIS* to the bundle branches. The bundle branches start in the ventricular septum with the *Septal bundle (SB)*, then split into the *left bundle branch (LBB)* and *right bundle branch (RBB)*.

From the LBB and RBB impulses are spread through the ventricular muscles themselves by another set of specialized conduction fibers called *Purkinje Fibers* (not shown in Figure 5.3).

Shortly after the ventricular muscles are depolarized they contract in concert. The wave of depolarization travels through the ventricular muscles in a very specific pattern. After a *short latency* period, the atrial and ventricular muscle cells spontaneously repolarize by the mechanisms described above. *Repolarization* leads to relaxation of the atria and ventricles.

5.2.1.1 Standard Limb Leads (I, II, and III)

The waves of depolarization and repolarization can be recorded by electrodes placed on the skin surface in a roughly triangular pattern first described by *Einthoven* (see [1]). Surface electrodes are placed on the right arm (RA), left arm (LA), and left leg (LL). An additional electrode is placed on the right leg to act as a ground. In a *bipolar* configuration, they record the difference in surface potential between two of the three electrodes. By convention, the positive and negative ends for each lead are taken as shown in Figure 5.4. The difference in potential between the RA and LA electrodes is designated as Lead I. Similarly, differences in potential between LA and LL are designated as Lead II and differences in potential between RA and LL are designated as Lead III. Leads I, II, and III are called the *standard limb leads* and are shown schematically in Figure 5.4. These leads measure electrical activity of the heart in the frontal plane, with the heart positioned at the center of an equilateral triangle as shown in Figure 5.4.

The electrical patterns recorded by these surface electrodes are called *ECG*. A typical Lead II ECG is illustrated.

Since these electrodes only record electrical activity in one plane (the frontal plane), other electrodes positions were added over time to record electrical activity in other planes and on the chest.

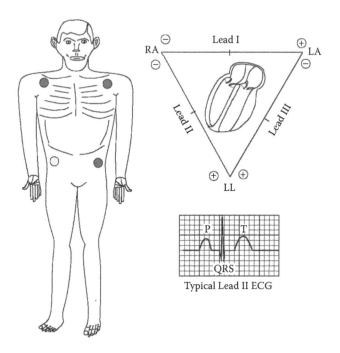

FIGURE 5.4
Placement of standard limb leads. Einthoven's triangle and a typical Lead II ECG signal.

5.2.1.2 Augmented Limb Leads (aVR, aVL, and aVF)

In an attempt to increase the quality and quantity of information obtained from ECG recordings, electrocardiographers have proposed various enhancements to the standard limb leads. The two limb leads that are in common use today are the V leads and the augmented V (aV) Leads. Both the V and aV leads compare an exploring electrode (+) with a reference voltage (−). In the aV system, the reference voltage is obtained by dropping the measured surface voltage at each of the standard limb leads through a 5000 Ω resistor. The exploring electrode (+) (RA, LA, or LL) is disconnected from its resistor. The two remaining voltages are brought together at a common point and act as the reference (−). For example, the aVR lead is formed by the voltage difference between the electrode at the RA (+) and the sum of the voltages from the LA and LL, each dropped through a 5000 Ω resistor (−). Currently, aVR (for the RA), aVL (for the LA), and aVF (for the left foot) are commonly used.

5.2.1.3 Chest Leads (V1–V6)

The chest leads use the difference between the surface potential of an exploring electrode (+) placed in various positions on the chest (see Figure 5.5) and a reference electrode (−) that consists of the sum of all three standard limb electrodes (RA, LA, and LL) each dropped through a 5000 Ω resistor.

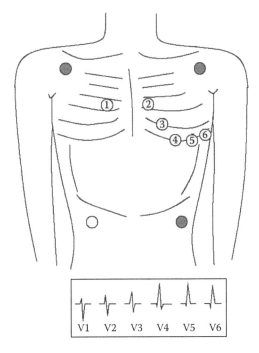

FIGURE 5.5
Placement of electrodes for standard 12-lead ECG. Typical QRS complexes recorded in Leads V1–V6.

These are designated as Leads V1–V6. Because the reference electrode is the electrical sum of the three standard limb leads dropped through resistors, the position of the reference electrode is taken to be at the middle of the equilateral triangle (in the approximate region of the AV node as shown in Figure 5.4). Since the exploring electrodes are placed at various positions on the chest surface, the chest leads monitor electrical activity in the horizontal plane. Chest leads allow views of the electrical activity of discrete, small, regions of the heart from both the right and left sides. Therefore, electrical abnormalities that might not be detected by the limb leads may be detected on the chest leads.

Both the standard and augmented limb leads monitor electrical activity of the heart in the frontal plane at different positions with respect to a circle drawn around Einthoven's triangle. The heart is assumed to be oriented within the circle as shown in Figure 5.5. The positive pole of Lead I (RA) is arbitrarily set at 0°, with positive angles going clockwise and negative angles going counterclockwise starting at the RA. In that coordinate system, Lead II monitors electrical activity at +60° and Lead III at +120°. In the coordinate system, Lead aVR monitors electrical activity at −150°, aVL monitors electrical activity at −30° and aVF monitors electrical activity at +90°.

5.2.2 The 12-Lead ECG

A routine ECG uses all 12 leads discussed above (Leads I, II, and III; aVR; aVL; aVF; and V1–V6). The placement of the electrodes used to monitor the 12 leads is shown in Figure 5.5.

TABLE 5.3

Chest Lead Placement

Lead	Location on Chest
V1	Fourth intercostal space at right sternal boarder
V2	Fourth intercostal space at left sternal boarder
V3	Equidistant between V2 and V4
V4	Fifth intercostal space at the left midclavicular line. Leads V5 and V6 are taken in the same horizontal plane as V4 and V6 are taken in the same horizontal plane as V4
V5	Anterior axillary line
V6	Mid-axillary line

Table 5.3 indicates the location of the currently used chest electrodes.

The parts of the ECG associated with atrial depolarization (the P wave), ventricular depolarization (the QRS complex), and ventricular repolarization (the T wave) are shown in Figure 5.4. Please note that the electrical activity recorded by the surface electrodes represent events taking place in atrial and ventricular *muscle mass* only. The mass of tissue associated with the specialized conduction pathways is too small for those signals to be recorded by the surface electrodes. Also note that since *atrial repolarization* takes place during the QRS complex, its signal is overwhelmed by the signal generated by the depolarization of the ventricular muscle mass and is not recorded by the surface electrodes.

Characteristic changes in both the rate and pattern of the ECG have enabled cardiologists to diagnose many cardiac diseases and abnormalities including cardiac *arrhythmias* (changes in heart rate due to conduction problems), *ischemia* (low flow through the coronary arteries), and cardiac muscle damage *(myocardial infarction)*. All these abnormalities produce changes in the rate and/or pattern of the ECG that can aid in the diagnosis of these (and other) cardiac diseases.

5.2.3 Guides for Interpretation of ECG Strips

Table 5.4 presents a checklist that will help in the systematic interpretation of ECG strips. Table 5.5 will help in identifying the type of arrhythmia, when there is a positive finding according to the guidelines in Step 2 of Table 5.4.

5.3 The Cardiac Cycle

Figure 5.6 shows the pressure changes during a single cardiac cycle in the left atrium, left ventricle, and aorta. The *ECG* is included to emphasize the fact that mechanical events are *initiated and preceded by* electrical events.

TABLE 5.4

Checklist for Interpretation of ECG Strips

---

1. Calibration

   Check 1.0 mV calibration signal. Standard = 10 mm/mV

2. Identify presence of sinus rhythm:

   Each P wave is followed by QRS

   Each QRS is preceded by a P wave

   The P wave is upright in Leads I, II, and III

   The PR interval is >0.12 s

   If these criteria are not met, indicates arrhythmia (see Table 5.5).

3. Heart rate calculation

   HR (beats/min) = 60 (s/min)/x (s/beat)

   where x = number of seconds between successive QRS peaks. Generally average 4–6 beats

   Normal rate = 60–100 beats/min (bradycardia < 60, tachycardia > 100)

4. Check interval times

   Normal PR = 0.12–0.20 s

   Normal QRS ≤ 0.10 s

   Normal QT ≤ (R–R interval)/2, if heart rate is normal

5. Mean electrical axis

   Normal if QRS is upright in Leads I and II (+90° to −30°)

6. P wave abnormalities

   Inspect *P* in Leads II and V1 for left and right atrial enlargement (*P* wave higher than 2.5 mm in Lead II)

7. QRS wave abnormalities

   Look for left and right ventricular hypertrophy

   Right ventricular hypertrophy

      R > S in Lead V1

   Left ventricular hypertrophy

      S in Lead V1 positive and R in V5 or V6 ≥ 35 mm *or*

      R in aVl > 11 mm *or*

      R in Lead 1 > 15 mm

   Look for bundle branch blocks (BBB). QRS > 0.10 s: *plus*

      RBBB: no Q, plus RSR′ in Lead V1

      LBBB: R–R′ in Leads V4–V6

8. Myocardial infarctions (MI)

   Look for Q wave abnormalities: Q waves > 0.4 s and/or ≥ 25% of the total QRS height

   May be present in 1 or more leads depending on the site of the MI

9. ST segment/T wave abnormalities

   ST elevations: Indicates transmural MI or pericarditis

      ST depressions/T wave inversions:

         Ischemia (acute; accompanied by chest pain) or MI

         Also occur with ventricular hypertrophy or BBBs

         Metabolic abnormalities and/or chemical imbalances

            Hyper/hypokalemia

            Hypo/hypercalcemia

10. Compare with patient's previous ECGs where available

TABLE 5.5

Arrhythmias

	ECG Findings
A. Bradyarrhythmias: HR < 60 BPM	
1. Sinus bradycardia	Normal P waves, HR < 60 BPM.
2. Conduction blocks	
1°:	PR Interval is prolonged.
2°:	*(Wenchkebach)*-P-wave rate is
	1. Constant but PR interval is progressively lengthened until a QRS is completely blocked
	2. *(Mobitz II)*-P-wave rate is constant and a QRS is eventually blocked without lengthening of the PR interval
3°:	1. P-wave and QRS rates are independent of each other
	2. QRS complexes are widened
	3. Junctional escape rhythm
	4. Normal width QRS complexes are not preceded by P-waves and slow HR (40–60 BPM)
	5. Ventricular escape rhythm. Even slower rates (30–40 BPM) and a widened QRS complex, not preceded by P-waves
B. Tachyarrhythmias: HR > 100 BPM	1. Sinus tachycardia. Normal P-waves and QRS complexes, HR > 100 BPM
	2. Atrial premature beats. An earlier than expected P-wave. Normal P and QRS
	3. Atrial flutter rapid, "sawtooth" appearing atrial activity with rates of 250–300 BPM, well-defined QRS
	4. Paroxysmal tachycardias. Sudden onset and termination, atrial rates of 140–250 BPM, normal QRS (P-wave may be hidden or abnormal during onset)
	5. Ventricular premature beats (VPBs). VPBs are unexpected, widened QRS-waves. If they appear in the place of an expected P-wave they are usually benign. If they occur in the place of an expected T-wave they may be life threatening
	6. Ventricular tachycardia. A series of three or more VPBs in a row
	7. Ventricular fibrilation. The most life-threatening arrhythmia. Results in disordered rapid stimulation of the ventricles, preventing them from contracting in a coordinated way. Results in a severe drop in CO and death, if not treated quickly

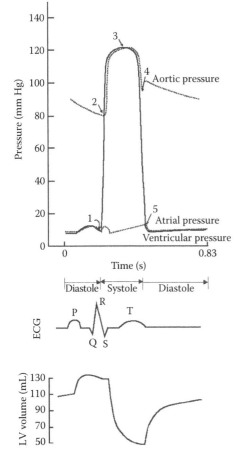

FIGURE 5.6
Pressure–volume changes during a single cardiac cycle. The ECG is included to show the relationship between the electrical and mechanical events.

The ventricular volume curve is included to illustrate the pulsatile nature of ventricular ejection.

The cardiac cycle is roughly divided into two phases: *systole and diastole*. Systolic events are associated with contraction of the *ventricles* and ejection of blood. Note that with this definition, *atrial contraction* is a *diastolic* event. Diastolic events are associated with relaxation of the ventricles and ventricular filling. There are two leaflet valves on each side of the heart that allow for pressure generation and one-way flow into and out of the ventricles. One of the valves is located between the atrium and the ventricle (the AV valve). The other valve is located between the ventricle and the aorta on the left side of the heart (the *aortic* valve) and between the ventricle and pulmonary artery (the *pulmonic* valve) on the right side of the heart.

The important events in the cardiac cycle are indicated in Figure 5.6. Since the cardiac cycle repeats itself every heartbeat, we can start anywhere. It is instructive to start at the P wave of the ECG in Figure 5.6. The P wave precedes atrial contraction (Point 1). At this point, the AV valve is still open and a small volume of blood is ejected into the ventricle. The volume in the ventricle is the *end-diastolic volume* (EDV), and the pressure in the ventricle is the *end-diastolic pressure* (EDP). EDV is a measure of ventricular stretch (or strain) before the ventricles contract. The passive stress in the ventricular wall at the end of diastole, as measured by either EDP (radial stress) or EDV (strain) is called **preload**. Shortly thereafter, the QRS complex, which initiates ventricular contraction, occurs. The initial contraction of the ventricles raises the ventricular pressure above the atrial pressure closing the AV valve. At this point, both the AV and aortic valves are closed and the volume in the ventricle is constant. The ventricle continues to contract against a fixed volume of an incompressible fluid (blood), raising the pressure in the ventricle. The pressure rise from Point 1 to 2 is called *isovolumic contraction*. At Point 2, the pressure in the ventricle is just greater than the aortic diastolic pressure (*afterload*) opening the aortic valve. With the opening of the aortic valve, the heart muscle has overcome the afterload and can now shorten, ejecting the stroke volume (SV) into the aorta. After Point 2, the pressures in the ventricle and aorta are very nearly equal until the aortic valve closes. Point 3 indicates aortic (and ventricular) systolic pressure. The middle of the T wave occurs just after Point 3. The T wave initiates ventricular repolarization and relaxation of the ventricular muscle, and pressure in the ventricle falls. At Point 4, the pressure in the ventricle falls below the pressure in the aorta, closing the aortic valve. The pressure in the aorta rises slightly (*the incisura*) as the elastic recoil of the aorta pushes blood against the closed aortic valve. Now, there is a constant volume of blood left in the ventricle (*end-systolic volume*). As the ventricular muscle continues to relax, ventricular pressure continues to fall until the pressure in the ventricle falls below the pressure in the atrium at Point 5 opening the AV valve. The phase from Point 4 to 5 is called *isovolumic relaxation*. Note that while the AV valve is closed, blood continues to return to the atrium from the pulmonary vein. This accounts for the continuous rise in pressure in the atrium during the time that the AV valve is closed. When the AV valve opens at Point 5, we begin ventricular filling. Ventricular filling continues until just after Point 1, when the QRS again initiates ventricular contraction, raising the pressure in the ventricle and closing the AV valve. The volume of blood ejected into the ventricle during the ejection phase of the cardiac cycle is the SV.

5.4 Propagation of the Pressure Wave

The pressure wave generated by left ventricular ejection is propagated throughout the systemic circulation. The pressure wave (or pressure pulse) is propagated through the blood-filled circulatory system with a characteristic

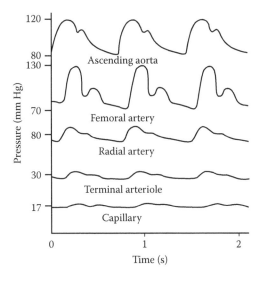

FIGURE 5.7
Pressure waves as they progress through the systemic circulation.

velocity that approaches the speed of sound in a viscous fluid. As with any other wave, the pressure wave is reflected at vessel branches and bends. At any position in the arterial tree, the forward pressure wave interacts with reflected waves from distal vessels to produce wave reinforcement (nodes) and cancellations (antinodes). Thus, the *morphology* (shape) of the pressure wave as recorded in various parts of the systemic circulation is different.

The resistive and capacitive elements of the arterial network also act as a **low-pass** filter. The high-frequency components of the pressure wave are filtered out as the pressure wave travels through the arterial network, resulting in an almost steady (nonpulsatile) pressure tracing in the capillaries.

Figure 5.7 shows the changes in amplitude (damping) and smoothing of the pressure pulse as it travels through the systemic circulation.

Figure 5.8 shows the average pressure at various positions in the systemic circulation from the aorta to the pulmonary venous system. The average pressure produced by the left ventricle (~100 mm Hg) is dissipated in the systemic circulation by the hemodynamic resistance of the vessels and blood and falls to ~0 mm Hg at the right atrium.

5.5 Impedance and Resistance

To better quantify flow, pressure, and volume relationships in the circulation, we introduce the concepts of *hemodynamic* impedance, *resistance, and compliance.* Since the pressure generated by the left ventricle is oscillatory and the aorta and arterial blood vessels are *distensible,* blood flow at any point in the arterial tree varies with time within a single cardiac cycle.

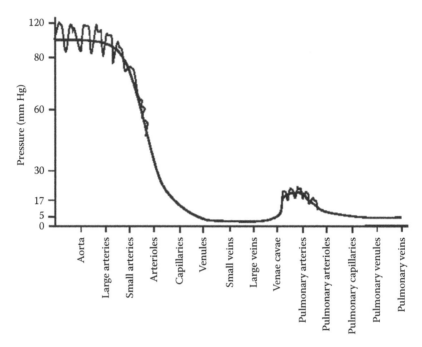

FIGURE 5.8
Pressure drop through the systemic and pulmonary circulations.

The term "cardiac output" (e.g., 5 L/min) is the volumetric output (flow) of the left ventricle (or the right ventricle) *averaged* over 1 min. Since pressure and flow in the large arteries vary with time, we use the concept of *impedance* to characterize the hemodynamic opposition of the vessels to flow. In general, the impedance is defined as the ratio of pressure (or pressure difference) to flow. In time-varying (AC) systems, impedance plays the same role as does *resistance* in steady state (DC) processes. The correspondence of hemodynamic impedance to its electrical analog is not exact. Whereas electrical impedance can contain resistive, capacitive, and inductive elements, hemodynamic impedance contains mainly resistive and capacitive (compliant) elements. The resistive elements correspond to vessel diameter, vessel length, and blood viscosity. Other factors such as vessel tapering, branching patterns, vasomotion, and the series–parallel arrangement of a particular vascular bed further influence equivalent resistance. Vessel compliance (capacitance) depends on the structure and material properties of the vessel wall. These, in turn, are influenced by the wall thickness, the ratio of elastin to collagen, the smooth muscle content, and the tertiary structure of the elastin and collagen fibers. Inductive contributions to impedance, where they are important, are related to the inertia of the blood (blood density) and vessel wall (wall density).

Impedance is a general concept that can be used to describe the opposition to flow of either a single vessel (aortic impedance) or an entire vascular bed (arterial impedance).

It is important to realize that impedance, as defined here, is a frequency-dependent quantity and *not* a time-dependent quantity.

Hemodynamic resistance is then impedance as measured at zero frequency (steady state).

Impedance is a *complex variable* and has both a *magnitude* and a *phase angle*. The form of the impedance relationship depends on the model we choose to represent the resistive and capacitive elements of the vessel or vascular bed and also on where the impedance measurement takes place. In general, the shapes of the pressure and flow curves (waves) at any point in the arterial system are determined by the superposition of two sets of waves.

- *Forward waves* generated by the ventricle and modified (damped) by the resistance and compliance of the arterial vessels in between the ventricle and the measurement site.

- *Reflected waves* returning from distal vessels.

Figure 5.9 shows the flow velocity in the femoral artery of a dog (upper curve) and the pressure gradient between the aorta and the femoral artery computed from simultaneous measurements of pressure at the two sites. The dog had a complete heart block and, in this experiment, the heart was paced at 2.85 Hz. This corresponds to a heart rate of 171 beats/min. Recognize that pulsatile flow in a vessel is not only a function of time but also of radial position at any time. Figure 5.10 shows the variation of flow velocity with radial position in the same set of experiments. Remember that volumetric flow at any axial position, z, is calculated by integrating the radial velocity profile over the cross-sectional area of the vessel.

Four different types of impedance have been defined [2].

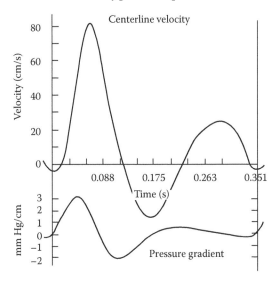

FIGURE 5.9
Centerline-flow velocity and pressure gradient as measured in the femoral artery of a dog.

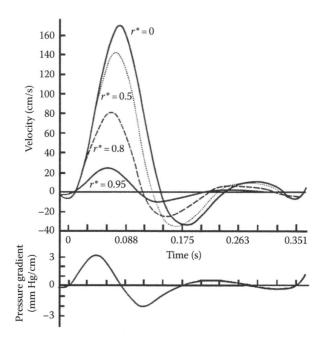

FIGURE 5.10
Flow velocity as a function of radial position.

5.5.1 Longitudinal Impedance

$$Z_L = \frac{(P_1 - P_2)}{Q} \tag{5.1}$$

This is the ratio of pressure gradient to flow and is the pulsatile analog of hemodynamic resistance. It is the opposition to flow between Points 1 and 2 along a vessel or vascular bed. Longitudinal impedance represents the impedance per unit length of a vessel and depends only on the local mechanical properties and geometry of the vessel and the blood. It does not depend on the properties of the vessels downstream. The longitudinal *resistance* of a vessel or vascular bed is the ratio of average pressure to flow and equals longitudinal impedance only at zero frequency.

At zero frequency, the longitudinal impedance reduces to the familiar *Ohms Law* relationship for steady-state hemodynamic resistance between any two points in the vascular system:

$$R_L = \frac{\Delta \bar{P}}{\bar{Q}} \tag{5.2}$$

where
  R_L is the hemodynamic resistance of the vessels
  $\Delta \bar{P} = \bar{P}_1 - P_2$ is the steady-state (time averaged) pressure drop
  Q is the steady blood flow between the two points

Units for hemodynamic resistance and impedance: When blood flow is measured in milliliters per second and pressure is measured in millimeters of mercury, one consistent set of units for hemodynamic resistance is mm Hg/(mL/s). The impedance modulus that is the modulus of pressure (or pressure gradient) divided by the modulus of flow is expressed in units of either dyn-s/cm⁵ for volumetric flow (mL/s) or dyn-s/cm³ for flow velocity (cm/s). The phase angle for impedance is usually reported in radians.

The equivalent hemodynamic resistance of all the vessels between the aorta and the right atrium is called the *total peripheral resistance*. It is used clinically as a measure of the progression of certain diseases such as renovascular hypertension and to determine the efficacy of some medications that are used to lower peripheral resistance in the treatment of systemic hypertension and heart failure.

Longitudinal impedance was introduced by Womersley [3,4] who studied flow in a thin-walled, fluid-filled tube. McDonald [5] further developed the concept to calculate pulsatile flow from the pulsatile pressure gradient recorded in an artery. We present Womersley's analysis in Chapter 4.

The longitudinal impedance, Z_L, can be interpreted on physical grounds by using Equations 4.6, 4.7, and 4.9 in Equation 5.1 and multiplying numerator and denominator by α^2 to give

$$Z_L = \frac{P_1 - P_2}{Q} = \frac{j\mu L \alpha^2 e^{-j\varepsilon_{10}}}{\pi R^4 M'_{10}} \tag{5.3}$$

We then use the Euler relation to replace $e^{-j\varepsilon_{10}}$ to give

$$Z_L = \frac{j\mu L \alpha^2}{\pi R^4 M'_{10}}\left[\cos \varepsilon_{10} - j \sin \varepsilon_{10}\right] = \frac{\mu L \alpha^2}{\pi R^4 M'_{10}}\left[\sin \varepsilon_{10} + j \cos \varepsilon_{10}\right]$$

and rearranging the above equation, we get

$$Z_L = \frac{\mu L}{\pi R^2 \left[\dfrac{M'_{10}}{\alpha^2}\right]}\left(\sin \varepsilon_{10} + j \cos \varepsilon_{10}\right) \tag{5.4}$$

as $\omega \to 0$, $(M'_{10}/\alpha^2) \to (1/8)$, and $\varepsilon_{10} \to (\pi/2)$ so that

$$Z_L \to \frac{8\mu L}{\pi R^4} \tag{5.5}$$

Equation 5.5 is the relationship for the hemodynamic resistance of a fluid in steady, laminar flow in a rigid tube. Analogous to the use in electrical terminology, the first term in Equation 5.4 is then a *resistance* to flow and the second term is a *reactance*.

$$\text{Reactance} = \frac{\mu L}{\pi R^4 \left(\dfrac{M'_{10}}{\alpha^2} \right)} \cos \varepsilon_{10}$$

The reactance term then approaches zero as $\omega \to 0$, which is true for a DC (steady-state) circuit.

5.5.2 Input Impedance

$$Z_i = \frac{P}{Q} \tag{5.6}$$

where P and Q are influenced by wave reflections. Input impedance is the relationship between pulsatile pressure and pulsatile flow as recorded in an artery feeding a particular vascular bed. For example, the ratio of pressure to flow in the ascending aorta is the *input impedance* to the systemic arterial system since the aorta feeds the systemic arterial system. In particular, it is the ratio of *harmonic** terms of pressure to the corresponding *harmonic* terms of flow. It is the most commonly used and also *the most clinically important* of all the impedance measurements.

A particular harmonic component of the pressure wave as measured at a particular point in the arterial system can be represented as a complex modulus and phase as

$$P = |P| e^{j(\omega t - \alpha)} \tag{5.7}$$

Similarly, the same harmonic component of flow can be represented as

$$Q = |Q| e^{j(\omega t - \beta)} \tag{5.8}$$

The input impedance for this harmonic term can then be written as

$$Z_i = \frac{P}{Q} = \frac{|P| e^{j(\omega t - \alpha)}}{|Q| e^{j(\omega t - \beta)}} = \frac{|P| e^{j(\beta - \alpha)}}{|Q|} \tag{5.9}$$

---

* Each of the frequencies is called a *harmonic* component of the periodic function. The total range of the frequencies that are used is called the frequency *spectrum*.

The real part of Equation 5.9 is then

$$\text{Re}\{Z_i\} = \frac{|P|}{|Q|}\cos(\beta - \alpha) = |Z|\cos(b - a) \tag{5.10}$$

where $|Z|$ is the modulus and $\theta = (\beta - \alpha)$ is the phase of this harmonic component of the input impedance.

It is important to note that when the term *impedance* or *vascular impedance* is used in the literature without further clarification, it is taken to mean *input impedance*.

5.5.3 Characteristic Impedance

$$Z_0 = \frac{P}{Q} \tag{5.11}$$

The characteristic impedance is the relationship between pressure and flow in a vessel under circumstances such that the pressure and flow waves are not influenced by wave reflection. This would require, for example, an infinitely long, straight vessel, a situation that never occurs in real life. Pressure and flow waves originating at the left ventricle with every heartbeat travel through the circulatory system at different characteristic speeds. In any vessel, the pressure wave travels at a speed approaching the speed of sound in a viscous fluid, while the flow wave travels at a speed of approximately the ratio of volumetric flow to cross-sectional area of the vessel (the average flow velocity). Since both pressure and flow are waves, they are reflected at vessel branches, bends, and in particular, at the level of the arterioles where there is a large increase in resistance to flow. At any position along the vascular network, the forward pressure and flow waves interact with the return waves reflected from downstream vessels. As in any wave pattern, the interactions of these waves produce wave reinforcement (nodes) and wave cancellation (antinodes) at different positions. This is a simplified explanation since the pressure and flow waves interact with the vessel wall and with each other. Moreover, each harmonic component of the pressure and flow waves has a different characteristic velocity and different reflection pattern, making the problem of accounting for these effects of reflected waves very complex.

Taylor [6–9] was the first to introduce the concepts of characteristic impedance and terminal impedance.

Recall that the longitudinal impedance, Z_L, is the ratio of the pressure difference between any two points in the circulation to the flow between those two points while the input impedance, Z_0, is the ratio of the pressure to the flow at any point in the circulation. Since pressure is a wave, the pressure difference between any two points is related to the absolute pressure through the complex wave velocity ($c/j\omega$), where c is the wave speed. This is the speed at which information propagates along the characteristics (eigenvalues) of the coupled set of equations of motion for the fluid and the vessel wall. Taylor [6–9] pointed out that input impedance is also related to longitudinal impedance through the complex wave velocity. Since the characteristic impedance, Z_0, is the input impedance, Z_i, in the absence of reflected waves, we first consider wave velocity, c, in the absence of reflected waves. This gives us a relationship between Z_0 and Z_L as

$$Z_0 = Z_L \left(\frac{c}{j\omega} \right) \tag{5.12}$$

We now substitute Equation 5.3 for Z_L and the definition of α^2 to obtain

$$Z_0 = \frac{c\mu L\alpha^2 \, e^{-j\varepsilon_{10}}}{\omega\pi R^4 M'_{10}} = \frac{c\rho e^{-j\varepsilon_{10}}}{\pi R^2 M'_{10}} \tag{5.13}$$

Since c varies directly with the elastic modulus of the vessel wall and inversely with its cross-sectional area [10], the characteristic impedance exhibits the same functional dependence on elastic modulus and cross-sectional area. That is, the stiffer the vessel and/or the smaller the radius, the larger the characteristic impedance.

5.5.4 Terminal Impedance

$$Z_T = \left(\frac{P}{Q} \right) \tag{5.14}$$

where P and Q are measured at the termination of the system immediately upstream from the reflecting site. The assumption is that the reflecting sites are the high-resistance arterioles, since in these vessels the resistance part of impedance dominates the reactance part. Terminal impedance is thus reported only in terms of its modulus as the mean pressure divided by the mean flow. While this definition is similar to that of peripheral resistance, it is not exactly the same for two reasons. First, there is a small (but significant) pressure drop between the large arteries and the arterioles, and, second, the pressure in the capillaries and beyond is not zero, even though there is very little pulsatile flow in these vessels. Therefore, terminal impedance is always less than peripheral resistance (as measured by mean arteriovenous pressure difference) between the same two points. This difference becomes important when venous pressure is high as is true for the pulmonary circulation under normal physiological conditions and for the systemic circulation in heart failure.

Assumptions involved in applying the concepts of vascular impedance

The concept of vascular impedance has been used to define quantitative relationships between pulsatile pressure and pulsatile flow in the arterial circulation. In that sense, the concept of impedance allows a more general description of the pressure–flow relationship than the concept of resistance, which is based on a time-averaged (steady state) ratio of pressure to flow. Strictly speaking, the flow in the arterial circulation is pulsatile so that the concept of a "resistance" to flow in the steady-state sense does not really apply anywhere. And yet, the steady-state analysis as represented by the Hagen–Poiseuille equation has proven to be very useful in practice. The Hagen–Poiseuille equation predicts that the hemodynamic resistance of a blood vessel varies inversely as the fourth power of the internal diameter of the vessel. The fourth power relationship between hemodynamic resistance and vessel diameter has been confirmed experimentally in both large caliper vessels and in the microcirculation. It has also provided a framework for understanding the major mechanisms for reflex regulation of arterial blood pressure and local (myogenic) regulation of blood flow (autoregulation).

The concept of vascular impedance originally arose from Womersley's [4,11] linearized solutions of the equations of motion of a fluid in a thin-walled, fluid-filled, elastic tube. Womersley's analysis then provided a framework for interpreting the interactions of pulsatile pressure and flow waves with the arterial wall. In particular, the smoothing and damping effects of the elastic or viscoelastic wall on the pulsatile pressure and flow waves. Just as the steady-state analysis has its utility and also its limitations, so too does Womersley's analysis. In particular, the analysis assumes simple harmonic motion, linear pressure–flow relationships, small deformations, the absence of residual stresses in the wall, and an isotropic vascular wall. Some of these assumptions are not true even under normal physiological conditions. Therefore, one should be aware of these limitations and apply the concepts appropriately.

McDonald [2, Ch. 11] presents experimental data on vascular impedance as measured at the femoral artery and the aorta as well as various tube models for simulating the impedance of the vascular system.

5.6 Elastic Modulus, Distensibility, and Compliance (Capacitance)

Hook's law states that, for many materials over a limited range of stress, strain is proportional to stress. A generalization of this law states that each of the six independent components of strain can be expressed as a linear function of the six independent components of stress. The proportionality constants in this case, as in any linear rate law, are *material properties*.

The proportionality constant that equates stress and strain is the *elastic modulus*. We will discuss these in more detail in the chapter on biomechanics (Chapter 10). To characterize an *anisotropic* elastic material, we must *independently* evaluate the following five elastic constants:

The modulus in the longitudinal direction, which is the ratio of longitudinal force per unit area to extension per unit length. This modulus is commonly called *Young's modulus*.

The *shear modulus*, or modulus of rigidity, which is the ratio of shear stress to angular strain.

The *bulk modulus*, which is the ratio of compressive stress to strain.

A *generalized longitudinal modulus*, which is the ratio of the longitudinal load per unit area to longitudinal strain.

The *Poisson's* ratio, which is the ratio of transverse to longitudinal strain.

An *anisotropic material* is one whose material properties (stress–strain relationships) vary with direction. If the material can be assumed to be *isotropic* (material properties are the same in all directions), then only two of these elastic constants need to be evaluated independently, since they all may be expressed in terms of any two of them. The two elastic constants that are most often used to characterize isotropic materials are Young's Modulus and the Poisson ratio.

5.6.1 The Bulk Modulus and Distensibility

Physiologists have used the concepts of *distensibility* and *compliance* (capacitance) to characterize the material properties of vessels and organ systems. The distensibility (also called the specific compliance) of a vessel is the *inverse of* the bulk modulus. Using the definition of bulk modulus, we can write

$$M_B = \frac{\text{Compressive stress}}{\text{Compressive strain}} \tag{5.15}$$

where M_B is the bulk modulus.

Consider a cube at initial volume V_0 and initial pressure P_0. We now increase the external pressure to P, which causes a change in volume, $\Delta V = V_1 - V_0$. The average compressive stress on the cube can be written as

$$\bar{\tau} = -\left[\frac{\tau_{xx} + \tau_{yy} + \tau_{zz}}{3} \right] \tag{5.16}$$

where the τ_{ij} are the *normal* stresses in the x-, y-, and z-directions, respectively. Pascal's law allows us to consider Equation 5.16 as defining a pressure, P, which will exert the same force in all directions. That is,

$$P = \tau_{xx} = \tau_{yy} = \tau_{zz} \tag{5.17}$$

If the same cube were now subjected to a compressive force in the x-direction, such that a point on the cube moved from position X_0, Y_0, Z_0 to position X_1, Y_1, Z_1, the relative decrease in length $(X_1 - X_0)/X_0$ is called the *longitudinal strain* denoted as $-\varepsilon_{XX}$. The relative decrease in length $(Y_1 - Y_0)/Y_0$ is called a *transverse strain* denoted as $-\varepsilon_{YY}$ and the relative decrease in length $(Z_1 - Z_0)/Z_0$ is an additional *transverse strain* denoted as $-\varepsilon_{ZZ}$. The convention is that an increase in length is a positive strain, while a decrease in length is a negative strain. Consider only longitudinal strain: For small deformations, one can show that the *compressive* strain is related to the change in volume as follows:

$$\frac{\Delta V}{V_0} = -(\varepsilon_{XX} + \varepsilon_{YY} + \varepsilon_{ZZ}) \tag{5.18}$$

where $-(\Delta V/V_0)$ is called the *volumetric strain*.

We can now define the bulk modulus in terms of pressure and volume using Equations 5.15 through 5.17 as

$$M_B = \frac{(\tau_{XX} + \tau_{YY} + \tau_{ZZ})}{3(\varepsilon_{XX} + \varepsilon_{YY} + \varepsilon_{ZZ})} = \frac{\Delta P}{(\Delta V/V_0)} \tag{5.19}$$

where $\Delta P = P - P_0$. Rearranging Equation 5.19 gives

$$\Delta P = M_B \left(\frac{\Delta V}{V_0} \right) \tag{5.20}$$

Equation 5.20 tells us that if we plot ΔP on the y-axis and $(\Delta V/V_0)$ on the x-axis, for small deformations, we will have a straight line whose slope is a material property, the bulk modulus of elasticity. In fact, the stiffer (or more elastic) the vessel, the *higher* the slope. That is, higher slopes indicate greater elasticity.

Physiologists commonly use the inverse of the bulk modulus, called the *distensibility* or, sometimes, the *specific compliance*. Therefore,

$$D = \frac{1}{M_B} = \left(\frac{(\Delta V/V_0)}{\Delta P} \right) \tag{5.21}$$

Equation 5.21 tells us that if we now plot $(\Delta V/V_0)$ on the y-axis and ΔP on the x-axis, for small deformations, we will get a straight line whose slope is the distensibility, D. For these plots, the stiffer (or more elastic) the vessel, the *lower* the slope. That is, higher slopes indicate greater distensibility. One should be aware that the bulk modulus (and the distensibility) is a function not only of the pressure change but also of both the pressure level and the degree of *vasoconstriction* of the vessel. This can be inferred from Equation 5.20. The bulk modulus is defined in terms of V_0, the initial volume of the vessel before the pressure change. As pressure increases and arterial vessels expand, the tertiary structure of the walls, namely the helical windings of elastic elements, start to contribute significantly to the mechanical properties of the wall and provide additional stiffness to the structure in a nonlinear fashion. Therefore, in arterial vessels, changes in pressure produce different changes in volume *at different initial pressures and volumes.*

Veins and venules on the other hand are much more distensible than arteries and arterioles of comparable diameter. These vessels do not show an increase in stiffness until they are quite full. At that point, there is a steep increase in stiffness as the veins are stretched further. Venoconstriction reduces the volume at which the steep increase occurs.

5.6.2 Distensibility and Compliance

The compliance (also called capacitance) of a vessel, organ, or part of the circulatory system is defined as distensibility times volume.

$$C = V_0 D = \frac{\Delta V}{\Delta P} \tag{5.22}$$

A subscript is usually used to further identify the compliance. For example, C_a would indicate arterial compliance, if we were to apply Equation 5.22 to the arterial system. Note that the definition of compliance can be applied to a single vessel (as in compliance of an artery or vein), an organ system (as in left ventricular compliance and pulmonary compliance), or a part of the circulatory system (as in venous compliance and arterial compliance). Equation 5.22 tells us that if the compliance of a vessel is constant, a given change in pressure will produce a proportional change in volume. Another way of looking at Equation 5.22 is that the compliance of a vessel or compartment is a measure of the ability of that vessel or compartment to store volume for a given change in pressure. In that sense, compliance is the mechanical analog of an electrical capacitance. For this reason, compliance is sometimes referred to as *capacitance* in the physiological literature. Since compliance contains distensibility in its definition, it is also used as a measure of stiffness. *A less compliant vessel is stiffer.*

Example 5.1 Venous Compliance

The venous system holds five to six times the volume of the arterial system (see Table 5.2) and veins are three to four times as distensible as arteries of comparable size. Since compliance is distensibility times volume, the venous system is 18–20 times as compliant as the arterial system over their respective ranges of physiological pressure. This compliance difference helps to explain the fact that mean central venous pressure (5–7 mm Hg) is about 5% of mean arterial pressure (80–100 mm Hg).

Example 5.2 Arterial Compliance

The *pulse pressure* is the difference between *systolic* and *diastolic* pressure as measured in an arterial vessel. The pulse pressure is a measure of the amplitude of the pressure wave. An approximate equation for the pulse pressure is

$$PP = P_{sys} - P_{dias} \approx \left(\frac{SV}{C_a} \right) \tag{5.23}$$

where

P_{sys} is the systolic pressure
P_{dias} the diastolic pressure
SV the *stroke volume* (the volume, in mL/beat, ejected from the left ventricle with each heartbeat)
C_a is the compliance of the arterial vessel

While many factors affect the SV (e.g., *preload, afterload, contractility, venous return, circulating drugs, and hormones*), there are only two major factors that affect arterial compliance; mean arterial pressure and the *aging process*. As we age, our arterial vessels get less compliant (stiffer), a process known as *arteriosclerosis*. The decreased arterial compliance with age is reflected in increased systolic and diastolic pressures as well as a gradual increase in pulse pressure. There is also a change in the shape of the arterial pressure wave (resulting from changes in reflection points from distal vessels) with age.

Throughout life, cyclic stresses and strains are produced in the walls of the aorta and large arteries by the pulsatile nature of the pressure and flow waves produced by the left ventricle. As with any material, the mechanical properties of the walls of the arterial vessels would deteriorate with time under the cyclic load. Fortunately, in almost all organs (central neurons are a major exception), cells are constantly being replaced by new cells that adapt to the local chemical and mechanical environment. This process is called *remodeling*. As part of the remodeling process, the aortic smooth muscle and endothelial cells synthesize and secrete the proteins that are responsible for the integrity and mechanical properties of the arterial wall. As we age, the remodeling process (by a mechanism that is not clearly understood) gradually alters the ratio of *elastin* to *collagen* in the walls of arterial vessels. This has two consequences. First, there is a decrease in compliance, and second, there is *dilation* of the large vessels to accommodate the increased wall stress at normal CO.

Because arterial vessels act much like a series of springs and shock absorbers to dampen and attenuate the pressure and flow waves as they travel away from the left ventricle, the remodeling process is not uniform. The aorta exhibits the greatest decrease in compliance (approximately 50%–80% change in 50 years). The compliance change decreases with distance from the left ventricle, so that changes in the brachial and femoral arteries are 10%–15% over 50 years. The smaller arteries and arterioles are virtually unaffected. The nonuniform nature of the compliance changes produces characteristic changes in the shape of the pressure and flow waves with age, particularly in the aorta. The nonuniform decreases in compliance along the arterial system changes the characteristic wave speed and also shifts the locations of the nodes and antinodes produced in the pressure wave by reflected waves. These characteristic changes in input impedance can be measured and used clinically to assess the progression of disease states such as hypertension, atherosclerosis, and diabetes (also, the aging process). For a more detailed treatment of input impedance, see McDonald [2, Ch. 11 and 12].

Example 5.3 Regulation of Mean Arterial Pressure and Blood Volume

Mean arterial pressure is maintained remarkably constant by two major reflex mechanisms. Beat-to-beat (short-term) regulation is maintained by a negative feedback neural mechanism that adjusts heart rate, SV, and peripheral resistance. Long-term regulation involves a second negative feedback mechanism that involves the kidneys and adjusts blood volume through the renin–angiotensin–aldosterone system (see Figure 1.6).

5.7 Aortic and Carotid Baroreceptors, Our Defense against the Effects of Gravity

Specialized cells whose processes terminate in a region of the medulla called the *cardiovascular control center* are present in the arch of the aorta and at the bifurcation of the common carotid artery (the carotid sinuses). They are called the *aortic-carotid baroreceptors*. These cells maintain an active train of impulses to the cardiovascular control center. They respond to stretch (changes in arterial pressure) by modulating the frequency of their impulse train. *A higher arterial pressure increases firing rate and a lower arterial pressure decreases firing rate.* The cardiovascular control center integrates this signal with other neural inputs and those originating from higher centers of the brain. The reflex arc then sends out neural impulses to adjust heart rate (through the sympathetic and vagal pathways), SV (through direct sympathetic connections to the myocardium), and peripheral resistance (through the sympathetic neuromuscular connections at the smooth muscle cells in peripheral arterioles). These adjustments occur within 1–2 heartbeats and are our major defense mechanism against gravity. If not for the baroreceptors, consider what would occur when you tried to get out of bed in the morning. As you rise to a sitting position, the blood in your head and neck would try to pool in your feet due to the action of gravity. This would reduce arterial pressure in your head and neck and decrease the oxygen being transported to your brain. The pooling is delayed somewhat by the one-way valves in the veins, which prevents blood from moving away from the heart. However, if this reduced arterial pressure were to persist for more than a few seconds, the lack of oxygen would cause you to feel dizzy and eventually you would lose consciousness. The baroreceptor reflex arc acts very quickly to raise arterial pressure and prevent these occurrences.

Equation 1.1 shows the relationship between mean arterial pressure and heart rate, SV, and peripheral resistance.

$$CO = \frac{\Delta P}{R_{TPR}} \qquad (1.1a)$$

where R_{TPR} is the total peripheral resistance and $\Delta P = \bar{P}_{arterial} - \bar{P}_{right\,atrium}$. The bars above the symbols indicate time-averaged values. However, $\bar{P}_{right\,atrium} \approx 0$ and $CO = HR \times SV$. Rearranging and solving for $P_{arterial}$ gives

$$\bar{P}_{arterial} = HR \times SV \times R_{TPR} \qquad (5.24)$$

We see that any factor that changes HR, SV, or R_{TPR} will change the mean arterial pressure as well.

The heart and the systemic circulation can be thought as two coupled mechanical systems, the heart being a pressure generator and the systemic circulation being a pressure dissipater. These two mechanical systems are not independent, but are coupled. In a later analysis, we shall demonstrate that the *ejection fraction* (ratio of SV:EDV) is the factor that couples the two mechanical systems.

5.8 Atriovenous Baroreceptors: Long-Term Regulation of Mean Arterial Pressure

A second set of baroreceptors are located in the atria, particularly at the junctions with the venae cavae and the pulmonary veins. These receptors are sensitive to stretch and participate in the regulation of arterial pressure through regulation of blood volume. These baroreceptors also have processes that terminate in the cardiopulmonary control center in the medulla and also at the thirst center of the hypothalamus. Because of their location, these stretch receptors respond to venous return. When venous return increases (e.g., from a blood transfusion or long-term peripheral edema), two mechanisms come into play, one short term and one long term. The short-term mechanism, called the *Frank–Starling* mechanism does not involve neural control. You can cut all the neural connections to the myocardium (as in a heart transplant) and the Frank–Starling mechanism would still operate. *We discuss the*

Frank–Starling mechanism in the following section. The long-term response to increased venous return involves the atriovenous baroreceptors. Increased venous return increases the firing rate of the atriovenous baroreceptors and the frequency of impulses reaching the cardiopulmonary control center in the medulla and the thirst center of the hypothalamus. These signals initiate several events, all designed to decrease blood volume.

1. A signal is sent from the hypothalamus to the anterior pituitary gland to decrease production of *antidiuretic hormone* (ADH), also called *vasopressin*. ADH is a powerful vasoconstrictor, but more importantly it acts on the collecting tubules of the kidney to increase water absorption. A decrease in ADH secretion allows less water to be reabsorbed in the collecting tubules and more water to pass through the kidneys and be excreted in the urine.

2. The increased venous return leads to an increased secretion of *atrial natriuretic hormone* (ANH). One of the functions of ANH is to inhibit the action of ADH. ANH also acts on arterioles to cause vasodilation, decreasing vascular resistance.

3. Output from the cardiopulmonary control center in the medulla decreases sympathetic stimulation of all arterioles, including the renal arterioles. The decreased renal arteriole pressure signals the kidney to decrease production of *renin* and *aldosterone*. Decreased renin production directly reduces production of angiotensin II, a powerful vasoconstrictor. This reduces blood pressure by reducing peripheral resistance. Aldosterone acts on the distal tubules of the kidney to conserve sodium. Decreased aldosterone production allows more sodium to be excreted in the urine, increasing urine osmolarity. The increased osmolarity pulls more water into the urine and also acts to reduce blood volume.

4. The actions of ADH, ANH, and aldosterone, all increase urine output and reduce blood volume. The actions of atrial natriuretic peptide (ANP) and reduced renin production reduce peripheral resistance. As indicated in the previous section, reduced peripheral resistance directly reduces mean arterial pressure. Reduced blood volume further contributes to reduced mean arterial pressure through the compliance effect. Remember that arterial compliance is defined as (Equation 1.2):

$$C_{art} = \frac{\Delta V}{\Delta P}, \quad \text{so that } \Delta P = \frac{\Delta V}{C_{art}} \tag{5.25}$$

Over the period of time that these adjustments take place, the arterial compliance, C_{art}, is approximately constant. Therefore, a decrease in the volume (ΔV) of the arterial system leads directly to a decrease in arterial pressure (ΔP).

Figure 1.6 is a schematic diagram of both the short-term (beat-to-beat) and long-term negative feedback reflex system that provides regulation of mean arterial pressure.

5.9 The Frank–Starling Mechanism (Heterometric Autoregulation)

An increase in venous return (e.g., as happens in a blood transfusion or by infusion of a liter of ringer's lactate) stimulates the atriovenous barorereceptors and initiates events that lead to long-term reduction in blood volume (and also venous return). These events, however, are not fast enough to maintain short-term homeostasis. The increased venous return activates another important reflex mechanism that *acts immediately* and *does not require neural intervention*. This mechanism, called the Frank–Starling mechanism, Heterometric Autoregulation, or Starling's Law of the Heart, depends only on the ability of a muscle to react with a stronger force of contraction when it is stretched (see Chapter 3). An increase in venous return increases EDV and stretches the ventricular muscle (increased preload). When the stretched ventricular muscle is stimulated to contract by the wave of depolarization (*P* wave of the ECG), the strength of the contraction increases with the degree of stretch over the physiological range. This happens because of a more favorable alignment of the actin–myosin crossbridges in the cardiac muscle. Thus, an increase in venous return leads to an increased SV. The Frank–Starling mechanism is independent of neural influence. You can cut all the neural connections to the myocardium (as in a heart transplant) and the mechanism still operates. This allows the heart to accommodate short-term changes in venous return without activating a neural reflex arc.

Figure 5.11 illustrates the Frank–Starling response. The important points to be recognized are that the mechanism only operates over a limited range and that the response depends on the *inotropic* state of the myocardium, that is, on the inherent ability of the cardiac muscle to develop force, also called *contractility*. Therefore, for any heart, there are a family of possible Frank–Starling curves depending on its inotropic state or contractility. The inotropic state of a muscle can be increased by treatment with positive inotropic agents such as norepinephrine and certain drugs. The inotropic state of a muscle can be decreased by negative inotropic agents such as a myocardial infarction or heart failure. The heart only operates on one Frank–Starling curve at a time.

Point a on the middle curve represents a normal left ventricular SV at a normal left ventricular end-diastolic volume (preload). After a mild myocardial infarction, the inherent ability of the ventricle to develop force (muscle contractility) would decrease and the heart would then be forced to operate

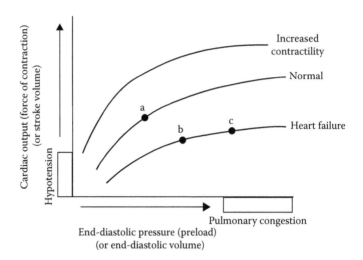

FIGURE 5.11

The Frank–Starling mechanism. As the cardiac muscle is stretched (increased preload), it responds with a greater force of contraction (cardiac output) over the physiological range. The heart only operates on one Frank–Starling curve at a time, depending on its inherent ability to develop force (contractility).

on the lower Frank–Starling curve. Point b on the lower curve represents a decrease in left ventricular SV that might occur after a mild myocardial infarction. If the right ventricle is still operating normally, there will then be an increase in left ventricular end-diastolic volume and because the pulmonary veins feed the left ventricle, an increase in pulmonary venous pressure. The Frank–Starling mechanism would attempt to increase left ventricular SV in response to the increase in left ventricular preload. However, at Point c, the backup of blood in the pulmonary system increases pulmonary capillary pressure, increasing water flow from the pulmonary capillaries to the alveoli, causing *pulmonary congestion or edema.*

5.10 Modeling the Ventricles: Time-Varying Elastance Left Ventricular Work Loops

The pressure–volume events of the left ventricle are presented in Figure 5.1. An alternative way of presenting these data is by means of pressure–volume or "work" loops. At any time during the cardiac cycle, there exists a pair of pressure–volume values for the left ventricle. We can eliminate time by plotting pressure against volume over the course of a single cardiac cycle. If we do, we get a closed figure as shown in Figure 5.12. The shape of the work loop changes with preload, afterload, and the inotropic state (contractility) of the

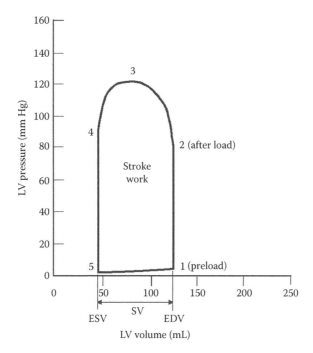

FIGURE 5.12
Left ventricular work loop.

ventricle. All the ventricular events associated with the cardiac cycle can be identified on the work loop diagram.

The numbers in Figure 5.12 correspond to the same events described in the text following Figure 5.1. Consider Figure 5.12. Venous return increases central venous pressure (ventricular filling pressure) and ventricular filling (preload). Just before position 1, the QRS complex of the ECG records the wave of depolarization that spreads through the ventricular muscle mass causing ventricular contraction. This raises the pressure in the ventricles above atrial pressure and closes the AV valve. With both the AV and aortic valves closed, the ventricles are contracting against a fixed volume of incompressible fluid. This raises ventricular pressure until the pressure in the ventricle is above aortic diastolic pressure (afterload). The constant volume pressure rise from Point 1 to 2 is called isovolumic contraction. At Point 2, the pressure in the ventricle is sufficient to open the aortic valve. Ventricular muscle can then shorten and we have the beginning of ejection. Ejection of the SV occurs from Point 2 to 4. Point 3 represents aortic (and ventricular) systolic pressure. Also at Point 3, the wave of depolarization (as recorded by the T wave of the ECG) starts to relax the ventricular muscle and pressure begins to fall. At Point 4, the pressure in the ventricle falls below the pressure in the aorta and the aortic valve closes. Between Points 4 and 5, we have both the aortic and AV valves closed and this segment of the work loop corresponds to isovolumic relaxation. At Point 5, the pressure in the ventricle falls below the pressure in the atrium, the AV valve opens and filling begins. Filling continues from Point 5 to 1, where the cycle repeats itself.

5.10.1 Cardiac Work

The heart derives its energy almost entirely from aerobic metabolism (ATP production). For this reason, myocardial oxygen consumption is directly related to myocardial energy use (ATP consumption). Table 5.6 lists the major determinants of myocardial oxygen consumption in a typical adult at rest.

We recognize two types of energy expenditure in the heart during contraction.

1. *External or useful work*. This is pressure–volume work called *stroke* or *pumping* work. It accounts for approximately 10%–15% of total energy consumption during contraction.

   The pressure–volume work performed by the left ventricle can be calculated from the area under the *P–V* curve. This is the mechanical work of pumping blood through the systemic circulation and can be calculated from the work loop as

   $$W(\text{stroke}) = \int p \, dV \tag{5.26}$$

   We recognize the integral as the area under the work loop. An approximate calculation of the pumping work is to treat the work loop as a rectangle with height as the mean arterial pressure, $\bar{P}$. The mean arterial pressure can be approximated as

   $$\bar{P} = P_{\text{diastolic}} + \frac{1}{3}\left(P_{\text{systolic}} - P_{\text{diastolic}}\right) \tag{5.27}$$

TABLE 5.6

Determinants of Myocardial Oxygen Consumption

Process	ATP Use (%) (O_2 Consumption)
1. Basal metabolism	12
Energy consumed in myocardial cellular processes other than contraction	
2. Muscle contraction. Reflects ATP use associated with:	75
a. Crossbridge cycling during isovolumic contraction and ejection phases of cardiac cycle	
b. ATP utilization for Ca^{2+} sequestration at the termination of each contraction	
3. Pumping or SW	13
Total	100

so that

$$W = \bar{P}\text{SV} \tag{5.28}$$

and

SV = stroke volume = end diastolic volume – end systolic volume.

2. *Tension or "Pressure" work.* This is the ATP consumed in maintaining the high wall stresses (tensions) during contraction. This represents 85%–90% of the total energy used by the muscle during contraction.

A heart will consume more ATP when it has to work at higher pressures, independent of $\int p\,dV$ (external work). Since energy supplies are limited, there is a tradeoff. The heart will be forced to give up external or useful work when it is forced to pump against higher pressures (higher afterloads).

Example 5.4 Cardiac Power and Efficiency

Cardiac power
The metabolic energy required to develop the pressure in isovolumic contraction at rest is about 10 W. Since the mean pulmonary artery pressure is about 1/7 of the aortic pressure and the outputs of the two ventricles are equal, the work of the right ventricle is about 1/7 that of the left ventricle.

LV: 8.75 W (7 kg m)
RV: 1.25 W (1 kg m)
Total: 10.0 W (8 kg m)

The mechanical power required to pump blood is the area under the work loop (*P–V* work) divided by the time per heartbeat. This is a measure of the ejection or pumping energy, not the total power. The stroke work (SW) is then given as

$$W = \bar{P}\,\text{SV} \tag{5.29}$$

$$\text{Power (or minute work)} = \frac{\text{SW}}{\text{time}} \tag{5.30}$$

Assuming a mean arterial pressure of 100 mm Hg and a SV of 75 mL at a heart rate of 75 beats/min (0.8 s/beat) gives

$$\text{Power} = \left[\frac{(75\,\text{mL})(100\,\text{mm Hg})}{0.8\,\text{s}}\right]\left[\frac{1330\,\text{dyn/cm}^2}{\text{mm Hg}}\right]\left[\frac{10^{-7}\,\text{W}}{\text{erg/s}}\right] = 1\,\text{W erg} = \text{dyn cm}$$

A comparison of pressure work (10 W) verses pumping work (1 W) demonstrates the high energy costs associated with developing wall tension for the myocardium.

Cardiac efficiency

$$\text{Efficiency} = \frac{(100)\text{Power}}{O_2 \text{ consumed}} \tag{5.31}$$

The volume of oxygen consumed by the heart depends on the amount and type of activity the heart performs. Under resting conditions, myocardial oxygen consumption is about 8–10 mL/min/100 g of heart muscle. This could increase several-fold during exercise and decrease moderately during conditions of hypotension and hypothermia. We assume resting conditions and a 300 g heart. We also assume that oxygen consumption is at 9 mL/min/100 g of heart muscle:

$$O_2 \text{ consumption} = \left[\frac{9\,\text{mL/min}}{100\,\text{g}}\right][300\,\text{g}] = 27\,\text{mL/min}$$

This is equivalent to 130 cal/min at a respiratory quotient* of 0.82.

Together, the ventricles deliver about 1(1/7) = 1.14 W. The ventricular energy efficiency is then given as

$$\text{Efficiency} = \left[\frac{1.14}{9.06}\right][100] = 13\%$$

The net efficiency is slightly higher since the requirements of the resting cardiac muscle should be subtracted (about 2 mL/min/100 g). The net contraction energy efficiency is then about 17% or 18%.

Time-varying elastance

Based on experiments in dogs, Suga and Sugawa [12,13] proposed that the pressure–volume relationship during the cardiac cycle could be modeled using a time-varying elastance. That is, the ventricular volume, $V(t)$, at any time during the cardiac cycle is proportional to the instantaneous pressure, $P(t)$, through a time-varying elastance, $E(t)$. Changes in filling pressure of the ventricle move the diastolic pressure through a characteristic curve called the diastolic pressure–volume relationship. The pressure at the end of diastole (Point 1 in Figure 5.14) is called the *EDP and is a measure of preload*. The entire filling curve between Points 5 and 1 is often referred as the end-diastolic pressure–volume relationship (EDPVR). A more complex relationship between

---

* The RQ is the ratio of O_2 consumed to CO_2 produced during aerobic metabolism. This depends on diet since the moles of oxygen consumed for oxidation of carbohydrates, proteins, and fats are different and the average value will vary with an individual's diet. The average adult respiratory quotient is about 0.82.

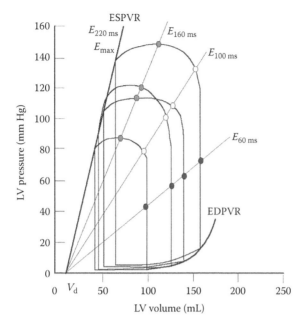

FIGURE 5.13
Work loops with regular change in preload and afterload time-varying elastance changes during a single cardiac cycle.

pressure and volume exists during systole. Suga and Sugawa made the following observations, which hold for *isovolumetric* (constant EDV) beats as well as varying pressure (*auxobaric*) beats in the left ventricle of dogs:

1. If *preload and afterload* are changed in a regular manner, the work loops change in a regular way as shown in Figure 5.13.

2. If the end-systolic pressure–volume points (at the closure of the aortic valve) are connected, they form a straight line as shown in Figure 5.13. This linear relationship is called the *end-systolic pressure–volume relationship (ESPVR)*. **The slope of the ESPVR, E_{max}, is independent of preload and afterload and only depends on the inotropic state (contractile state) of the ventricle**. Positive inotropic agents such as norepinephrine increase E_{max} and negative inotropic agents such as heart failure decrease E_{max}. *The slope of the ESPVR is used as an index of ventricular contractility.*

   a. The time to E_{max} from the onset of systole, T_{max}, is not affected by EDV (preload) but is affected by afterload.

   b. Positive inotropic agents shorten T_{max}.

   c. Heart rate increases (by pacing) have little effect on E_{max}. However, T_{max} shortens significantly with increased heart rate.

3. Connecting corresponding pressure–volume points at corresponding times during systole also results in a linear relationship (at each time point) with the intercept on the volume axis (V_d) being almost constant as shown in Figure 5.13. V_d is an experimentally determined constant.

4. The relationship between pressure and volume at any time during systole can be approximated by a time-varying elastance as

$$P(t) = E(t)\left[V(t) - V_d\right] \tag{5.32}$$

5. The time-varying elastance (*P–V* ratio) can be normalized by dividing $E(t)$ by E_{max}. If time during a cardiac cycle is divided by T_{max}, the normalized *P–V* ratio curve is approximately Gaussian shaped (Figure 5.14) and for any loading condition and any inotropic state can be represented by two parameters, E_{max} and T_{max}.

The time-varying elastance can then be expressed by the relationship:

$$E(t) = E_{max}E_n\left(\frac{t}{T_{max}}\right) \tag{5.33}$$

Here E_{max} is the maximum value of $E(t)$ and $E_n(t/T_{max})$ is the normalized elastance (Figure 5.14). The maximum value of $E_n(t/T_{max})$ is $E_n(1)$, which equals 1.

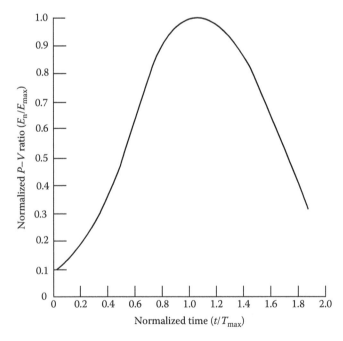

FIGURE 5.14
Normalized time-varying elastance, $E_n(t/T_{max})$.

Shroff et al. [14] and Barnea et al. [15] used a third-order polynomial to approximate $E_n(t)$ as follows:

$$E_n(t) = a_1 t + a_2 t^2 + a_3 t^3 \tag{5.34}$$

The coefficients, a_i, were obtained from a least squares fit of experimental data. Typical values for the a_i are

$$a_1 = 0.158, \quad a_2 = 2.685, \quad \text{and} \quad a_3 = -1.841$$

T_{max} is the time at which $E(t)$ reaches its maximum value, E_{max} starting from the beginning of systole. T_{max}, can be calculated as follows [16]:

$$T_{max} = \frac{(413 - 1.7 HR)}{1000} \tag{5.35}$$

HR is the heart rate in beats per minute and T_{max} is in seconds.

The right ventricle can be modeled in a similar manner, with the elastance values being about 20% of their corresponding values for the left ventricle.

A model for the left ventricle

A mass balance over the left ventricle (assuming constant density for blood) is given as

$$\frac{dV_{lv}}{dt} = F_{pv} - F_{lv} \tag{5.36}$$

Using Ohm's law for the flow rates into and out of the left ventricle (ignoring the contribution due to atrial contraction) is given as

$$F_{pv} = \left(P_{pv} - P_{lv}\right) G_{lv} \tag{5.37}$$

and

$$F_{lv} = \left(P_{lv} - P_{la}\right) G_{ao} \tag{5.38}$$

Using the time-varying elastance,

$$P_{lv} = E_{lv}(t)\left[V_{lv} - V_{lv0}\right] \tag{5.39}$$

and

$$E_{lv}(t) = E_{lv,max} E_n\left(\frac{t}{T_{max}}\right) \tag{5.40}$$

To simulate the AV and aortic valves, the following conditions are used:

$$F_{pv} = 0, \quad \text{when} \quad P_{lv} > P_{pv} \tag{5.41}$$

and

$$F_{lv} = 0, \quad \text{when } P_{ao} > P_{lv} \tag{5.42}$$

where

   V is the volume (L)
   F is the flow rate (L/min)
   G is the hemodynamic conductance (1/hemodynamic resistance) [(L/min)/mm Hg]
   $E(t)$ is the time-varying elastance
   lv is the left ventricle
   pv is the pulmonary veins
   ao is the aorta
   V_{lv0} is the volume axis intercept of the line connecting the maximum left ventricular elastance P/V points for different loaded beats

Figure 4.16 shows that, experimentally, V_{lv0} (V_d in Figure 5.13) is a constant. The model does not take into account the variation of hemodynamic resistance within each heartbeat.

Example 5.5 Model Results

The following pressure–time curve and elastance–time curve were generated using this model and the following parameters:

   HR = 75 beats/min, $E_{lv,max}$ = 2 mm Hg/mL (in systole), E_{lv} = 0.051 (in diastole), $E_{rv,max}$ = 0.45 mm Hg/mL, V_{lv0} = 10 mL, V_{rv0} = 0 mL, G_{lv} = 50 (mL/s)/mm Hg, and G_{ao} = 50 (mL/s)/mm Hg.
   T_{max} is calculated from Equation 5.35.
   E_{lv} is calculated from Equations 5.34 and 5.40.

The simulation was carried out using MATLAB® and Simulink®* software (Figures 5.15 and 5.16).

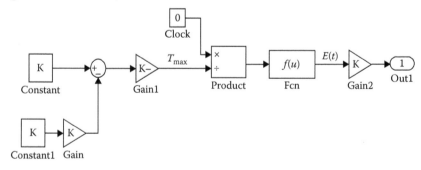

FIGURE 5.15
Simulink® block diagram for calculating time-varying elastance for left and right ventricles.

---

* MATLAB® and Simulink® are copywrites of The Mathworks, Inc.

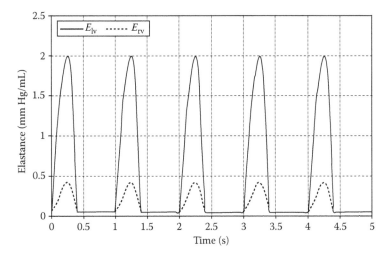

FIGURE 5.16
Model output of time-varying elastance.

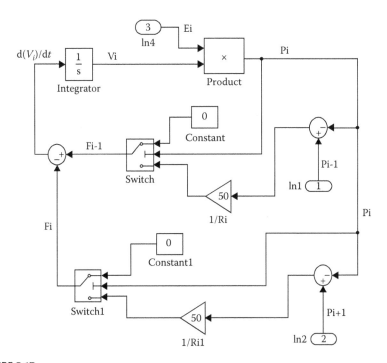

FIGURE 5.17
Simulink® block diagram for calculation of left ventricular pressures and volumes. The limit switches act as one-way valves.

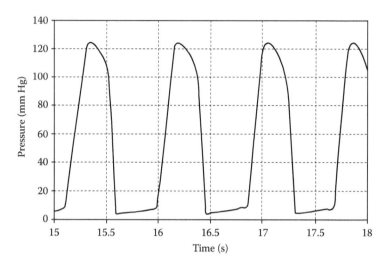

FIGURE 5.18
Left ventricular pressure versus time as calculated from the model.

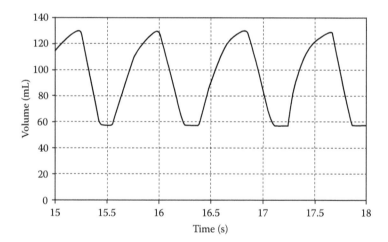

FIGURE 5.19
Left ventricular volume versus time as calculated from the model.

The Simulink block diagram for calculation of pressure and volume in the left ventricle is shown in Figure 5.17.

Figures 5.18 and 5.19 show the pressure–time and volume–time curves for the left ventricle produced by the model.

If we combine these two graphs so as to eliminate time, we produce a left ventricular work loop that is very similar to Figure 5.14. Figure 5.20 illustrates the model output of the LV work loop.

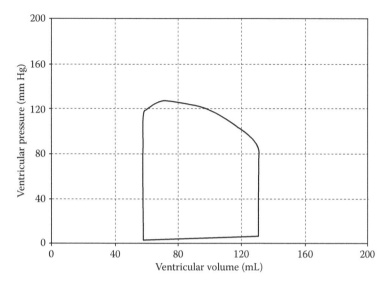

FIGURE 5.20
Left ventricular work loop produced by combining the outputs in Figures 5.20 and 5.21 in such a way as to eliminate time. Each P and V point on the graph represents a single time point in the cardiac cycle.

The ventricular model is useful to study the effects of changes in cardiac contractility ($E_{lv,max}$), preload (EDV), and afterload (aortic diastolic pressure) on CO, cardiac work, and pumping work.

Because the ventricles are mechanical force generators that exist in a closed loop with circulatory systems (systemic and pulmonary) that are force dissipaters, we would like to couple the model of the ventricles with a model of the circulatory vessels to close the loop.

5.11 A Compartmental Model for the Circulatory System

The systemic and pulmonary circulations can be thought of as series–parallel networks of compartments in which pressures and flows change with time. A large number of compartments may be required to closely simulate normal physiological flows and volumes. Each compartment is represented by a particular three-element Windkessel model. Each compartment is then joined to its neighboring compartments to construct a series–parallel arrangement that represents the closed-loop circulatory system.

A mass balance over any compartment (assuming constant blood density) gives

$$\frac{dV_i}{dt} = F_{i+1} - F_i \tag{5.43}$$

Using Ohm's law

$$F_i = (P_i - P_{i+1})G_i \tag{5.44}$$

From the definition of compliance

$$P_i = \left(\frac{(V_i - V_{i0})}{C_i}\right) \tag{5.45}$$

The model does not take into account the variation of either conductance or compliance with pressure.

Here, V_{i0} is the unstressed volume of the compartment in milliliter and C_i is the compliance of the compartment. The use of compliance, rather than impedance in this model assumes that the resistance contribution of the compartment is small relative to the capacitance contribution over the physiological range. In spite of these limitations, the approximate Windkessel model does a good job simulating flows and pressures in the systemic and pulmonary compartments.

Example 5.6 Aortic Pressure Wave

The pressure–time curves for the aorta were simulated using Equations 5.43 through 5.45. The parameter values used were

$$G = 10.59; \quad V_{i0} = 60.0; \quad \text{and} \quad C = 0.625.$$

Figure 5.21 is the Simulink block diagram for the aorta.

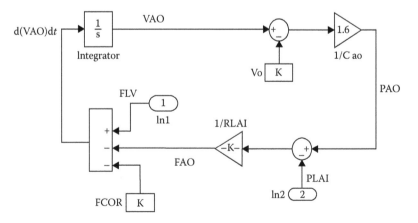

FIGURE 5.21
Simulink® block diagram for simulation of aortic pressure, volume, and flow.

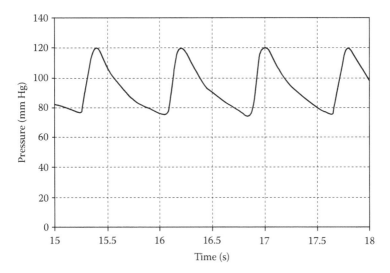

FIGURE 5.22
Model simulation of aortic pressure wave.

Figure 5.22 shows the model output. Note that the shape of the aortic pressure–time curve is very similar to that measured experimentally (see Figure 5.7). However, the dicrotic notch (closure of the aortic valve) is not reproduced well.

The complete compartmental model contains 20 series–parallel compartments corresponding to the major anatomical and organ systems [17]. The model can reproduce many of the major physiological mechanisms such as the Frank–Starling effect (heterometric autoregulation) and the almost linear relationship between SV and heart rate over the physiological range of heart rates. Patel [17] has also added reflex regulation of arterial pressure and local regulation of blood flow (autoregulation) by several organs and the coronary arteries.

5.11.1 Coupling of the Ventricles with the Systemic Circulation

Sunangawa et al. [18] used the left ventricular pressure–volume relationship and the hemodynamic impedance to analyze the mechanical coupling of the left ventricle and the arterial system. The idea was to analytically predict the working SV using the mechanical properties of the left ventricle and the arterial system.

Starting with the experimental observations of Suga et al. [19] that the ESPVR is approximately linear over the physiological range. Also, although E_{max}, the slope of the ESPVR, changes with changes in ventricular contractility (inotropic state of the ventricle), it is relatively independent of preload and afterload (see Figure 5.16).

Equation 4.33 can be written at end systole as

$$P_{es} = E_{max} = (V_{es} - V_0) \tag{5.46}$$

and

$$SV = V_{ed} - V_{es} \tag{5.47}$$

so that

$$V_{es} = V_{ed} - SV \tag{5.48}$$

and

$$P_{es} = E_{max}(V_{ed} - SV - V_0) \tag{5.49}$$

Equation 5.49 tells us that for a given V_{ed} (preload), P_{es} varies inversely with SV. This is shown schematically in panel b of Figure 5.23.

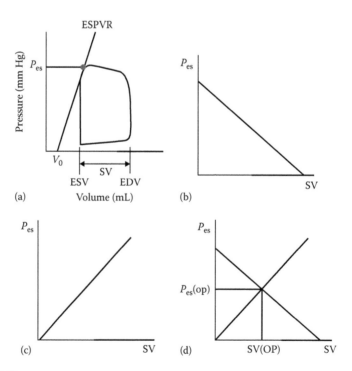

FIGURE 5.23
(a) Left ventricular work loop. (b) Ventricular end-systolic pressure as a function of SV. (c) Arterial end-systolic pressure as a function of SV. (d) SV of the cardiovascular system at mechanical equilibrium.

The properties of the arterial system are now approximated by the three-element Windkessel model. Using Ohm's law, the mean arterial flow rate (averaged over the time of one cardiac cycle, T) is

$$\langle F_{art} \rangle = \frac{\langle P_{art} \rangle}{(R_c + R_a)} \tag{5.50}$$

where $\langle P_{art} \rangle$ is the average arterial pressure over one cardiac cycle.

Figure 5.26 shows a typical arterial pressure curve. We can define two areas under the curve; A_s, the area under the pressure–time curve during systole and A_d, the area under the pressure–time curve during diastole. Here, t_s is the ejection time, t_d the diastolic filling time, and T is the time for one cardiac cycle. The area A_s is approximated by the rectangle* $P_{es} t_s$.

The mean arterial flow, $\langle F_{art} \rangle$, is just the SV divided by the cycle time, T. From Figure 5.23, the mean arterial pressure, $\langle P_{art} \rangle$, is approximated by the area $(A_s + A_d)/T$, so that Equation 5.50 is then approximately

$$\frac{SV}{T} = \frac{\dfrac{(A_s + A_d)}{T}}{R_c + R_a} \tag{5.51}$$

Solving for the SV gives

$$SV = \frac{(A_s + A_d)}{R_c + R_a} \tag{5.52}$$

The diastolic area, A_d, in Figure 5.23 is approximated by

$$A_d = P_{es} t \left[1.0 - \exp\left(\frac{-t}{t}\right) \right] \tag{5.53}$$

Here, t_d is the duration of arterial diastole and τ is the time constant of diastolic pressure decay, which equals the product of the arterial resistance, R_s, and arterial compliance, C_a.

Then, approximately

$$A_s + A_d = P_{es} 1.0 + t \left\{ \left[1.0 - \exp\left(\frac{-t_d}{t}\right) \right] \right\} \tag{5.54}$$

Substituting Equation 5.54 in Equation 5.52 and solving for P_{es} gives

$$P_{es} = \frac{R_c + R_a}{t_S + \tau\left[1.0 - \exp(-t_d/\tau) \right]} SV \tag{5.55}$$

---

* In general, the end-systolic pressure, P_{es}, is not the same as the mean systolic pressure, $\langle P_s \rangle$. This difference is ignored in this analysis.

Equation 5.55 states that P_{es}* varies linearly with SV. This is shown in panel c of Figure 5.23.

If the vascular parameters (R_a, R_c, and C_a) remain constant and t_s and t_d also remain constant, we can denote the slope of the P_{es}–SV relationship as E_a (effective arterial elastance), so that

$$E_a = \frac{R_c + R_a}{t_s + \tau\left[1.0 - \exp(-t_d/\tau)\right]} \tag{5.56}$$

Using Equation 5.56, we can write Equation 5.55 as

$$P_{es} = E_a SV \tag{5.55a}$$

The effective arterial elastance, E_a, can then be estimated as the ratio P_{es}/SV.

Although these relationships have been deduced from experimental measurements in dogs, Kelly et al. [20] and Cohen-Solal et al. [21] have validated these results in humans.

Now the analytical relationship between P_{es} and SV for the ventricle is given by Equation 5.49 and the relationship between P_{es} and SV for the arterial system is given by Equation 5.55, we can estimate the equilibrium SV for the coupled left ventricle–arterial system by solving the two equations simultaneously as

$$SV = \frac{V_{ed} - V_d}{1.0 + E_a/E_{max}} \tag{5.57}$$

The graphical solution is shown schematically in panel d of Figure 5.23. Cohen-Solal et al. extended this analysis and was able to demonstrate that for $V_0/SV \ll 1.0$, the ratio E_a/E_{max}, which is an index of the mechanical coupling of the left ventricle with the arterial system, is approximately

$$\frac{E_a}{E_{max}} \approx \frac{1.0}{EF} - 1.0 \tag{5.58}$$

where EF is the ejection fraction. The ejection fraction is the fraction of the ventricular volume at end-diastole that is ejected by the ventricle with each stroke. It is defined by

$$EF = \frac{SV}{V_{ed}} \tag{5.59}$$

---

* See footnote on previous page.

From Equations 5.58 and 5.59, it is clear that the ejection fraction reflects the coupling between the left ventricle and the arterial system. It is directly related to the systolic ejection fraction of the left ventricular pump and inversely proportional to the mechanical properties of the arterial system [21]. This result illustrates the important role that the mechanical properties of the arterial system play in determining the overall pumping performance of the left ventricle.

Cohen-Solal et al. [21] compared a group of normal males with a similar group of hypertensive males without heart failure measuring hemodynamic and angiographic variables. The study data supported several important conclusions:

- Normal hearts operate at an E_a/E_{max} of about 0.6, corresponding to maximum efficiency of SW.

- The left ventricle and the arterial system remain correctly coupled in hypertensive condition. However, there are marked differences in the adaptation of individuals to hypertension. These differences mainly involve the remodeling of the ventricle and changes in wall thickness and geometry (and impaired muscle function).

- EF was similar in hypertensives and controls. E_{max} (ventricular stiffness) was significantly increased in hypertensives as compared with controls, E_a was increased as well, making the ratio E_a/E_{max}, not significant overall.

5.12 The Failing Heart

When cardiologists speak of *heart failure* what they are generally referring to is the failure of the left side of the heart (left ventricular failure). While failure of the right side of the heart occurs, it does so with far less frequency than failure of the left side. Since the cardiac cycle consists of two distinct phases, diastolic filling and systolic pumping, the heart can fail in two distinct modes. These are termed *systolic dysfunction* and *diastolic dysfunction*. Figure 5.24 is a schematic diagram that illustrates the most common causes of heart failure [22].

5.12.1 Systolic Dysfunction

Several compensatory mechanisms are called into action in heart failure that serve to buffer the fall in CO and help to maintain sufficient blood pressure to adequately perfuse the vital organs.

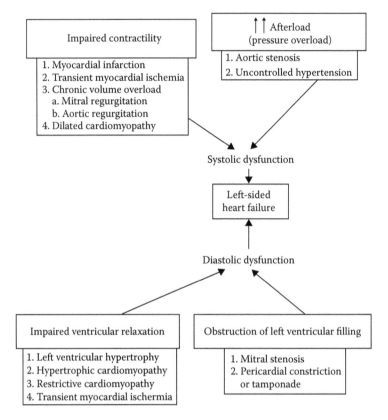

FIGURE 5.24
Some common causes of left heart dysfunction.

- The Frank–Starling mechanism (see Figure 5.11).
- Neurohormonal activation (baroreceptors, the renin–angiotensin–aldosterone system).
- Development of myocardial hypertrophy (thickening) of the left ventricle and an increase in ventricular mass and rearrangement of ventricular geometry (called *remodeling*) reduces systolic wall stress (see Figure 5.25).

Ventricular remodeling can also occur when the aortic or mitral (AV) valves are incompetent, either through disease or as a result of drug abuse. An incompetent aortic valve (one whose leaflets do not form a tight seal at closure) allows backflow of blood (from the aorta to the ventricle) during diastole. Such a condition is called aortic (or mitral) *regurgitation*. Because of the backflow of blood during diastole, the ventricle dilates (grows larger) to accommodate the larger EDV. The ventricle must also increase SV to maintain a normal CO, since some of the SV returns to the ventricle rather than moving forward through the aorta.

Figure 5.25 and the table below compare some of the geometric changes in ventricular mass, wall thickness, and pressures in both pressure and volume overload.

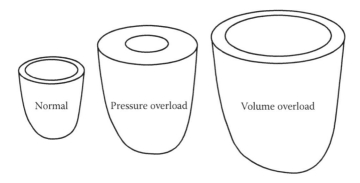

FIGURE 5.25
Left ventricular (LV) pressure, Left ventricular mass index (LVMI),* wall thickness, ratio of wall thickness to ventricular radius at diastole (h/R), and longitudinal wall stress (σ_z) in patients with normal hearts compared with well-compensated patients with pressure and volume overload. (Data from Grossman, W. *Am. J. Med.* 69, 576, 1980.)

LV pressure (mm Hg)	$117/10 \pm 7/1$	$226/23 \pm 6/3$	$138/23 \pm 7/2$
*LVMI** (g/m²)	71 ± 8	206 ± 17	196 ± 17
LV wall thickness (mm)	8.2 ± 0.6	15.2 ± 0.9	10.6 ± 0.5
h/R	0.34 ± 0.03	0.56 ± 0.05	0.33 ± 0.02
Left ventricular longitudinal stress (dyn/cm⁻² × 10³), σ_z			
Peak systolic	151 ± 4	161 ± 24	175 ± 7
End diastolic	17 ± 2	23 ± 3	41 ± 3

Since the vascular and pulmonary systems form closed circuits that are in series, systolic failure of the left side of the heart (isolated systolic dysfunction) affects the pulmonary circulation and leads to pulmonary congestion and pulmonary edema. If the right heart is functioning normally, and the left heart fails to keep up, blood volume backs up in the pulmonary venous system. The increased volume in the pulmonary veins leads to an increased pressure in the pulmonary veins, through the pulmonary compliance effect. The increased pulmonary venous pressure is reflected back to the pulmonary capillaries since the venous vessels are highly compliant and the pulmonary circuit is a low-pressure system. The increased pulmonary capillary pressure forces fluid into the interstitial space by filtration. As filtration rate increases, the capacity of the pulmonary lymphatics is exceeded and fluid builds up in the interstitial space. This leads at first to pulmonary

---

* LVMI is the mass of a patient's left ventricle (as estimated from echocardiography), in grams, divided by the patients body surface area in meter square.

congestion and then progresses to pulmonary edema as the condition worsens. Eventually, the excess fluid in the pulmonary alveoli compromises blood–alveolar gas exchange.

5.12.1.1 The Laplace Equation for Ventricular Stress

Since both volume and pressure overload lead to ventricular remodeling, including *dilation* (enlargement) and *hypertrophy* (increased wall thickness), it is important to be able to estimate the changes in wall stress. One simple estimate of the average wall *tension is given by the Laplace* equation. This was originally derived for vessels where the wall was thin enough so that the stress distribution through the wall could be neglected.

The Laplace equation overestimates the *average* circumferential wall stress, since it assumes that the entire pressure gradient is felt by the wall.

We now define the "transmural" pressure gradient. That is, the difference between the internal pressure and the external pressure across the wall of a hollow vessel.

$$P_{TM} = P_{in} - P_{out}.$$

The Laplace wall stress per unit length (wall tension) for a thin-walled cylindrical vessel can be calculated as

$$\tau = \frac{(P_{TM}R)}{2h}. \tag{5.59a}$$

where
  h is the wall thickness = $R_o - R_i$
  R is the average wall thickness (see Figure 5.27a)

A more rigorous biomechanical calculation of the distribution of wall stresses in the left ventricle during systole is given by Chaudhry et al. [23].

Equation 5.59a tells us that the wall stress is directly proportional to transmural pressure gradient and vessel radius and inversely proportional to wall thickness, which means that transmural pressure increases (e.g., hypertension), and ventricular dilation (increased vessel radius) both increase wall stress, while hypertrophy is the remodeling process that attempts to *compensate* for these two factors.

5.12.2 Diastolic Dysfunction

Approximately 30% of patients with heart failure have normal systolic function [22]. These patients exhibit diastolic abnormalities that cause an increase in diastolic filling pressure. Two mechanisms are involved: either an increase in diastolic ventricular relaxation time (an active, metabolically dependent process), an increased left ventricular stiffness (decreased ventricular compliance, a passive process), or both. Patients with diastolic dysfunction often present with vascular congestion since the higher diastolic filling pressures are reflected back to the pulmonary and vascular venous systems. In particular, there is noticeable distension of the jugular veins of the neck.

5.12.3 Right Heart Failure

The right side of the heart has a relatively thin-walled, compliant ventricle and feeds the pulmonary circulation, which is a low-pressure, highly compliant system. Because of these factors, the right heart easily adapts to changes in venous return (preload) without large increases in filling pressure. Diastolic dysfunction of the right heart is rare. Also because of these factors, the right ventricle is very sensitive to increases in afterload (pulmonary artery pressure). The major cause of right-sided failure is left ventricular dysfunction. Failure of the left ventricle not only increases pulmonary capillary pressure and causes pulmonary edema, but the pressure increase can be reflected back to the pulmonary artery.

Right-sided failure in the absence of left-sided failure is very rare and is always caused by a primary pulmonary disease such as interstitial lung disease (e.g., asbestosis) or chronic obstructive pulmonary disease (COPD).

Common causes of right heart failure

- Cardiac causes: pulmonic valve stenosis
- Diseases of the lung tissue: adult respiratory distress syndrome
- Diseases of the lung vasculature: primary pulmonary hypertension

In the right heart failure, the increased diastolic pressure is reflected back to the right atrium, resulting in distension of the large systemic veins (observed in the neck). If the right heart fails and the left heart is normal (isolated right heart failure), the reduced pulmonary venous flow will result in a lower left ventricular preload and a reduced CO.

5.12.4 Compensatory Mechanisms

Several physical, genetic, and neurohormonal compensatory mechanisms come into play during left ventricular systolic dysfunction. They are briefly discussed below.

- The Frank–Starling mechanism: The heart shifts to a lower Frank–Starling curve. Venous return (preload) is increased (the right heart is functioning normally), stretches the left ventricle, and it responds with a greater force of contraction, improving SV, but at a higher central venous pressure.
- Ventricular hypertrophy (increased ventricular mass): Increased wall stress due to pressure or volume overload stimulates cellular and genetic changes that lead to ventricular remodeling in an

attempt to reduce the wall stress. The mass (but not the number) of cardiac myocytes is increased as is the mass of the interstitial matrix (ventricular hypertrophy). Increased muscle mass helps to maintain force of contraction and, if wall thickness increases, leads to a reduction in wall stress. These changes also lead to reduced ventricular compliance and an increase in diastolic filling pressure. The hypertrophy that develops depends on whether the ventricle is subjected to pressure overload or volume overload. In pressure overload (e.g., hypertension or aortic stenosis), the hypertrophy occurs due to the addition of sarcomeres in parallel with existing units and is termed *concentric* hypertrophy. The wall thickness increases and there is little dilation of the ventricular chamber. In volume overload (mitral or aortic regurgitation), sarcomere units are added is series and the ventricular chamber dilates so that the ratio of wall thickness to chamber radius is approximately normal. This is termed *eccentric* hypertrophy. In this case, wall stress will increase.

- Neurohormonal changes: The changes induced by left ventricular systolic failure, increased ventricular stress, and cyclic strain also include activation of the adrenergic nervous system, the renin–angiotensin–aldosterone system, the release of vasopressin (ADH), and ANH.

- The baroreceptor reflexes stimulate afferent sympathetic and vagal reflex activity from the cardiovascular control center in the medulla. These reflexes increase heart rate, ventricular contractility (from the spillover of norepinephrine at the neuromuscular junctions in the intact myocardium), and total peripheral resistance by vasoconstriction of systemic arterioles. These factors increase CO and arterial pressure. The renin–angiotensin–aldosterone system and ADH increase blood pressure by two mechanisms: directly by vasoconstriction of systemic arterioles and indirectly by conservation of water and sodium in the renal system. The increase in circulating volume increases arterial pressure through the capacitance of the arterial system. ANH is also a powerful vasoconstrictor. However, the initial beneficial effects of the increased circulating volume in restoring blood pressure are often detrimental in the long run, since the increased venous return to the heart exacerbates the backup of fluid in the lungs and accelerates pulmonary congestion and edema.

- Cellular and intracellular changes: The changes in pressure, cyclic stress, and strain in the myocardium also initiate genetic changes in expression of contractile proteins, stretch-activated ion channels, catalytic enzymes, surface receptors, and second messengers in myocytes. There is experimental evidence that changes occur in calcium homeostasis, decreased responsiveness of the myofilaments to calcium and changes in excitation–contraction coupling and cellular energy utilization [22].

5.13 Summary

This chapter has reviewed the working of the cardiovascular system. Flow, pressure, and volume relationships were reviewed, principles of the ECG discussed and the mechanical events of the cardiac cycle presented. The mechanical properties of the vascular system were defined in terms of impedance, elastic modulus, distensibility, and compliance. The mechanisms whereby blood pressure and blood volume are regulated and the Frank–Starling mechanism for regulation of SV were discussed. Cardiac work and efficiency were defined and a model for ventricular elastance was introduced. The ventricular model was coupled to a windkessel model of the circulatory system to synthesize a compartmental model for the cardiovascular system. Finally, the failing heart was discussed along with the compensatory mechanisms that come into play in pressure and volume overload hypertrophy.

Problems

5.1 You are asked to do a preliminary design for a new portable ECG machine suitable for a physicians' office. The design objectives are portable (120 VAC + battery operated), light weight, LED display plus standard ECG 12-lead output on thermal paper, built-in diagnostics (according to rules in handout), using standard Ag–AgCl electrodes, easy to set up and operate.

  a. List all the design considerations you can think of for the hardware associated with the design of the instrument. Include materials of construction, the calibration of the signal, measuring and conditioning the signal, displaying the results in real time, and printing a selected set of ECG strips. Since this instrument will be used on humans, address (electrical) safety considerations that need to be incorporated into the design.

  b. List all the design considerations for the software necessary to display the results in real time, enter patient data, give the physician a choice of which set of signals to print, which to save (to a recoverable file), and how to interpret the ECG signals in such a way as to aid the physician in his/her diagnosis.

5.2 Assume that you have available conditioned and digitized ECG results for Lead II of a patient in an Excel spreadsheet. There are two columns of data, millivolts and time (in milliseconds). Further assume that there are 100 values per second and you have 10 worth of data to analyze, along with a calibration pulse.

a. Write a simple software routine (in any high level language) to calculate heart rate using the time between the QRS complex of 2 adjacent beats. You have to first think of a way to identify the QRS complex from the data.
b. Write a subroutine that will check the calibration.
c. Write a subroutine that will determine whether there is a normal sinus rhythm (see ECG interpretation table).
d. Write a subroutine that will allow calculation of PR, QRS, and QT intervals and check them against the guidelines in the ECG interpretation table.
e. Write a subroutine that will output these values to an Excel spreadsheet (or any formatted table).

5.3 Consider the data shown in Figure P5.3, which represents pressure–time and volume–time data for a patient with compensated moderate hypertension.

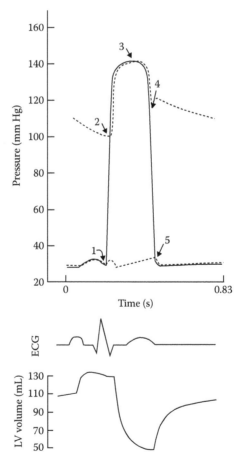

FIGURE P5.3

a. Construct a left ventricular work loop from these data. Label the Points 1, 2, 3, 4, and 5 on your diagram.

b. Describe the events that take place at Points 1, 2, 4, and 5.

c. Label preload and afterload on your diagram.

d. Calculate LV pumping energy in Watts for this patient. Comparing this patient to a person with normal blood pressure, explain why the calculated pumping energy is not a good measure of the total metabolic energy expended by this person's left ventricle.

e. In an echocardiographic study of this patient's heart, the left ventricular wall thickness was measured as 12 mm. Assuming that the pressure exerted by the pericardial sac (outside the heart) is 2 mm Hg, and using a cylindrical model of the ventricle, calculate the ventricular stress at Point 3. Report the stress in dyn/cm² or N/m². What is the importance of this stress?

5.4 A patient with heart problems was given a positive inotropic agent (dobutamine) just prior to surgery to augment his ventricular function during surgical trauma. The following diagrams represent his left ventricular work loops measured before (A) and after (B) treatment with dobutamine (Figure P5.4). The dashed lines represent his ventricles response to an increase in aortic diastolic pressure in each case.

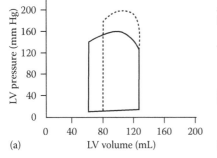 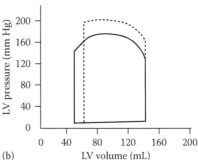

FIGURE P5.4

a. Calculate his change in ejection fraction at his normal afterload.

b. If the patient's heart rate remained constant at 60 beats/min, calculate his CO in liters per minute before and after treatment.

c. What percent change occurred in the contractility of his left ventricle after treatment?

d. Calculate his pumping work in watts before and after treatment.

5.5 A left ventricular assist device (LVAD) is a blood pump that can partially or completely take over the work of the left ventricle, allowing the damaged heart muscle to rest and potentially recover. In the case of slow or inadequate recovery, the LVAD can aid the patient to regain normal physiological conditions without progressive degradation while they are waiting for a donor heart to become available.

An LVAD consists of the blood-circulating device (pump), energy converter that drives the pump (driver), prime mover or source of energy, control system that governs the functioning of the whole system, and other necessary accessories. The devices are designed as either implantable or as extracorporeal (usually placed in the abdomen). Figure P5.5 is a schematic diagram illustrates the functional relationships of an extracorporeal (external) LVAD.

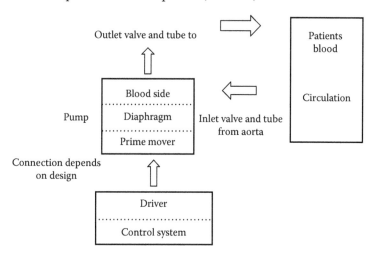

FIGURE P5.5

You are asked to do a preliminary design for a new LVAD.
a. For each of the elements described above, circulating device, energy converter, prime mover, and control system, list the design options you would explore. In particular, discuss possible types of circulating devices, their pros and cons for this application, possible driver-prime mover combinations and their energy sources, and the elements that have to go into design of an appropriate control system for this application.
b. Discuss the materials selection for each of these units as well as materials for the inlet and outlet valves and tubes.
c. What are the major issues that must be addressed for the design of the valves on the inlet and outlet tubes? Why?

5.6 Given the following diagram of an experimental setup to measure changes in blood flow with changing aortic compliance (stiffness) (Figure P5.6).

Here, $Q_i(t)$ is the time-varying output of a pump that simulates the pressure-flow characteristics of the left ventricle. Assume that $P(t)/[V(t) - V_0] = E(t)$, a time-varying elastance. Also, assume that $E(t)$ is a linear function of time during systole, such that $E(t) = Kt$ and constant at E_d during diastole. Here, V_0, K, and E_d are all constants.

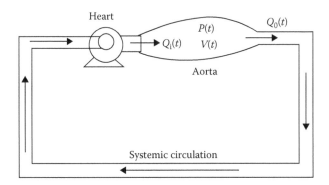

FIGURE P5.6

Set up a model that will allow you to compute $Q_0(t)$, given $Q_i(t)$, V_0, K, and E_d, during both systole and diastole. Include appropriate initial conditions. *Note, you do not have to solve the model*

5.7 Describe the propagation of the action potential in the heart, starting at the SA node. What nerve acts as the normal pacemaker of the heart?

5.8 Draw Einthoven's triangle and identify the three standard ECG Leads on the triangle. By convention, what causes a positive deflection in any lead?

5.9 Use Figure P5.9 to answer the next question.

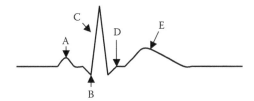

FIGURE P5.9

a. Which one of the labels identifies the part of the ECG that corresponds to ventricular *repolarization*?

b. Which of the labels identifies the signal that precedes atrial *depolarization*?

5.10 How can you tell from an ECG that a patient has a normal sinus rhythm?

5.11 How do you compute heart rate from an ECG strip?

5.12 Describe in sequence the events that occur in the heart during the cardiac cycle.

5.13 Draw a left ventricular work loop. Identify the events of the cardiac cycle on the work loop. Identify preload and afterload.

5.14 How is contractility defined? How is contractility determined from pressure–volume diagrams?

5.15 List the factors that are responsible for mean arterial pressure. Describe the sequence of events that occurs within 1–2 heartbeats when you stand up after lying in bed all night.

5.16 Describe autoregulation of blood flow (use a graph of flow versus time for simplicity). List the sequence of events that occurs during autoregulation using either the myogenic hypothesis or the metabolic hypothesis.

5.17 List the two ways in which the heart can fail and at least three causes of each type of heart failure.

a. The heart can compensate (partially) for both pressure and volume overload in different ways. List 2 compensatory mechanisms for heart failure.

b. Laplace's law relates wall stress to transmural pressure, vessel radius, and wall thickness. According to Laplace's law, an increase in which factors will increase wall stress and an increase in which factor(s) will decrease wall stress? Why are each of these factors important in cardiac performance?

5.18 Interpret the following ECG in terms of the criterion in Tables 5.2 and 5.4. Be sure to follow items 1–4 and 6–8 in your interpretations. The leads are the standard limb leads (I, II, and III). The square waves at the beginning of each tracing are the 1 mV calibration signal and the chart speed is 25 mm/s. Each small division on the chart is 1 mm and each large division is 5 mm (Figure P5.18).

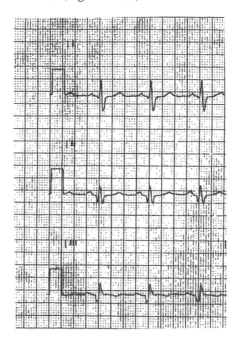

FIGURE P5.18

5.19 The graph in Figure P5.19 is a left ventricular work loop.

    a. Calculate the left ventricular pumping energy in Watts. The patient's heart rate is 60 beats/min.

    b. If pumping energy represents only 10% of the total metabolic energy of the heart, and the right ventricle only expends 1/7 of the metabolic energy of the left ventricle, estimate the total metabolic energy expended by the heart of this patient.

    c. Identify preload and afterload in the diagram. Calculate the stress in the left ventricle at preload, using Laplace's equation, if the patient's left ventricular chamber measures 1.0 cm in radius, and the external pressure on the epicardial surface is 1 mm Hg.

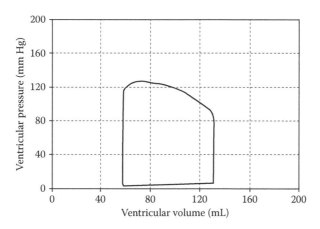

FIGURE P5.19

References

1. Guyton, A.C. and Hall, J.E. *Textbook of Medical Physiology*, W.B. Saunders Co., New York, 2000.

2. McDonald, D.A. *Blood Flow in Arteries*, 3rd ed. Edward Arnold, London, Chapter 11, 1990.

3. Womersley, J.R. Method for the calculation of velocity, rate of flow and viscous drag in arteries when the pressure gradient is known. *J. Physiol.* 127: 553–563, 1955.

4. Womersley, J.R. An elastic tube theory of pulse transmission and oscillatory flow in mammalian arteries. WADC Technical Report TR 56-614, 1957.

5. Bergel, D.H. The static elastic properties of the arterial wall. *J. Physiol.* 156: 445–457, 1961.

6. Taylor, M.G. An approach to the analysis of the arterial pulse wave. I. Oscillations in an attenuating line. *Phys. Med. Biol.* 1: 258–269, 1957.

7. Taylor, M.G. An approach to the analysis of the arterial pulse wave. II. Fluid oscillations in an elastic pipe. *Phys. Med. Biol.* 2: 321–329, 1957.

8. Taylor, M.G. The influence of the anomalous viscosity of blood upon its oscillatory flow. *Phys. Med. Biol.* 3: 273–280, 1959.
9. Taylor, M.G. An experimental determination of the propagation of fluid oscillations in a tube with a viscoelastic wall; together with an analysis of the characteristics required in an electrical analogue. *Phys. Med. Biol.* 4: 63–82, 1959.
10. Bergel, D.H. The dynamic elastic properties of the arterial wall. *J. Physiol* 156: 458–467, 1961.
11. Womersley, J.R. Method for the calculation of velocity, rate of flow and viscous drag in arteries when the pressure gradient is known. *J. Physiol.* 127: 553–563, 1955.
12. Suga, H. and Sagawa, K. Instantaneous pressure-volume relationships and their ratio in the excised, supported canine left ventricle. *Circ. Res.* 35: 117–126, 1974.
13. Suga, H., Hayashi, T., and Shirahata, M. Ventricular systolic pressure-volume area as a predictor of cardiac oxygen consumption. *Am. J. Physiol.* 240: H39–H44, 1981.
14. Shroff, S.G., Janicki, J.S., and Weber, K.T. Left ventricular systolic dynamics in terms of its chamber mechanical properties. *Am. J. Physiol.* 245: H110–H124, 1983.
15. Barnea, O. Mathematical analysis of coronary autoregulation and vascular reserve in closed-loop circulation. *Comput. Biomed. Res.* 27: 263–275, 1994.
16. Rubanyi, G.M., Romero, J.C., and van Houtte, P.M. Flow-induced release of endothelium-derived relaxing factor. *Am. J. Physiol.* 250: H1145–H1149, 1986.
17. Patel, T. Quantitative assessment of reflex blood pressure regulation using a dynamic model of the cardiovascular system. MS Thesis. New Jersey Institute of Technology, Newark, NJ, 2002.
18. Sunagawa, K., Maughan, W.L., Burkhoff, D., and Sagawa, K. Left ventricular interaction with arterial load studied in isolated canine ventricle. *Am. J. Physiol.* 245: H773–H780, 1983.
19. Suga, H. and Sagawa, K. Instantaneous pressure-volume relationships and their ratio in the excised, supported canine left ventricle. *Circ. Res.* 35: 117–126, 1974.
20. Kelly, R.P., et al. Effective arterial elastance as an index of arterial vascular load in humans. *Circulation* 86: 513–517, 1992.
21. Cohen-Solil, A., et al. Left ventricular-arterial coupling in systemic hypertension: Analysis by means of arterial effective and left ventricular elastances. *J. Hypertension* 12: 591–598, 1994.
22. Lilly, L.S., ed. *Heart Failure: Pathophysiology of Heart Disease.* Williams and Wilkins, Baltimore, MD, p. 201. Chap. 9, 1998.
23. Chaudhry, H.R., Bukiet, B., Findley, T.W., and Ritter, A.B. Stresses and strains in the passive left ventricle. *J. Biol. Syst.* 4: 535–554, 1996.
24. Grossman, W. Cardiac hypertrophy: Useful adaptation or pathologic process? *Am. J. Med.* 69: 576–580, 1980.
25. Ferrari, G., De Lazzari, C., Guaragno, M., Tosti, G., Mancini, A. A simple method for E(max) evaluation: in vitro results. *Int J Artif Organs* 27(2): 149–156, 2004.
26. Ferrari, G., Kozarski, M., De Lazzari, C., Gorczynska, K., Mimmo, R., Guaragno, M. Tosti, G., and Darowski, M. Modeling of cardiovascular system: Development of a hybrid (numerical-physical) model. *Int. J. Artif. Organs.* 26(12): 1104–1114, 2003.

27. Wang, J.J., O'Brien, A.B., Shrive, N.G., Parker, K.H., Tyberg, J.V. Time-domain representation of ventricular-arterial coupling as a windkessel and wave system. *Am. J. Physiol. Heart Circ. Physiol.* 284(4): H1358–H1368, 2003.

28. Laffon, E., Galy-Lacour, C., Laurent, F., Ducassou, D., and Marthan, R. MRI quantification of the role of the reflected pressure wave on coronary and ascending aortic blood flow. *Physiol. Meas.* 24(3): 681–692, 2003.

29. Papaioannou, T.G., Mathioulakis, D.S., and Tsangaris, S.G. Simulation of systolic and diastolic left ventricular dysfunction in a mock circulation: The effect of arterial compliance. *J. Med. Eng. Technol.* 27(2): 85–89, 2003.

30. Prisant, L.M., Resnick, L.M., and Hollenberg, S.M. Arterial elasticity among normotensive subjects and treated and untreated hypertensive subjects. *Blood Press. Monit.* 6(5): 233–237, 2001.

31. Ferreira, A.S., Santos, M.A., Barbosa, F.J., Cordovil, I., and Souza, M.N. Determination of radial artery compliance can increase the diagnostic power of pulse wave velocity measurement. *Physiol. Meas.* 25(1): 37–50, 2004.

32. Frankel, S.K. and Fifer, M.A. Heart failure, in Lilly, L.S., ed., *Pathophysiology of Heart Disease.* 2nd ed., Williams and Wilkens, Baltimore, 1998.

Further Readings

Adams, D., Barakeh, J., Laskey, R., and Breemen, C.V. Ion channels and regulation of intracellular calcium in vascular endothelial cells. *FASEB J.* 3: 2389–2400, 1989.

Asakura, T. and Karino, T. Flow patterns and spatial distribution of atherosclerotic lesions in human coronary arteries. *Circ. Res.* 66: 1045–1066, 1990.

Bayliss, W. On the local reaction of the arterial wall to changes in internal pressure. *J. Physiol. (Lond.)* 28: 220–231, 1992.

Beltran, A., McVeigh, G., Morgan, D., Glasser, S.P., Neutel, J.M., Weber, M., Finkelstein, S.M., and Cohn, J.N. Arterial compliance abnormalities in isolated systolic hypertension. *Am. J. Hypertens. Oct.* 14(10): 1007–1011, 2001.

Berridge, M.J. Inositol triphosphate and diacylglycerol: Two interacting second messengers. *Annu. Rev. Biochem.* 56: 159–193, 1987.

Curi, M.A., Skelly, C.L., Quint, C., Meyerson, S.L., Farmer, A.J., Shakur, U.M., Loth, F., and Schwartz, L.B. Longitudinal impedance is independent of outflow resistance. *J. Surg. Res.* 108(2): 191–197, 2002.

Ferris, C.D. and Snyder, S.H. Inositol 1,4,5,-triphosphate-activated calcium channels. *Annu. Rev. Physiol.* 54: 469–488, 1992.

Glasser, S.P. On arterial physiology, pathophysiology of vascular compliance, and cardiovascular disease. *Heart Dis.* 2(5): 375–379, 2000.

Grey, E., Bratteli, C., Glasser, S.P., Alinder, C., Finkelstein, S.M., Lindgren, B.R., and Cohn, J.N. Reduced small artery but not large artery elasticity is an independent risk marker for cardiovascular events. *Am. J. Hypertens.* 16(4): 265–269, 2003.

Kirber, M.T., Walsh, J.V., and Singer, J.J. Stretch-activated ion channels in smooth muscle a mechanism for initiation of stretch-induced contraction. *Pflugers Arch.* 412: 339–345, 1988.

Lambermont, B., Kolh, P., Ghuysen, A., Moonen, M., Morimont, P., Gerard, P., Tchana-Sato, V., Rorive, G., and D'Orio, V. Effect of hemodiafiltration on pulmonary hemodynamics in endotoxic shock. *Artif Organs.* 27(12): 1128–1133, 2003.

Langille, B.L. and Adamson, S.L. Relationship between blood flow direction and endothelial cell orientation: Arterial branch sites in rabbits and mice. *Circ. Res.* 48: 481–488, 1981.

Letsou, G., Rosales, O., Maitz, S., Vogt, A., and Sumpio, B. Stimulation of adenylate cyclase activity in cultured endothelial cells subjected to stretch. *J. Cardiovasc. Surg.* 31: 634–639, 1990.

Mo, M., Eskin, S.G., and Schilling, W.P. Flow-induced changes in Ca signaling of vascular endothelial cells: Effect of shear stress and ATP. *Am. J. Physiol.* 260: H1698–H1707, 1991.

Moore, J.E., Ku, D.N., Zarin, C.K., and Glagov, S. Pulsatile flow visualization in the abdominal aorta under differing physiological conditions: Implications for increased susceptibility to atherosclerosis. *J. Biomech. Eng.* 114: 391–397, 1992.

Morris, C.E. Mechanosensitive ion channels. *J. Membr. Biol.* 113: 93–107, 1990.

Nollert, M.U., Hall, E.R., Eskin, S.G., and McIntire, L.V. Effect of flow on arachidonic acid metabolism in human endothelial cells. *Biochim. Biophys. Acta.* 1005: 72–78, 1989.

Nollert, M.U., Panaro, N.J., and McIntire, L.V. Regulation of genetic expression in shear stress-stimulated endothelial cells. *Ann. N.Y. Acad. Sci.* 665: 94–104, 1992.

Okano, M. and Yoshida, Y. Endothelial cell morphometry of atherosclerotic lesions and flow profiles at aortic bifurcations in cholesterol fed rabbits. *J. Biomech. Eng.* 114: 301–308, 1992.

Olufsen, M.S., Nadim, A., and Lipsitz, L.A. Dynamics of cerebral blood flow regulation explained using a lumped parameter model. *Am. J. Physiol. Regul. Integr. Comp. Physiol.* 282(2): R611–R622, 2002.

O'Rourke, M.F., Staessen, J.A., Vlachopoulos, C., Duprez, D., Plante, G.E. Clinical applications of arterial stiffness: Definitions and reference values. *Am. J. Hypertens.* 15(5): 426–444, 2002.

Papaioannou, T.G., Mathioulakis, D.S., Nanas, J.N., Tsangaris, S.G., Stamatelopoulos, S.F., and Moulopoulos, S.D. Arterial compliance is a main variable determining the effectiveness of intra-aortic balloon counterpulsation: Quantitative data from an in vitro study. *Med. Eng. Phys.* 24(4): 279–284, 2002.

Rosales, O. and Sumpio, B. Changes in cyclic strain increase inositol triphosphate and diacylglycerol in endothelial cells. *Am. J. Physiol. Heart. Circ. Physiol.* 262: C956–C962, 1992.

Schwartz, L.B., O'Donohue, M.K., and Purut, C.M. Myointimal thickening in experimental vein grafts is dependent on wall tension. *J. Vasc. Surg.* 15: 176–186, 1992.

Sumpio, B., Banes, A., and Levin, L. Mechanical Stress stimulates aortic endothelial cells to proliferate. *J. Vasc. Surg.* 6: 252–256, 1987.

Sumpio, B. and Widmann, M. Enhanced production of an endothelium derived contracting factor by endothelial cells subject to pulsatile stretch. *Surgery* 108: 277–282, 1990.

White, G.E. and Fujiwara, K. Expression and intracellular distribution of stress fibers in aortic endothelium. *J. Cell. Biol.* 103: 63–70, 1986.

Part II

Signal Processing

6

Biomedical Signals Processing

6.1 Introduction

The main objective of this chapter is to introduce the fundamentals of discrete-time signals and systems and to provide a working knowledge necessary for performing signal processing of common signals derived from the human body. Ultimately, any engineering analysis must be compared with experimental data both for validation and for estimating the sensitivity of the system to small changes in parameter values. Modern biomedical instrumentation predominantly produces electrical signals as outputs, no matter what physical variable is being measured. Also, data acquisition is invariably linked to the digital computer. This section, therefore, presents the elements of signal processing necessary for a biomedical engineer, both from a theoretical as well as from a practical point of view. The chapter starts with the major concept of time to frequency domain transformation. It then introduces the basic building blocks required to perform this analysis in the digital domain. This includes digital signals and their mathematical representation. Once this is accomplished actual biomedical examples are given to demonstrate major concepts such as aliasing and quantization.

6.2 The Time to Frequency Domain Transform

Let us begin with a major digital signal processing (DSP) concept: signals can be represented either in the time or frequency domain, the selection is a matter of which representation serves a more useful purpose. This is possible because DSP can take signals traditionally analyzed in the time domain and transform them into the frequency domain. The benefit of transforming data between the time and frequency domains is that a biomedical student looking to better understand medical signals can gain more insight into the signals of interest.

Before we venture into the realm of DSP, let's demonstrate the benefits of a frequency domain view with a simple mathematical example. Figure 6.1

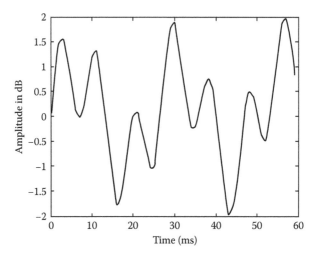

FIGURE 6.1
Two combined sinusoids in the time domain.

represents the summation of two sinusoids, $y = \sin(2\pi 40t) + \sin(2\pi 110t)$. Lost in the time domain, Figure 6.1, two combined sinusoids, clearly gives the student little insight as to the composition of this signal.

However, once the compound sinusoid is translated into the frequency domain, the result is a clear representation as to the composition of the signal. This is made evident by Figure 6.2, two combined sinusoids in the frequency domain.

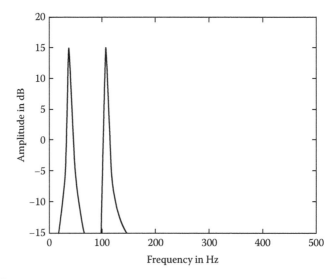

FIGURE 6.2
Two combined sinusoids in the frequency domain.

In this example, the frequency spectrum representation is clearly more informative as to the signal content, than the time-domain representation. This transformation was done with the assistance of MATLAB® and advanced fast Fourier transform (FFT) processing. At this point, the student is not ready to perform these operations. However, the concept of a complex waveform being composed of elementary sine and cosine waves is not beyond the student. Later, we will explore this in depth, when we cover the Fourier transform. For the time being, we merely need to be concerned with Euler's formula, which allows us to represent any complex number $re^{j\theta}$ as the sum of $r\cos\theta + r\sin\theta$. This identity permits us to express a complex exponential signal in Cartesian form. Thus, given a complex signal that is a function of time, such as $Re^{j(wt+\theta)}$, it can be expressed as $R\cos(wt + \theta) + jR\sin(wt + \theta)$. Euler's formula and the two inverse Euler identities (defined in equations 6.1 to 6.3) are all one needs to generate a spectrum representation of a complex signal.

$$\cos\theta = \frac{e^{j\theta} + e^{-j\theta}}{2} \quad \text{Inverse Euler-1} \tag{6.1}$$

$$\sin\theta = \frac{e^{j\theta} - e^{-j\theta}}{2j} \quad \text{Inverse Euler-2} \tag{6.2}$$

$$e^{j\theta} = \cos\theta + j\sin\theta \quad \text{Euler's identity} \tag{6.3}$$

As an example, let's take a signal that consists of two sinusoids.

$$x(t) = 4\cos\left(100\pi t - \frac{\pi}{4}\right) + 10\cos\left(50\pi t - \frac{\pi}{2}\right) \tag{6.4}$$

Once we apply the inverse Euler identity the following equation results:

$$x(t) = 2e^{-j\pi/4}e^{j2\pi 50t} + 2e^{j\pi/4}e^{-j2\pi 50t} + 5e^{-j\pi/2}e^{j2\pi 25t} + 5e^{j\pi/2}e^{-j2\pi 25t} \tag{6.5}$$

From Equation 6.5, the double-sided line spectrum of the signal can be directly plotted, as shown in Figure 6.3, double-sided line spectrum of two sinusoids. Notice that this is considered a double-sided line spectrum because both positive and negative frequencies have been identified. It is a direct result of Euler's inverse identity, which introduces the existence of both positive and negative frequencies and therefore halves the amplitude of each frequency, adhering to the preservation of energy concept. Had we derived the single-sided line spectrum, the amplitude of the positive frequencies would have been twice as large and there would have been no negative frequencies.

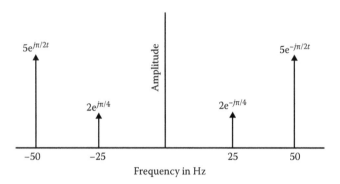

FIGURE 6.3
Double-sided line spectrum of two sinusoids.

More will be discussed about negative frequencies when we consider the Fourier series.

Figure 6.3 highlights a few points that should be noted. The amplitude, phase, and frequency of each tone that constitute the signal are all clearly identified. The phase of the 50 Hz signal is $-\pi/4$ and it appears in the form of a constant phase shift, $e^{-j\pi/4}$. The frequency of this component is also identified in the plot as 50 Hz and in Equation 6.5, it appears as $e^{j2\pi50t}$.

Two additional points need to be noted in Figure 6.3. Before we do this recall, a function satisfying the condition $f(t) = f(-t)$ is called an *even* function. Similarly, if a function satisfies the condition $f(t) = -f(-t)$, it is an *odd* function. With this in mind, the amplitudes of the positive and negative frequencies are identical, thus we see magnitude is an even function, as it reflects about the y-axis directly. However, the phase of the positive and negative frequencies are opposite in sign because phase is an odd function and a reflection about the y-axis results in a change of the sign of the phase.

From Euler's inverse identity, all of the characteristics of the double-sided line spectrum have been identified, so it is now easy to plot the input signal as a frequency spectrum, labeling the amplitude, phase, and frequency.

Working merely in the analog domain, without the use of any DSP tools, we have identified a number of points within this section. We now need to extend these concepts to the digital realm. Before that can be accomplished, we need to first understand what a digital signal looks like and how one converts an analog signal into a digital signal.

6.3 Introduction to Digital Signals

Biological signals, such as the electrocardiogram (ECG), electroencephalogram (EEG), and electromyogram (EMG), are produced in analog form. In this form, the signal is defined over a continuum of both time and

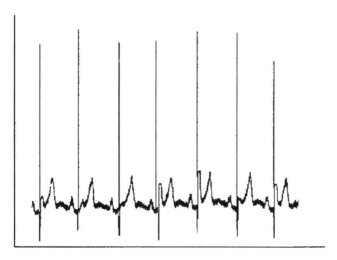

FIGURE 6.4
Analog ECG signal.

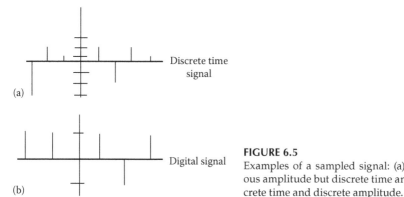

Discrete time
signal

(a)

Digital signal

(b)

FIGURE 6.5
Examples of a sampled signal: (a) continuous amplitude but discrete time and (b) discrete time and discrete amplitude.

amplitude. Figure 6.4 is an example of an analog ECG signal. In order to analyze such signals on a digital computer, the signal must be transformed into a signal that is defined only at a particular set of values of time and amplitude. Such a signal is called a digital signal. Figure 6.5 shows examples of a signal defined for continuous amplitude but discrete time (discrete-time signal) and a signal defined for both discrete time and amplitude (digital signal). In this chapter, we will present the basic concepts of analog and digital signals. The operations necessary to convert an analog signal into a discrete-time (sampling) and discrete-amplitude (quantization) signal will be presented later in this chapter. We will also discuss in this chapter the origins and characteristics of three common bioelectric signals, the ECG, EEG, and EMG.

6.4 Sequences

We will now consider signals that are defined only at discrete values of the independent variable, which we will assume is time. We will also assume that the dependent variable is continuous.

Such a signal can be obtained by sampling a continuous signal, such as an ECG. Assume that the sampling is performed at uniform intervals $t = nT$, where T is the interval between time samples and n is an integer. Such a sequence of numbers can be denoted as

$$x(nT) = \frac{x(t)}{t} = nT, \quad 0 < n < \infty \tag{6.6}$$

Figure 6.5 is an example of such a sequence. The sequence may also be written as

$$\{x(nT)\} = \{x(0), x(T), x(2T), \ldots\} \tag{6.7}$$

The process of generating the values in Figure 6.6a is as shown in Figure 6.6b, where we assume that the sampling device is the exact value of the input at the sampling instant.

It is common practice to abbreviate the notation $x(nT)$ by $x(n)$, where it is assumed that the sampling is performed every T seconds. Such abbreviations are common in signal processing and should not be confusing once the reader has become sufficiently familiar with the material.

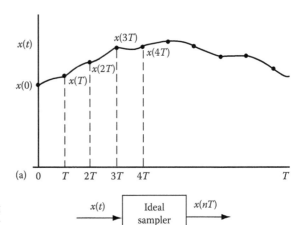

FIGURE 6.6
Illustration of the sampling process: (a) signal to be sampled and (b) sampler.

The following are examples of several common sequences used in signal processing theory.

a. *Unit impulse sequence.* This is the most fundamental sequence and is defined by

$$\delta(n) = \begin{cases} 1, & n = 0 \\ 0, & n \neq 0 \end{cases} \qquad (6.8)$$

Figure 6.7 is an example of a unit impulse sequence.

b. *Constant sequence.* This sequence has the same real value for all values of n and is defined by

$$x(n) = A, \quad -\infty \leq n \leq \infty \qquad (6.9)$$

with the graphical description in Figure 6.8.

c. *Unit step sequence.* The unit step sequence is defined by

$$u(n) = \begin{cases} 1, & n \geq 0 \\ 0, & n < 0 \end{cases} \qquad (6.10)$$

and is shown in Figure 6.9. Notice that $u(n)$ is defined to be 1 at $t = 0$, whereas the continuous function $u(t)$ is not defined at $t = 0$.

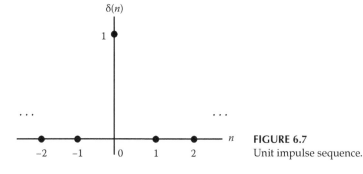

FIGURE 6.7
Unit impulse sequence.

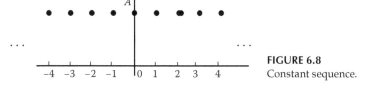

FIGURE 6.8
Constant sequence.

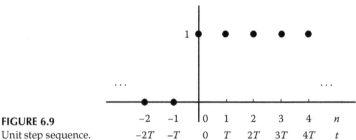

FIGURE 6.9
Unit step sequence.

d. *Shifted sequences.* Just as continuous signals can be shifted in time, so sequences can be shifted. For example, a shifted unit impulse sequence is given by

$$\delta(n-n_0) = \begin{cases} 1, & n = n_0 \\ 0, & n \neq n_0 \end{cases} \tag{6.11}$$

and is shown in Figure 6.10. Similarly, a shifted unit step sequence is given by

$$u(n-n_0) = \begin{cases} 1, & n \geq n_0 \\ 0, & n < n_0 \end{cases} \tag{6.12}$$

and is shown in Figure 6.11.

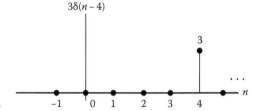

FIGURE 6.10
Shifted unit impulse sequence.

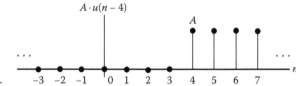

FIGURE 6.11
Shifted unit step sequence.

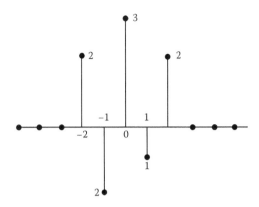

FIGURE 6.12
General example of a sequence.

e. *General description of any sequence.* Since each point in a sequence, $x(n)$, can be considered as a shifted impulse sequence, any sequence can be written as a weighted sum of shifted unit impulses. The relationship is given by

$$x(n) = \sum_{m=-\infty}^{m=+\infty} x(m)\delta(n-m) \qquad (6.13)$$

where $x(m)$ gives the weight (amplitude) of the sample located at $n = m$ and the location specified by the shifted impulse $\delta(n - m)$. As an example, the sequence shown in Figure 6.12 may be expressed as

$$x(n) = 2\delta(n+2) - 2\delta(n+1) + 3\delta(n) - \delta(n-1) + 2\delta(n-2) \qquad (6.14)$$

6.5 Bioelectric Signals

Throughout this chapter, we will be using three signals as examples to illustrate the concepts of signal processing. In this section, we will briefly discuss the production of these signals as well as their characteristics. We will concentrate on the time-domain representation of the signals as well as their frequency characteristics. For a more complete discussion of the physiology relative to the production of these signals, consult any text on human physiology.

The three signals, the ECG, the EEG, and the EMG, are all produced as electrical signals and can be acquired at the body surface through the use of surface electrodes. The instrumentation used to acquire the signals is similar for the three except for changes in required amplification and filter settings. Figure 6.4 shows a normal ECG, which is produced at the body surface to the contraction of the heart. When the heart (cardiac muscle) contracts and relaxes, ionic currents are produced at the surface of the heart, which in turn produce a voltage at the body surface. The peaks (or waves) in the signal can be related to the contraction and relaxation of the atria and ventricles. The amplitude of the largest peak (R wave) is approximately 1 mV, and the accepted frequency range for the ECG is between 0.05 and 100 Hz.

The EEG is produced at the surface of the scalp due to the nerve cell activity in the brain and shows a continuous oscillating electrical activity. The EEG has an amplitude of approximately 100 μV and a frequency range of 0.5–100 Hz. The state of brain activity can be assessed by examining the energy in the EEG in four frequency ranges: the beta range (13–30 Hz), the alpha range (8–13 Hz), the theta range (4–7 Hz), and the delta range (below 3.5 Hz). This will be discussed in more detail later in this chapter. Figure 6.13 is an example of a normal EEG.

The EMG is produced at the body surface due to the electrical activity of contracting muscles immediately beneath the surface. The EMG amplitude can be as high as 5 mV and can have frequency content as high as 10,000 Hz. Figure 6.14 is an example of an EMG signal.

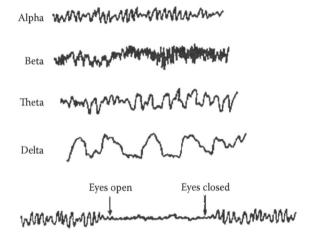

FIGURE 6.13
Example of a normal EEG showing the components in the alpha, beta, theta, and delta frequency bands.

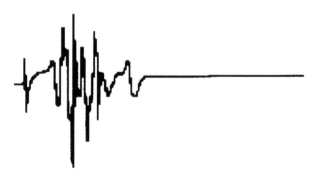

FIGURE 6.14
Normal EMG.

6.6 Pulse Decomposition: An Important DSP Concept

At this point, we have introduced both analog and digital versions of biomedical signals. Even before we proceed further, one very important concept of DSP is introduced—this is pulse decomposition. We have seen that an analog signal can be broken down into a series of pulses, which we can analyze in the digital domain. This is important because once we realize that every pulse looks the same, merely different in amplitude, we can determine how any linear time-invariant system will respond to a series of pulses, based on how the system responds to one basic pulse. Before we proceed too far with this concept, let us review two important concepts required before we can produce a digital signal from an analog one, that would be signal quantization and sampling.

6.7 Signal Quantization

An integrated circuit called an analog-to-digital converter (A/D converter) actually performs the sampling operation. The A/D converter is controlled by the computer that signals the A/D converter at a sampling rate determined by the user, which conforms to the sampling theorem. The resulting samples are carried from the A/D converter to the computer and are usually stored in either the random access memory (RAM) or on a disk (the computer's hard drive or another storage medium). For details on the operation of an A/D converter, the reader is invited to examine the following text [1].

The A/D converter actually performs two functions on the incoming continuous-time signal. One function is the sampling operation described above. The second operation is *quantization* of the signal. The quantization operation can be explained with the aid of Figure 6.14. The figure shows a sine wave that has been sampled at equal intervals. The computer cannot store the amplitude of a signal as a number of volts but instead stores data as 1's and 0's (binary storage). Therefore, it is necessary to convert all incoming data into this binary format. In Figure 6.14, the range (vertical axis) of the signal is divided into four equal intervals. (As we will see, this division into four intervals is not acceptable for proper computer storage but just serves as a simple illustration of the procedure.) The computer can refer to these four regions by using two binary numbers or *bits*. If a sample occurs in the bottom-most region, it will be stored in the computer as 00. The other three regions, from bottom to top, will be stored in the computer as 01, 10, and 11, respectively. Therefore, in Figure 6.15, the first sample would be stored as 10, the second and third as 11, and so on. As we can see, the second and third samples would both be stored in the computer as the

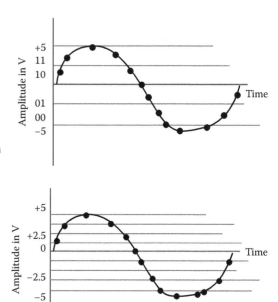

FIGURE 6.15
Illustration of four quantization levels.

FIGURE 6.16
Illustration of eight quantization levels.

same value even though their actual analog values are different. This is called *quantization error* and, although it can be reduced, it cannot be totally eliminated.

The quantization error defined above can be reduced in two ways. The first is to increase the number of bits (or regions) into which the signal is divided. In Figure 6.16, three bits (eight regions) are used to characterize the signal, rather than the two bits (four regions) used above. Now, sample points two and three fall into different regions and would be stored as different values. It is clear that the larger the number of bits used to store the signal, the lower will be the quantization error. The price that is paid for increasing the number of bits is increased storage required for the data. Therefore, a compromise is usually found that has an acceptable quantization error and an acceptable storage requirement. Typical signal quantization uses 8 bits (256 regions), 12 bits (4,096 regions), or 16 bits (65,536 regions).

The second way to reduce the quantization error is to utilize the maximum available range for the signal. In Figures 6.15 and 6.16, we are assuming that the A/D converter can accept a signal that lies between −5 and +5 V. If the signal goes outside of this range, the converted digital value would revert to the value of the maximum allowable amplitude. This has the effect of *clipping* off the top of the signal. In order to reduce the quantization error, the signal should be as close as possible to the maximum permissible values without exceeding the values. In Figure 6.17, the signal amplitude is too small. Notice, how all the samples are crowded into a portion of the possible quantization regions with the others unused. This causes an increase in quantization error

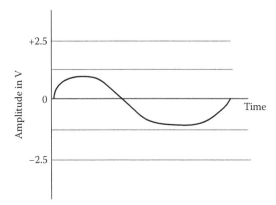

FIGURE 6.17
Incorrect signal amplitude producing large quantization error.

since samples of different amplitudes will be stored in the computer with the same digital value. The relationship between the range of the signal amplitude and the range of the A/D converter can be optimized by adjusting either the signal amplitude (by amplifying or attenuating the signal) or the maximum range of the A/D converter.

6.8 Sampling Rate

Once quantization error is addressed, the next question arises. How fast must the A/D converter operate? The sampling rate of the A/D converter determines what frequencies can and cannot be resolved. For example, consider Figure 6.18. The vertical dashed lines show where the signal is sampled. A frequency higher than the sampling rate is clearly present. However, because we have sampled at a much slower rate, we can no longer

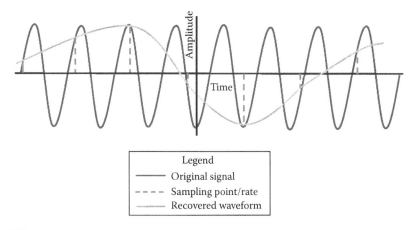

Legend
——— Original signal
- - - - Sampling point/rate
········ Recovered waveform

FIGURE 6.18
Aliased frequency.

determine that the higher rate is present. Instead, we are left reporting a frequency that is clearly less than the actual frequency. This frequency is that of the dotted line.

From the example above, it is clear that the sampling rate must be established with knowledge of the signal we are trying to recover. We must ensure that we can recover the highest frequency component found within the analog signal. Now, we need a way to relate the sampling frequency to the highest signal component. That relationship has already been determined by the Nyquist sampling theorem. This theorem states that to unambiguously recover a given frequency component, we need to sample at a frequency greater than twice the highest frequency component we desire to recover. Once this criterion is met, the false frequency report is avoided. It turns out that the reason for the false frequency report is related to how the sampling frequency interacts with the frequencies present. The aliased frequencies that are produced are equal to the modulo of the sampling frequency with the signal frequencies present, or as shown here:

$$F_{\text{Aliased}} = f_{\text{Present}} \text{ Mod } F_{\text{Sampling frequency}} \tag{6.15}$$

As an example of aliased frequencies, take a 10 Hz signal that is sampled at a rate of 12 Hz. What aliased frequencies will the sampled signal appear to contain? We must first perform the modulo arithmetic and then determine which of these aliased frequencies fall within the detection range of the system. The frequencies that meet the detection range are the frequencies that meet the Nyquist criteria, meaning they must fall below half the sampling frequency. The answer is as shown in the following.

- Apply $F_a = f \bmod (f_s)$
  - Frequencies that appear $= F \pm (n) \times F_s$
  - $(n = 1)$: $10 + 12 = 22$ $10 - 12 = -2$
  - $(n = 2)$: $10 + 24 = 34$ $10 - 24 = -14$
  - $(n = 3)$: $10 + 36 = 46$ $10 - 36 = -26$
- Which of these frequencies enters the Nyquist Frequency range of -6 to $+6$ Hz?
  - Answer: -2 Hz

A spectral diagram of the aliased frequency appears in Figure 6.19, Nyquist plot.

As another example of aliasing, consider an ECG signal, which we assume has no frequency components above 40 Hz. We, however, did not realize there was a 60 Hz noise component present. We, therefore, sampled the signal at 90 Hz; that would satisfy the sampling theorem if the 60 Hz noise component were not present. If the original spectrum is as shown in Figure 6.20a,

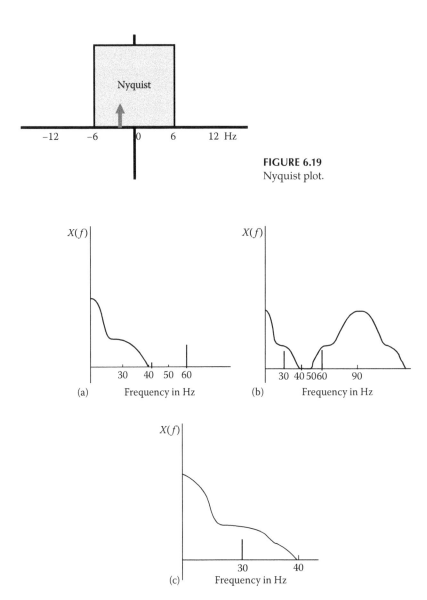

FIGURE 6.19
Nyquist plot.

FIGURE 6.20
(a) Spectrum of an ECG signal with 60 Hz noise present. (b) Spectrum after sampling at 90 Hz.
(c) Spectrum after low-pass filtering at 50 Hz.

the spectrum of the sampled signal is shown in Figure 6.20b. If we now try to recover the original signal by low-pass filtering at 50 Hz, which would be acceptable if the noise were not present, the resulting spectrum is as shown in Figure 6.20c. Notice how the error in the sampling rate produces an incorrect spectrum of the recovered signal.

It is beneficial to understand mathematically what happens when a signal is sampled at too low a rate. To this point, let's examine what happens when the signal below is sampled with a sampling rate that does not meet the Nyquist criteria. Given the 25 Hz signal defined in Equation 6.16, let us sample it at 25 Hz.

$$x(t) = 2\cos(50\pi t) \tag{6.16}$$

In order to meet the Nyquist criteria, we should sample at a frequency above 50 Hz. However, we have chosen to sample at 25 Hz. In order to mathematically sample this signal, we merely replace the variable t in Equation 6.16 with n/f_s, which produces the following:

$$x(n) = 2\cos\left(\frac{50\pi n}{f_s}\right) \tag{6.17}$$

Replacing f_s with 25 Hz results in Equation 6.18.

$$x(n) = 2\cos\left(\frac{50\pi n}{25}\right) \tag{6.18}$$

Reducing the terms results in Equation 6.19.

$$x(n) = 2\cos(2\pi n) \tag{6.19}$$

Recall that a cosine function always equals 1 at multiples of 2π. Thus, Equation 6.19 simply reduces to Equation 6.20.

$$x(n) = 2 \tag{6.20}$$

Apparently, we have sampled at too slow a rate, and if the sampling frequency stays perfectly in sync with the input signal, we will catch the input signal at exactly the same level every time. To translate this back to the time domain, we would now substitute for n with $f_s t$. However, n has dropped out, thus we are left with merely a DC bias, as shown in Equation 6.21.

$$x(t) = 2 \tag{6.21}$$

We could elaborate on the topic of aliased frequencies, however, we will hold off on this discussion until after the student has a better understanding of the Fourier series. Now that the student has an understanding of sampling and the value of a spectral plot, we can proceed with an investigation into the discrete Fourier transform (DFT).

Problems

6.1 A signal is made up of the following sinusoidal components:

$$x(t) = 4\cos\left(200\pi t + \frac{\pi}{2}\right) + 3\cos\left(600\pi t - \frac{\pi}{3}\right)$$

a. Determine the two-sided line spectrum, displaying the magnitude and phase of each signal component.
b. State whether or not the signal is periodic. If periodic, determine the period.

6.2 A signal composed of sinusoids is defined as follows:

$$x(t) = 10 + 20\cos\left(400\pi t + \frac{\pi}{4}\right) + 8\cos\left(800\pi t - \frac{\pi}{3}\right)$$

a. Sketch the two-sided line spectrum of this signal.
b. Determine the period.
c. Determine the amplitude and phase of all harmonic content.

6.3 Given the two-sided line spectrum shown in Figure P6.3, find the following:
a. Equation for $x(t)$ as a sum of cosines.
b. Whether or not the function is periodic. If periodic determine the period.
c. What is the fundamental frequency ω_0 and the amplitude and phase of all harmonic components.

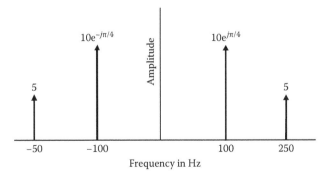

FIGURE P6.3
Line spectrum.

6.4 Given the two-sided line spectrum shown in Figure P6.4, find the following:
   a. An equation that describes $x(t)$ as a sum of cosines.
   b. Whether or not the function is periodic. If it is periodic, determine the period.

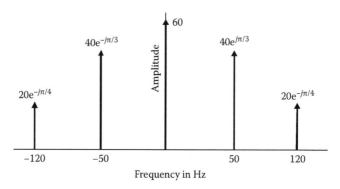

FIGURE P6.4
Line spectrum.

6.5 A signal $x(t)$ is defined as follows:

$$x(t) = 10 + 5\cos\left(100\pi t + \frac{\pi}{2}\right) - 3\cos\left(2000\pi t - \frac{\pi}{3}\right)$$

What condition must the sampling rate satisfy so that the signal can be correctly recovered without frequency aliasing?

6.6 The signal defined in the following is sampled at a set rate.

$$x(t) = 4\cos\left(40\pi t - \frac{\pi}{2}\right) - 2\cos(50\pi t)$$

   a. What must the sampling rate be in order to correctly recover the original frequencies without aliasing?
   b. What is recovered if the sampling rate is 25 Hz?

6.7 A signal to be sampled is described as follows:

$$x(t) = 20\cos\left(20\pi t + \frac{\pi}{3}\right)$$

Determine what the signal above looks like in the digital domain when the sampling rate is
   a. 8 Hz.
   b. 40 Hz.

6.8 A discrete-time signal is defined as follows:

$$x(n) = 5\cos\left(0.2\pi n - \frac{\pi}{6}\right)$$

If this signal above was obtained from a single tone in the time domain that was sampled at $f_s = 4000$ samples/s, then determine three different tones that could have produced this signal $x[n]$.

6.9 How can the quantization error of a system be reduced?

6.10 Describe the unit impulse sequence with both equations and a graph.

6.11 Describe the unit step sequence with both equations and a graph.

Reference

1. Tomkins, W.J. and Webster, J.G. *Interfacing Sensors to the IBM PC*. Prentice-Hall, Englewood Cliffs, NJ, 1988.

Further Readings

McClellan, J., Schafer, R. and Yoder, M. *DSP First: A Multimedia Approach*. Prentice-Hall, Upper Saddle River, NJ, 1999.

Oppenheim, A. and Willsky, A. *Signals and Systems*. Prentice-Hall, Englewood Cliffs, NJ, 1997.

Phillips, C. and Parr, J. *Signals, Systems and Transforms*. Prentice-Hall, Englewood Cliffs, NJ, 1995.

Proakis, J.G. and Manolakis, D.G. *Digital Signal Processing*. Macmillan, New York, 1992.

7

Fourier Series

In 1822, the French mathematician J.B. Fourier, while studying problems in the flow of heat, showed that any arbitrary periodic function could be represented by an infinite sum of sinusoids of harmonically related frequencies. Several words in this sentence need clarification at this point. A continuous function $f(t)$ is said to be periodic with period T if $f(t) = f(t + T)$ for any T. Several functions that fulfill this requirement are shown in Figure 7.1.

In (a) we have a train of pulses, in (b) we have a train of half wave rectified sine waves, and in (c) an arbitrary continuous but periodic function. Of special interest to us are the sinusoids

$$f_1(t) = \cos \frac{2\pi n}{T} t = \cos n\omega_0 t \qquad (7.1)$$

and

$$f_2(t) = \sin \frac{2\pi n}{T} t = \sin n\omega_0 t \qquad (7.2)$$

where n is any integer (or zero). Each frequency of the sinusoids, $n(\omega_0) = 2\pi n/T$, is said to be the nth *harmonic* of the *fundamental* frequency ω_0. Thus a periodic wave will be described in terms of its fundamental frequency, its second harmonic, third harmonic, etc., where each of these frequencies is simply related to the period T. Note that the fundamental frequency and period are related as follows:

$$\omega_0 = 2\pi f_0 = \frac{2\pi}{T} \qquad (7.3)$$

where

f_0 is the fundamental frequency in cycles per second or Hz
ω_0 is the fundamental frequency in radians per second
T is the period in seconds per cycle

Figure 7.2 shows how a square wave can be built up from a sum of harmonically related sine waves. Notice how the sum begins to approach a square wave as more terms are added. Higher frequency terms are needed in order

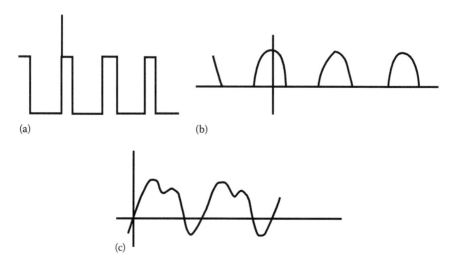

FIGURE 7.1
Examples of periodic functions.

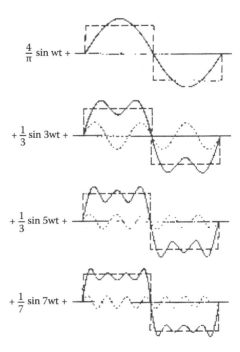

$\frac{4}{\pi}$ sin wt +

$+\frac{1}{3}$ sin 3wt +

$+\frac{1}{3}$ sin 5wt +

$+\frac{1}{7}$ sin 7wt +

FIGURE 7.2
Synthesis of a square wave from harmonically related sine waves.

to reproduce the sharp corners of the square wave. The mathematical expression of the Fourier series of a square wave will be derived in the following.

If $f(t)$ is periodic, the Fourier series is

$$f(t) = a_0 + a_1 \cos \omega_0 t + a_2 \cos 2\omega_0 t + \cdots + a_n \cos n\omega_0 t$$

$$+ \cdots + b_1 \sin \omega_0 t + \cdots + b_n \sin n\omega_0 t + \cdots \tag{7.4}$$

The series in Equation 7.4 may be written in a number of equivalent forms, one of which is obtained by recognizing that for all n

$$a_n \cos n\omega_0 t + b_n \sin n\omega_0 t = c_n \cos(n\omega_0 t + \theta_n) \tag{7.5}$$

where

$$c_n = \sqrt{a_n^2 + b_n^2} \quad \text{and} \quad \theta_n = -\tan^{-1} \frac{b_n}{a_n} \tag{7.6}$$

Combining pairs of terms in Equation 7.4 gives the equivalent form of the Fourier series

$$f(t) = c_0 + c_1 \cos(\omega_0 t + \theta_1) + \cdots + c_n \cos(n\omega_0 t + \theta_n) + \cdots \tag{7.7}$$

with $c_0 = a_0$ and all other c_n and θ_n defined by Equation 7.6.

Observe that if we know that a Fourier series is to be constructed in the form of Equation 7.7, then the set of numbers c_n and θ_n contains all the needed information. Plots by which the information may be displayed are shown in Figure 7.3. The plot of c_n as a function of n or $n\omega_0$ is called an *amplitude spectrum* (Figure 7.3a); the plot of n or $n\omega_0$ is the *phase spectrum* (Figure 7.3b). Later, we will distinguish between the line (or discrete) spectrum we have just discussed and the continuous spectrum defined for all frequencies.

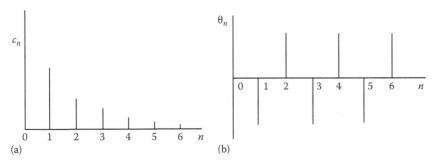

FIGURE 7.3
Examples of an (a) amplitude spectrum and (b) phase spectrum.

7.1 Evaluation of Fourier Coefficients

The evaluation of the a and b coefficients in (7.5) is accomplished by using the orthogonality property of the sin and cos functions. The derivation of the equations to evaluate the coefficients can be found in many books on the subject [1,2]. In this book, we are concerned only with the results and applications. Equations 7.8 through 7.10 are the resulting formulas for evaluation of a_0, a_n, and b_n.

$$a_0 = \frac{1}{T} \int_0^T f(t)\, dt \tag{7.8}$$

$$a_n = \frac{2}{T} \int_0^T f(t) \cos n\omega_0 t\, dt \tag{7.9}$$

$$b_n = \frac{2}{T} \int_0^T f(t) \sin n\omega_0 t\, dt \tag{7.10}$$

The following example illustrates the method.

Example 7.1 Evaluation of Fourier Coefficients

Figure 7.4a shows a square wave voltage that we wish to represent by a Fourier series. The waveform is written as

$$v(t) = \begin{cases} V, & 0 < t < T/4 \\ -V, & T/4 < t < 3T/4 \\ V, & 3T/4 < t < T \end{cases} \tag{7.11}$$

To calculate a_0, we can use two methods. One method involves noticing by inspection that the average value over one period is zero so that $a(0) = 0$ is obtained without using Equation 7.8. This is the preferred method since it results in the answer with the least amount of work. If it is desired to use Equation 7.8 we can proceed as follows:

$$a_0 = \frac{1}{T} \int_0^T v(t)\, dt = \frac{1}{T} \left(V \int_0^{T/4} dt - V \int_{T/4}^{3T/4} dt + V \int_{3T/4}^T dt \right) = \frac{V}{T}\left(\frac{T}{4} - \frac{T}{2} + \frac{T}{4} \right) = 0 \tag{7.12}$$

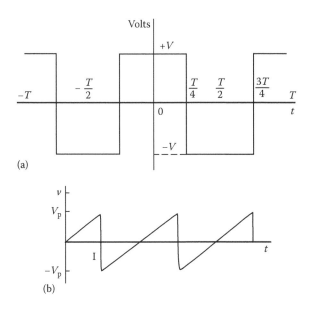

FIGURE 7.4
(a) Square wave for Example 7.1. (b) Sawtooth wave.

The value of a_1 may be obtained with Equation 7.12 setting $n = 1$ as follows:

$$a_1 = \frac{2}{T}\left(V \int_0^{T/4} \cos \omega_0 t \, dt - V \int_{T/4}^{3T/4} \cos \omega_0 t \, dt + V \int_{3T/4}^{T} \cos \omega_0 t \, dt \right) \tag{7.13}$$

$$= \frac{2V}{\omega_0 T}\left[\sin \frac{\omega_0 T}{4} - \left(\sin \frac{3\omega_0 T}{4} - \sin \frac{\omega_0 T}{4} \right) + \left(\sin \omega_0 T - \sin \frac{3\omega_0 T}{4} \right) \right] \tag{7.14}$$

Since $\omega_0 T = 2\pi$ (a useful result that is worth remembering), we obtain

$$a_1 = \frac{V}{\pi}(1 + 2 + 1) = \frac{4V}{\pi} \tag{7.15}$$

Applying the same procedure for all n, we obtain

$$a_n = \begin{cases} \dfrac{+4V}{n\pi}, & n = 1, 5, 9, \ldots \\[2mm] \dfrac{-4V}{n\pi}, & n = 3, 7, 11, \ldots \\[2mm] 0, & n = \text{even} \end{cases} \tag{7.16}$$

$$b_n = 0, \quad \text{all } n \tag{7.17}$$

Thus the Fourier series is

$$v(t) = \frac{4V}{\pi}\left(\cos \omega_0 t - \frac{1}{3}\cos 3\omega_0 t + \frac{1}{5}\cos 5\omega_0 t - \frac{1}{7}\cos 7\omega_0 t + \cdots\right) \tag{7.18}$$

This sum of harmonically related sinusoids is equal to the square wave of Figure 7.4a.

Figure 7.4b shows a sawtooth wave. The reader should verify that the Fourier series expression for the sawtooth wave of Figure 7.4b is given by Equation 7.19.

$$v(t) = \frac{2V_p}{\pi}\left(\sin \omega t - \frac{1}{2}\sin 2\omega t + \frac{1}{3}\sin 3\omega t - \frac{1}{4}\sin 4\omega t + \cdots\right) \tag{7.19}$$

7.2 Waveform Symmetries

In this section, we study a number of interesting properties of periodic signals, and rules that will simplify the evaluation of Fourier coefficients. A function satisfying the condition

$$f(t) = f(-t) \tag{7.20}$$

is called an *even* function. The function in this group of most importance to us is $\cos \omega_0 t$. Similarly, if a function satisfies the condition

$$f(t) = -f(-t) \tag{7.21}$$

it is an *odd* function. The function $\sin \omega_0 t$ is an important example of an odd function.

It can also be shown that

$$\text{Sum of even functions} = \text{even function} \tag{7.22}$$

$$\text{Sum of odd functions} = \text{odd function} \tag{7.23}$$

$$\text{Product of two odd functions} = \text{even function} \qquad (7.24)$$

$$\text{Product of two even functions} = \text{even function} \qquad (7.25)$$

$$\text{Product of an even and odd function} = \text{odd function} \qquad (7.26)$$

and that for any even function f_e:

$$\int_{-t_0}^{t_0} f_e(t)\,dt = 2\int_{0}^{t_0} f_e(t)\,dt \qquad (7.27)$$

while for an odd function f_0:

$$\int_{-t_0}^{t_0} f_0(t)\,dt = 0 \qquad (7.28)$$

It can then be shown that, for an even function $f(t)$, $b_n = 0$ and the Fourier series coefficients become

$$a_n = \frac{4}{T}\int_{0}^{T/2} f(t)\cos n\omega_0 t\,dt \qquad (7.29)$$

$$a_0 = \frac{2}{T}\int_{0}^{T/2} f(t)\,dt \qquad (7.30)$$

and the Fourier series contains only cosine terms and a constant.

For an odd function $f(t)$, $a_0 = 0$ and $a_n = 0$ and the Fourier series coefficients become

$$b_n = \frac{4}{T}\int_{0}^{T/2} f(t)\sin n\omega_0 t\,dt \qquad (7.31)$$

and the Fourier series contains only sine terms.

Figure 7.5 is an example of an even function while Figure 7.6 is an example of an odd function. The reader should show as an exercise that the Fourier series for the odd function of Figure 7.6 is given by

$$f(t) = \frac{8}{\pi^2}\left(\sin\omega_0 t - \frac{1}{9}\sin 3\omega_0 t + \frac{1}{25}\sin 5\omega_0 t - \frac{1}{49}\sin 7\omega_0 t + \cdots\right) \qquad (7.32)$$

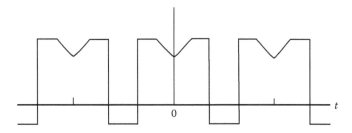

FIGURE 7.5
Example of an even function.

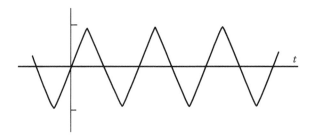

FIGURE 7.6
Example of an odd function.

7.3 Exponential Form of the Fourier Series

The Fourier series can be expressed in an equivalent form in terms of exponentials, $e^{\pm jn\omega_0 t}$. This form will be particularly important when we make the transition to the Fourier transform. Suppose we group the terms in the series together as

$$f(t) = a_0 + \sum_{n=1}^{\infty} \left(a_n \cos n\omega_0 t + b_n \sin n\omega_0 t \right) \tag{7.33}$$

However, as we have seen previously through the Euler identities, the cosine and sine can be expressed in terms of exponentials as

$$\cos n\omega_0 t = \frac{1}{2} \left(e^{jn\omega_0 t} + e^{-jn\omega_0 t} \right) \tag{7.34}$$

$$\sin n\omega_0 t = \frac{1}{2j} \left(e^{jn\omega_0 t} - e^{-jn\omega_0 t} \right) \tag{7.35}$$

Substituting these expressions into Equation 7.33, we obtain

$$f(t) = a_0 + \sum_{n=1}^{\infty} \left(a_n \frac{e^{jn\omega_0 t} + e^{-jn\omega_0 t}}{2} + b_n \frac{e^{jn\omega_0 t} - e^{-jn\omega_0 t}}{2j} \right) \tag{7.36}$$

Grouping the exponential terms and noting that $1/j = -j$, we obtain

$$f(t) = a_0 + \sum_{n=1}^{\infty} \left[\left(\frac{a_n - jb_n}{2} \right) e^{jn\omega_0 t} + \left(\frac{a_n + jb_n}{2} \right) e^{-jn\omega_0 t} \right] \tag{7.37}$$

We next introduce a new coefficient to replace the a and b coefficients. We define

$$c_n = \frac{a_n - jb_n}{2}, \quad c_{-n} = \frac{a_n + jb_n}{2}, \quad \text{and} \quad c_0 = a_0 \tag{7.38}$$

The new form of Equation 7.36 is

$$f(t) = c_0 + \sum_{n=1}^{\infty} \left(c_n e^{jn\omega_0 t} + c_{-n} e^{-jn\omega_0 t} \right) \tag{7.39}$$

Equation 7.39 represents the double sided line spectrum we introduced in the previous chapter when we discussed Euler identities. It represents both positive and negative frequency components. Allowing n to range through the values from 1 to ∞ in this equation is equivalent to allowing n to range from $-\infty$ to $+\infty$ (including zero) in the equation:

$$f(t) = \sum_{n=-\infty}^{\infty} c_n e^{jn\omega_0 t} \tag{7.40}$$

where the coefficients c_n can be obtained from

$$c_n = \frac{1}{T} \int_0^T f(t) e^{-jn\omega_0 t} \, dt \tag{7.41}$$

This is the complex or exponential form of the Fourier series. Note that the coefficients c_n are complex and are related to a_n and b_n by

$$c_n = |c_n| e^{j\phi_n} \tag{7.42}$$

$$|c_n| = \frac{1}{2}\sqrt{a_n^2 + b_n^2} \tag{7.43}$$

$$\phi_n = \tan^{-1} -\frac{b_n}{a_n} \tag{7.44}$$

for all n except $n = 0$ where $c_0 = a_0$.

Example 7.2 Complex Fourier Series

We will now find the complex Fourier series for the wave shown in Figure 7.7. Using $-\tau/2$ as the starting point for the integration, we have, from Equation 7.41 (note that points A and B in the figure correspond to $-\tau/2$ and $\tau/2$, respectively).

$$c_n = \frac{1}{T}\int_{-\tau/2}^{\tau/2} e^{-jn\omega_0 t}\, dt \tag{7.45}$$

$$c_n = \frac{j}{n\omega_0 T}\left(e^{\frac{-jn\omega_0\tau}{2}} - e^{\frac{jn\omega_0\tau}{2}}\right) \tag{7.46}$$

$$c_n = \frac{\tau}{T}\frac{\sin\left(n\omega_0\tau/2\right)}{n\omega_0\tau/2} \tag{7.47}$$

Therefore, the exponential series representation of $f(t)$ is

$$f(t) = \left(\frac{\tau}{T}\right)\sum_{n=-\infty}^{n=\infty} \frac{\sin\left(n\omega_0\tau/2\right)}{n\omega_0\tau/2} e^{jn\omega_0 t} \tag{7.48}$$

Note that since $f(t)$ has even symmetry, $b_n = 0$ for all n and hence we expect the coefficients to be real.

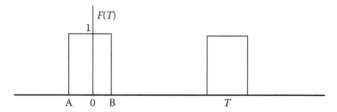

FIGURE 7.7
Waveform for Example 7.2.

7.4 Fourier Transforms and Continuous Spectra

In this section, we consider the case where the signal occurs once in some finite time interval. This is accomplished by letting the period T become infinite in the Fourier series. The line spectrum then becomes a continuous spectrum and the result is the Fourier transform.

One method of deriving the Fourier transform from the Fourier series is to consider a periodic pulse of magnitude V and duration a as shown in Figure 7.8. The function $f(t)$ is defined as

$$f(t) = \begin{cases} V, & -a/2 < t < a/2 \\ 0, & -T/2 < t < a/2,\ a/2 < t < T/2 \end{cases} \tag{7.49}$$

The Fourier coefficients can be determined from Equation 7.40 as

$$c_n = \frac{1}{T} \int_{-a/2}^{a/2} V e^{-jn\omega_0 t}\, dt \tag{7.50}$$

$$c_n = \frac{V}{n\pi} \left(\frac{e^{jn\omega_0 a/2} - e^{-jn\omega_0 a/2}}{2j} \right) = V \frac{\omega_0 a}{2\pi} \left(\frac{\sin\left(n\omega_0 a/2\right)}{n\omega_0 a/2} \right) \tag{7.51}$$

Now since $T = 2\pi/\omega_0$, Equation 7.51 is in the form $(\sin x)/x$ which is shown plotted in Figure 7.9.

Note from the plot that the function has the value 1 when $x = 0$ corresponding to the case $n = 0$ in Equation 7.50. Note also that c_n has both positive and negative values, corresponding to a phase ϕ_n of either $0°$ or $180°$.

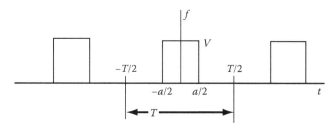

FIGURE 7.8
Periodic pulse.

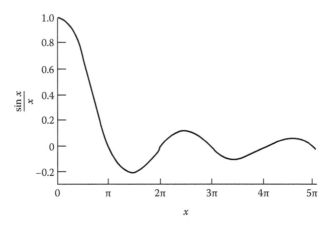

FIGURE 7.9
Plot of Fourier coefficients.

Equation 7.51 has values only for discrete frequencies $n\omega_0$. These are the values of frequencies at which there will be lines in the spectrum for c_n. The *envelope* of c_n is a *continuous* function found by replacing $n\omega_0$ with ω, and so,

$$\text{Envelope of } c_n = V\frac{a}{T}\frac{\sin(\omega a/2)}{(\omega a/2)} \tag{7.52}$$

This envelope will play an important role as our derivation progresses.

We next examine c_n as the ratio a/T changes. We will do this for $a/T = 1/2$, $1/5$ and finally generalize for $1/N$. For $a/T = 1/2$, we have

$$\text{Envelope of } c_n = \frac{V}{2}\frac{\sin(\omega a/2)}{\omega a/2} \tag{7.53}$$

which is shown in Figure 7.10. The envelope has zero values when

$$\frac{\omega a}{2} = \pm\pi, \pm 2\pi, \pm 3\pi, \dots \tag{7.54}$$

or at the frequencies

$$\omega = \frac{\pm 2\pi}{a}, \frac{\pm 4\pi}{a}, \frac{\pm 6\pi}{a}, \dots \tag{7.55}$$

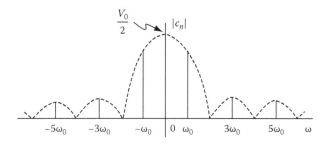

FIGURE 7.10
Envelope of c_n.

Now the fundamental frequency in $f(t)$ is $\omega_0 = 2\pi/T$, so that for $T = 2a$, c_n has values at

$$n\omega_0 = \frac{n\pi}{a} \tag{7.56}$$

Comparing Equations 7.54 and 7.55 we see that the even ordered harmonics have zero value except for c_0. This is illustrated in Figure 7.10.

Repeating next for the case $a/T = 1/5$, the envelope for the amplitude spectrum is

$$\text{Envelope of } c_n = \frac{V}{5} \frac{\sin(\omega a/2)}{\omega a/2} \tag{7.57}$$

Lines will be present in the amplitude spectrum when

$$n\omega_0 = \frac{2n\pi}{5a} \tag{7.58}$$

We now see that lines at $\pm 5\omega_0$, $\pm 10\omega_0$, etc., will have zero amplitude. The amplitude spectrum for this case is shown in Figure 7.11. In the general case, for $a/T = 1/N$, the spacing between the lines is still $\Delta\omega = \omega_0$ and the amplitude of the lines will be

$$|c_n| = \frac{V}{N} \left| \frac{\sin(n\omega_0 a/2)}{n\omega_0 a/2} \right| \tag{7.59}$$

where

$$\omega_0 = \frac{2\pi}{Na} \tag{7.60}$$

which is shown in Figure 7.11.

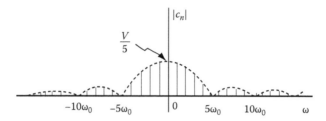

FIGURE 7.11
Amplitude of c_n.

We are now ready to consider the situation when $N \to \infty$ which is the same as $T \to \infty$ since $a/T = 1/N$. Thus in Figure 7.8, the pulse width a remains fixed and we have produced the equivalent of a single pulse. Let us consider the product of T and c_n which is given by

$$c_n T = \int_{-T/2}^{T/2} f(t) e^{-jn\omega_0 t}\, dt \tag{7.61}$$

Let this product be given the notation

$$F(jn\omega_0) = c_n T = \frac{2\pi}{\omega_0} c_n = \frac{2\pi}{\Delta\omega} c_n \tag{7.62}$$

since $\Delta\omega = \omega_0$ is the spacing between lines. Therefore,

$$c_n = \frac{F(jn\omega_0)\Delta\omega}{2\pi} \tag{7.63}$$

Returning to the general expression for the envelope of c_n (Equation 7.47), since $\Delta\omega = \omega_0 = 1/T$, Equation 7.47 becomes

$$\text{Envelope of } \frac{c_n}{\Delta\omega} = a\frac{V}{2\pi}\frac{\sin(\omega a/2)}{\omega a/2} \tag{7.64}$$

which is independent of N and of T, so that it will not change as $N \to \infty$. Since from Equation 7.63,

$$\frac{c_n}{\Delta\omega} = \frac{F(jn\omega_0)}{2\pi} \tag{7.65}$$

then

$$F(jn\omega_0) = a \frac{V}{2\pi} \frac{\sin(\omega a/2)}{\omega a/2} \tag{7.66}$$

and since from Equation 7.39

$$f(t) = \sum_{n=-\infty}^{\infty} c_n e^{jn\omega_0 t} = \sum_{n=-\infty}^{\infty} \frac{F(jn\omega_0)}{2\pi} e^{jn\omega_0 t} \Delta\omega \tag{7.67}$$

then as $T \to \infty$, $\Delta\omega \to d\omega$, $n\omega_0 = n\Delta\omega \to \omega$ and the summation becomes an integral. Therefore, as a final result, we have

$$f(t) = \frac{1}{2\pi} \int_{-\infty}^{\infty} F(j\omega) e^{j\omega t} dt \tag{7.68}$$

The expression for $F(j\omega)$ is found from Equation 7.61 by making the same substitutions as above.

$$F(j\omega) = \int_{-\infty}^{\infty} f(t) e^{-j\omega t} dt \tag{7.69}$$

Equations 7.68 and 7.69 constitute the Fourier transform pair. $F(j\omega)$ is the Fourier transform of $f(t)$, and $f(t)$ is the inverse Fourier transform of $F(j\omega)$. In general, $F(j\omega)$ is complex and therefore we can write $F(j\omega) = |F(j\omega)|e^{j\phi(\omega)}$ where $|F(j\omega)|$ is known as the *continuous amplitude spectrum* and $\phi(\omega)$ the *continuous phase spectrum* of $f(t)$.

7.5 Some Useful Fourier Transform Pairs

In this section, we will derive some transforms that are useful in biomedical signal processing. We will also try to give an intuitive discussion to lend insight into how to interpret the transform in terms of the frequency content of the signal. In general, we can say that if a time function $f(t)$ changes rapidly in time, the Fourier transform will have terms out at high values of frequency. On the other hand, signals that change slowly in time will have transforms with terms at low values of frequency.

1. *The unit impulse function*:
   The unit impulse function $\delta(t)$ is defined as follows:

   $$\delta(t) = \begin{cases} 0 & \text{for } t \neq 0 \\ \infty & \text{for } t = 0 \end{cases} \tag{7.70}$$

   and has unit area so that

   $$\int_{-\infty}^{\infty} \delta(t)\,dt = 1 \tag{7.71}$$

   Also the unit impulse has the property that, for any function $f(t)$,

   $$\int_{-\infty}^{\infty} f(t)\delta(t - t_0)\,dt = f(t_0) \tag{7.72}$$

   The impulse is shown in Figure 7.12. The Fourier transform of the unit impulse is easily derived as follows:

   $$F(j\omega) = \int_{-\infty}^{\infty} \delta(t)e^{-j\omega t}\,dt \tag{7.73}$$

   Utilizing the property in Equation 7.72, we get

   $$F(j\omega) = e^0 = 1 \tag{7.74}$$

   Thus the Fourier transform of a unit impulse is the constant. The magnitude spectrum is shown in Figure 7.13.

2. *The exponential function*:
   Consider the exponential defined for both positive and negative t as shown in Figure 7.14.

   $$f(t) = e^{-a|t|}, \quad \text{all } t \tag{7.75}$$

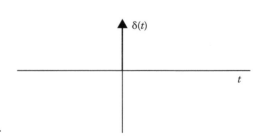

FIGURE 7.12
Impulse function.

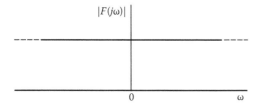

FIGURE 7.13
Fourier transform of the unit impulse.

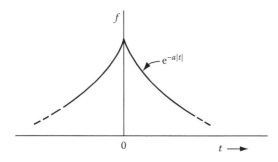

FIGURE 7.14
Exponential function.

The corresponding Fourier transform is found from defining the equation as

$$F(j\omega) = \int_{-\infty}^{0} e^{at} e^{-j\omega t}\, dt + \int_{0}^{\infty} e^{-at} e^{-j\omega t}\, dt = \frac{1}{a - j\omega} + \frac{1}{a + j\omega} = \frac{2a}{\omega^2 + a^2} \tag{7.76}$$

Note that $e^{-|a|t} \to 0$, as $t \to \pm\infty$. This is a real function whose magnitude spectrum is shown in Figure 7.15.

3. *The constant function*:
   We will now consider the Fourier transform of the constant function $f(t) = 1$. Because of the difficulty of deriving the transform, we will only present the result and then discuss the result intuitively. It can be shown that the Fourier transform of the function $f(t) = 1$ is

$$F(j\omega) = 2\pi\delta(\omega) \tag{7.77}$$

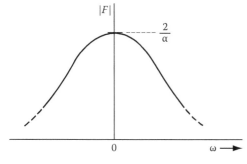

FIGURE 7.15
Fourier transform of an exponential function.

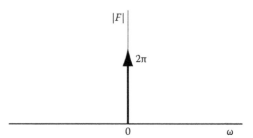

FIGURE 7.16
Fourier transform of a constant function.

as illustrated in Figure 7.16. Notice that a function which is constant for all time can be thought of as having zero frequency variation. In electrical engineering, such a function is called DC. The transform has a component only at zero frequency, which we would expect from the above statement. Therefore, the magnitude spectrum should agree with our intuition in terms of where in frequency the energy of the signal is located. We will further illustrate this concept in future examples.

4. *The complex exponential*:
We will now derive the Fourier transform of $f(t) = e^{j\omega_0 t}$, where ω_0 is a constant. Since we have just stated that

$$\int_{-\infty}^{\infty} 1 e^{-j\omega t}\, dt = 2\pi\delta(\omega) \tag{7.78}$$

it follows that for $f(t) = e^{j\omega_0 t}$,

$$F(j\omega) = \int_{-\infty}^{\infty} 1 e^{j\omega_0 t} e^{-j\omega t} dt = 2\pi\delta(\omega - \omega_0) \tag{7.79}$$

5. *Sine and cosine functions*:
Since the sine and cosine can be written in terms of complex exponentials as

$$\cos\omega_0 t = \frac{e^{j\omega_0 t} + e^{-j\omega_0 t}}{2} \quad \text{and} \quad \sin\omega_0 t = \frac{e^{j\omega_0 t} - e^{-j\omega_0 t}}{2j} \tag{7.80}$$

we may use Equations 7.78 and 7.79 to determine their Fourier transforms as

$$\cos\omega_0 t \Leftrightarrow \pi\left[\delta(\omega - \omega_0)\delta(\omega - \omega_0)\right] \tag{7.81}$$

$$\sin\omega_0 t \Leftrightarrow -j\pi\left[\delta(\omega - \omega_0) - \delta(\omega - \omega_0)\right] \tag{7.82}$$

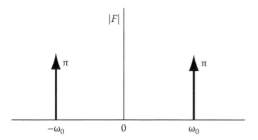

FIGURE 7.17
Fourier transform of a sinusoid.

where the symbol $\Leftrightarrow$ indicates a Fourier transform pair. The magnitudes of the transforms for both the sine and cosine are identical and are shown in Figure 7.17. Note that the transform indicates that all the energy in the signal is concentrated at the frequency ω_0 as we would expect. The component at frequency $-\omega_0$ is present due to the mathematics of the Fourier transform. We will discuss the significance of the "negative frequency" components later in this chapter.

6. *Generalized periodic functions*:
   Let us derive the expression for the Fourier transform of a periodic function. As we recall, any periodic function can be written as a Fourier series whose exponential form is repeated in the following:

$$f(t) = \sum_{n=-\infty}^{\infty} c_n e^{jn\omega_0 t} \tag{7.83}$$

The Fourier transform of this function $f(t)$ is given by

$$F(j\omega) = \Im\left[\sum_{n=-\infty}^{\infty} c_n e^{jn\omega_0 t} \right] = \sum_{n=-\infty}^{\infty} c_n \Im\left[e^{jn\omega_0 t} \right] \tag{7.84}$$

where $\Im$ indicates the Fourier transform operation. Now from Equation 7.79, we know the Fourier transform of the complex exponential. We can therefore substitute Equation 7.79 into Equation 7.84 and obtain

$$F(j\omega) = 2\pi \sum_{n=-\infty}^{\infty} c_n \delta(\omega - \omega_0) \tag{7.85}$$

Notice that the Fourier transform consists of energy only at the fundamental frequency and at multiples of the fundamental frequency, and that the amount of energy at each frequency is given by the Fourier series coefficient at that frequency. This important result relates the Fourier series and transform.

7.6 Discrete Fourier Transforms, Fast Fourier Transforms

Let us start by deriving the Fourier series expansion of periodic sampled signals.

Let $x_p(t)$ be a continuous time periodic signal whose period is L seconds, and is shown in Figure 7.18. We can represent $x_p(t)$ by a Fourier series using Equations 7.39 and 7.40 as follows:

$$x_p(t) = \sum_{n=-\infty}^{\infty} c_n \, e^{jn\omega_0 t} \tag{7.86}$$

where

$c_n = 1/L \displaystyle\int_0^L x_p(t)e^{-jn\omega_0 t}\,dt$ is the nth Fourier series coefficient

$\omega_0 = 2\pi/L$ is the fundamental radian frequency

Next we assume that $x_p(t)$ is band limited, that is, its Fourier series expansion has a finite number of harmonics. Thus,

$$c_n = 0 \quad |n| > M \tag{7.87}$$

where M is a positive integer. Therefore, the bandwidth B of $x_p(t)$ is given by

$$B = Mf_0 = \frac{M}{L} \tag{7.88}$$

where $f_0 = 1/L$ is the fundamental frequency in Hz.

We will now sample the signal $x_p(t)$ at the minimum allowable sampling frequency according to the sampling theorem. Therefore, the sampling frequency $f_s = 2B = 2M/L$ samples per second. This results in the periodic sequence

$$x(m) = x(mT), \quad 0 \le m \le 2M-1 \tag{7.89}$$

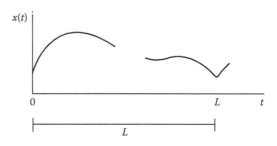

FIGURE 7.18
A continuous time signal with period L.

containing $2M$ terms and having length L. The corresponding sampled signal, $x_p^*(t)$ is

$$x_p^*(t) = \sum_{m=0}^{2M-1} x(m)\delta(t - mT) \tag{7.90}$$

We will now find the Fourier series expansion for the sampled signal $x_p^*(t)$. Thus using the expression in Equation 7.86, we have

$$x_p^*(t) = \frac{1}{L} \sum_{n=-M}^{M} c_n e^{jn\omega_0 t} \tag{7.91}$$

where $c_n = \int_0^L x_p^*(t)e^{-jn\omega_0 t}\,dt$, $|n| \le M$. To evaluate the coefficients c_n we substitute Equation 7.90 into the expression for c_n. We then obtain

$$c_n = \int_0^L \left[\sum_{m=0}^{2M-1} x(m)\delta(t - mT) \right] e^{-jn\omega_0 t}\,dt \tag{7.92}$$

Interchanging the order of integration and summation, there results

$$c_n = \sum_{m=0}^{2M-1} x(m) \left[\int_0^L e^{-jn\omega_0 t} \delta(t - mT)\,dt \right] \tag{7.93}$$

which yields

$$c_n = \sum_{m=0}^{2M-1} x(m)e^{-jnm\omega_0 T} \tag{7.94}$$

Since $L = 2MT$ and $\omega_0 = 2\pi/L$ Equation 7.93 becomes

$$c_n = \sum_{m=0}^{2M-1} x(m)e^{-jnm\pi/M}, \quad |n| \le M \tag{7.95}$$

We will now derive the *discrete Fourier transform* and relate it to the derivation that was just discussed above. The discrete Fourier transform of a sequence $x(m)$, $0 \leq m \leq N - 1$ is defined as

$$X(n) = \sum_{m=0}^{N-1} x(m)W^{nm}, \quad 0 \leq n \leq N-1 \tag{7.96}$$

where
  N is finite
  $W = e^{-2j\pi/N}$
  $X(n)$ is the nth Fourier coefficient

We note that W is the nth root of unity since $e^{2j\pi} = 1$. It can then be shown that the discrete Fourier transform is directly related to the Fourier series of the periodic sampled signal $x_p^*(t)$ in Equation 7.90 as follows:

$$\begin{aligned} c_0 &= X(0) \\ c_k &= X(k) \\ c_{-k} &= X(2M-k), \quad 1 \leq k \leq M \\ c_{\pm M} &= X(\pm M) \end{aligned} \tag{7.97}$$

where $N = 2M$.

It can also be shown that the sequence $x(m)$ can be recovered from $X(n)$ by Equation 7.98. This is the inverse discrete Fourier transform.

$$x(m) = \frac{1}{N} \sum_{n=0}^{N-1} X(n)W^{-nm}, \quad 0 \leq m \leq N-1 \tag{7.98}$$

We will now list some of the properties of the discrete Fourier transform. Derivations of these properties can be found in many books on the subject [1,2].

If $x(m)$, $0 \leq m \leq 2M - 1$ is a real sequence, then

1. $X(M+k) = \tilde{X}(M-k)$, $0 \leq k \leq M$, where $\tilde{X}(n)$ denotes the complex conjugate of $X(n)$. $\hspace{2cm}$ (7.99)

2. $X(0)$ and $X(M)$ are real numbers.

3. Parseval's theorem relating the power in the sequence $x(m)$ in both time and frequency domains can be stated as follows:

$$\sum_{m=0}^{2M-1} x^2(m) = \frac{1}{N}\left[X^2(0) + 2\sum_{n=1}^{M-1} |X(n)^2| + X^2(M) \right] \tag{7.100}$$

with $N = 2M$. If $x(m)$ is obtained by sampling a voltage or current across a one ohm resistor, then the left-hand side of Equation 7.100 represents the average power in the sequence. Thus, we define the discrete Fourier transform *power spectrum* of a real sequence $x(m)$ as

$$P_0 = \frac{X^2(0)}{N^2}$$

$$P_n = \frac{2|X(n)|^2}{N^2} \quad 1 \le n \le M \tag{7.101}$$

$$P_M = \frac{X^2(m)}{N^2}$$

where
   P_0 is the DC power
   P_M is the power in the Mth harmonic

Since $X(n)$ is a complex number, we can express it as $X(n) = |X(n)|e^{j\psi_n^x}$, which leads to the definitions

$$p_n^x = |X(n)| \tag{7.102}$$

is the discrete Fourier transform amplitude spectrum, and

$$\psi_n^x = \tan^{-1}\left\{ \frac{\text{Im}[X(n)]}{\text{Re}[X(n)]} \right\} \tag{7.103}$$

is the discrete Fourier transform phase spectrum.

We will now briefly discuss the *fast Fourier transform* (FFT). The FFT is a very efficient algorithm for computing the discrete Fourier transform. The efficiency is needed because it can be shown that to compute N discrete Fourier transform coefficients requires approximately N^2 complex multiplications and N^2 complex additions. By using the redundancy present in the values taken by W_N and its powers, the number of operations required can be considerably reduced. One restriction that ensues because of the reduction in numbers of operations is that N must be a power of 2. The FFT is described in detail in many texts [2,3] and will not be discussed further in this volume.

7.7 Frequency Resolution and Zero Padding

We will now address the following three questions. When we take a Fourier transform of a sampled time function,

1. How many points are there in the resulting transform?
2. How far apart in frequency are the points in the resulting transform (i.e., what is the frequency resolution of the resulting transform)?
3. How can the frequency resolution be changed?

The answers are discussed in the following:

1. *Number of points in the resulting transform*: We begin by referring to Equation 7.97 where it is shown that, when the discrete Fourier transform is taken of a sampled time function containing N samples, the resulting Fourier transform also contains N components. However, it is also shown in Equation 7.97 that the resulting Fourier components are situated on the negative frequency axis as well as on the positive frequency axis. There will be $N/2$ components on the positive frequency axis as well as a component at zero frequency. These $(N/2) + 1$ components are the only ones that are normally displayed by commercially available software. Therefore, we can state that a sampled signal with N samples produces a Fourier transform with $(N/2) + 1$ components where one component is at zero frequency.

2. *Frequency resolution*: Again refer to Equation 7.97 where we have shown a correspondence between the discrete Fourier transform and the discrete Fourier series of a periodic function. We have assumed that the original time signal was L seconds long, and had a fundamental frequency $f_0 = 1/L$. Since the components in a discrete Fourier series are separated by f_0 Hz, we conclude that the components in the Fourier transform are also separated by f_0 Hz, a spacing in Hz equal to the reciprocal of the length of the time signal in seconds. This spacing is referred to as the *frequency resolution* of the Fourier transform. For example, a signal which is 5 s long will produce a Fourier spectrum with components spaced at $1/5$ Hz intervals. Notice that the frequency resolution is independent of the sampling frequency of the time signal or of the time between samples.

3. *Changing the frequency resolution*: We can address this point by extending the above discussion. Since the frequency resolution is equal to the reciprocal of the length of the time signal, the only way we can improve the frequency resolution is by lengthening the time signal. Let us illustrate this with an example. Suppose we are interested

in examining the Fourier spectrum of a section of an electrocardio-gram (ECG) signal. We acquired 60 s of the signal that was sampled at 200 samples per second. We then take the Fourier transform of 10 s of the signal. This produces a spectrum with frequency resolution of 0.1 Hz. If this resolution is not satisfactory, we can improve the reso-lution by increasing the length of the signal which is transformed. Since we have acquired 60 s of the signal, we can improve the fre-quency resolution up to 1/60 Hz by using the entire signal.

Let us now consider another example where we acquired 10 s of an ECG signal. If we transform the entire signal, we will produce a spectrum with a frequency resolution of 0.1 Hz. If this resolution is not satisfactory, we can-not increase the length of the transformed signal since we have used the entire signal. Therefore, we resort to the method of *zero padding* to improve the frequency resolution. It can be shown that, if a time signal is extended by appending a string of zeros to the end of the signal at the same time spacing as the signal, the Fourier transform of the resulting time function will be the same as the Fourier transform of the original signal except that the frequency resolution will be the reciprocal of the length of the padded signal. As an example, consider the above instance where we acquired 10 s of an ECG signal which was sampled at 200 samples per second. If we now pad the signal with 10 s of zeros at $200 s^{-1}$ (a total of 2000 zeros) we improve the frequency resolution from 0.1 to 0.05 Hz. If we pad with 90 s of zeros at $200 s^{-1}$ (a total of 18,000 zeros) we increase the frequency resolution to 0.01 Hz.

We will now give an example to illustrate the above points. We acquire 120 s of an ECG signal and sample at 200 samples per second. This produces a total of 24,000 samples. We then request our FFT software to compute a Fourier transform of 2048 samples (note that we usually specify a power of two when computing an FFT to maximize the efficiency of the algo-rithm). This produces a frequency spectrum of 1025 points (one more than half of the number of points transformed). One point is at zero frequency. The spacing of the remaining points can be calculated by noting that 2048 samples occupy 10.24 s. Therefore the frequency resolution is 1/10.24 Hz or 0.0977 Hz. The spectrum therefore ranges from 0 to 100 Hz that is half of the sampling frequency. Now, suppose we would like to change the frequency resolution to 0.01 Hz, we have two choices since we have not used all of our acquired data. We need a signal length of 100 s to produce the desired frequency resolution. Since our total signal length is 120 s, we can use 100 s of the acquired signal and need not resort to zero padding. However, if we desire to use only 10.24 s of the signal (we would choose this option if we thought that the signal characteristics might not be con-stant over the needed 100 s time interval), we can zero pad with 17,952 zeros (20,000 total samples in 100 s minus 2048 samples for the signal) producing an equivalent result.

7.8 Power Spectrum

In this section, we will define the power spectrum and discuss some of its properties. The power spectrum, or power spectral density, is defined as the square of the magnitude of the Fourier transform of a signal:

$$P(\omega) = F(j\omega)F^*(j\omega) = |F(j\omega)|^2 \tag{7.104}$$

where
    $P(\omega)$ is the power spectrum
    $F(j\omega)$ is the Fourier transform

The amplitude of the power spectrum is always nonnegative for a real signal, and the power over a finite frequency interval from ω_1 to ω_2 can be obtained by integration.

$$P(\omega_1, \omega_2) = \int_{\omega_1}^{\omega_2} P(\omega)\,d\omega \tag{7.105}$$

If it were possible to record a signal of infinite duration, and to calculate its power spectrum, we would obtain a result that was totally reliable and with no uncertainties. However, in real situations we have data of limited duration and only finite time and computational resources to apply to the calculation of power spectra. We therefore, must make estimates of the power spectra using the notion that each finite length recording is a sample taken from an infinite duration recording. The periodogram is one such estimate of the signal power spectrum based on the amplitude squared Fourier transform.

The discrete Fourier transform of a discrete signal is given by Equation 7.96. The discrete frequency periodogram is given by

$$P(n) = \frac{1}{N}|X(n)|^2 \tag{7.106}$$

The effectiveness of the periodogram as an estimator of the underlying power spectrum is measured by the variance and bias of the estimates. The derivation of the variance is beyond the scope of this text. However, an important property of the variance can be stated and discussed. As the length of the time domain signal is increased, the variance of the periodogram does not approach zero but instead approaches a constant value. Therefore, the periodogram does not approach the true power spectrum as the length of the time domain signal increases. The periodogram therefore contains jagged curves with strong fluctuations about the true spectrum value. As a result, the periodograms become difficult to interpret. This problem can be overcome in part by controlling the variance by means of averaging procedures, which will be discussed in the following.

The bias in an estimate is the extent to which the estimate always yields an erroneous result irrespective of the extent that the variance is controlled. The bias arises from the error caused by the windowing of the data causing energy to leak from parts of the spectrum where it should be to frequencies on either side. (The topic of windowing will be further explored in Section 8.3 of this book. For this discussion, just consider windowing a truncation of the data, which produces problems that need to be analyzed.) The periodogram is thus a biased estimate of the underlying power spectrum and the bias is caused by the window process. Recall that the frequency resolution of the periodogram also depends on the length of the window. Therefore, to provide for low bias as well as good frequency resolution, the window must be narrow in the frequency domain or wide in the time domain. This means that a signal of long time duration must be transformed.

We will now discuss methods of reducing the variance of a spectral estimate, thus effecting a smoothing on the data. The simplest method would be to average adjacent samples in the periodogram according to some weighting function or window. A more complicated process would involve segmenting the time domain recording into sections, calculating a periodogram for each, and then forming the average of a number of segmental periodograms. We will first discuss the windowing method. The method is similar to time domain windowing already discussed except that the window is now constructed in the frequency domain. The new smoothed periodogram is then the convolution of the original periodogram with the window function. The problem with this method is that the averaging process will reduce the frequency resolution as it reduces the variance and that the resolution worsens as the window is made wider in frequency.

The second method is that of averaging the periodogram. The idea is that the time domain data is broken up into a number of equal length segments and the periodogram of each is calculated. The periodogram of the whole recording is then the average of all the segmental periodograms. The method assumes that the length of each segment of data is such that no component exists in the spectrum of low enough frequency to cause data in one segment to be correlated with data in another. This implies that successive segmental periodograms are independent of each other, which in turn implies that the variance of the averaged periodogram varies inversely with the number of periodograms in the average, and approaches zero as the number becomes large. As in the method discussed above, the price that is paid for the reduced variance is worsened frequency resolution due to the averaging process.

7.9 Coherence

We will now discuss the concept of coherence that is a frequency domain measure of the similarity between two signals. It is similar to the cross power spectrum that also indicates the frequency domain similarity of two signals, except that the coherence is normalized so that a value of zero indicates uncorrelated signals and a value of one indicates identical signals. The coherence is defined as

$$C_{xy}(\omega) = \frac{G_{xy}(\omega)}{\sqrt{G_{xx}(\omega)G_{yy}(\omega)}} \qquad (7.107)$$

where

$G_{xy}(\omega)$ is the cross power spectrum at frequency ω between signals $x(t)$ and $y(t)$

$G_{xx}(\omega)$ and $G_{yy}(\omega)$ are the power spectra of $x(t)$ and $y(t)$, respectively

More commonly used is the magnitude squared coherence (MSC) function, defined as

$$MSC = \left|C_{xy}(\omega)\right|^2 = \frac{\left|G_{xy}(\omega)\right|^2}{G_{xx}(\omega)G_{yy}(\omega)} \qquad (7.108)$$

This calculation provides us with several pieces of information about the relationship of the signals $x(t)$ and $y(t)$. For example, an obvious result would be the determination of any frequencies where there is joint presence of energy. However, the coherence function also provides information about the relative phase between frequency components in the two signals as well as mechanism of addition of new frequency components, whether by additive noise or by a nonlinear change in signal properties. We will consider the coherence function later in Chapter 9.

7.10 Properties of the Fourier Transform

A derivation of properties of the Fourier transform can be found in many textbooks (for example, see books by Nilsson [1] and Ziemer [2]). In this chapter we will briefly list the important properties.

If $f(t) \Leftrightarrow F(j\omega)$, then it follows that

1. *Superposition property*

$$af_1(t) + bf_2(t) \Leftrightarrow aF_1(j\omega) + bF_2(j\omega) \qquad (7.109)$$

2. *Time scaling property*

$$f(at) \Leftrightarrow \frac{1}{|a|} F\left(\frac{j\omega}{a}\right) \qquad (7.110)$$

3. *Differentiation property*

$$\frac{d^n}{dt^n} f(t) \Leftrightarrow (j\omega)^n F(j\omega) \qquad (7.111)$$

4. *Integration property*

$$\int_{-\infty}^{t} f(x)\,dx \Leftrightarrow \frac{1}{j\omega} F(j\omega) \qquad (7.112)$$

5. *Time shift property*

$$f(t-a) \Leftrightarrow F(j\omega)e^{-j\omega a} \qquad (7.113)$$

6. *Modulation property*

$$f(t)e^{jn\omega_0 t} \Leftrightarrow F\left[j(\omega - \omega_0)\right] \qquad (7.114)$$

$$f(t)\cos\omega_0 t \Leftrightarrow \frac{1}{2}\left\{F\left[j(\omega - \omega_0)\right] + F\left[j(\omega + \omega_0)\right]\right\} \qquad (7.115)$$

$$f(t)\sin\omega_0 t \Leftrightarrow \frac{1}{2}j\left\{F\left[j(\omega - \omega_0)\right] - F\left[j(\omega + \omega_0)\right]\right\} \qquad (7.116)$$

7.11 Filtering

In order to discuss examples of the use of the Fourier transform and the spectral density, we must first consider the process of filtering. Filtering allows us to remove portions of the Fourier spectrum and retain others. This allows us, for example, to remove unwanted noise, to enhance desirable characteristics

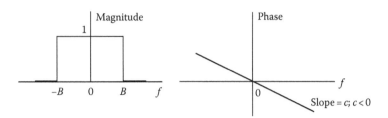

FIGURE 7.19
Characteristics of ideal low-pass filter.

of the signal, or to change the shape of the signal. We will consider in this section four *ideal* filters. We will not be concerned at this stage with practical implementation of the filters that will come later. We will define the filters in this section and discuss their use in succeeding sections.

1. *Ideal low-pass filter*: An ideal low-pass filter allows all frequencies below a specified "cutoff frequency" to remain and eliminates all others. The characteristic of the filter can be displayed as a function of frequency as shown in Figure 7.19. For frequencies below the cutoff frequency, f_c, the filter has a "gain" of 1 implying that components at those frequencies are not attenuated or amplified. For frequencies above f_c, the gain is zero implying that those frequency components are completely attenuated.

2. *Ideal high-pass filter*: An ideal high-pass filter allows all frequencies above the cutoff frequency f_c to remain and eliminates all others. The characteristic of the filter is shown in Figure 7.20.

3. *Ideal band-pass filter*: An ideal band-pass filter allows all frequencies in the band between frequencies f_1 and f_2 to remain and eliminates all others. The characteristic of the filter is shown in Figure 7.21.

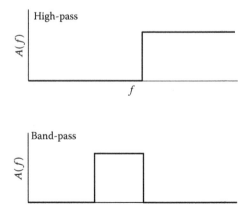

FIGURE 7.20
Characteristics of ideal high-pass filter.

FIGURE 7.21
Characteristics of ideal band-pass filter.

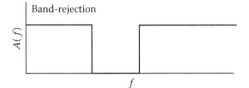

FIGURE 7.22
Characteristics of ideal band-reject filter.

4. *Ideal band-reject filter*: This filter is also called a notch filter, and eliminates all frequency components in the band between f_1 and f_2. The
characteristic of this filter is shown in Figure 7.22.

7.12 Examples

We will now give several examples of the use of the Fourier transform in
problems which relate to biomedical applications. The examples will also
serve to increase our intuitive feeling for the Fourier transform and spectral
density.

1. *Sum of two sinusoids*
   Figure 7.23 is a plot of the function $f(t) = \sin(t) + \sin(10t)$.
   Notice that the frequencies of the sinusoids are far apart and can
   be easily distinguished on the figure. The Fourier transform of $f(t)$
   is shown in Figure 7.24. Note that there are components of the spectrum at both frequencies 1 and 10 rad/s and at no other frequencies.

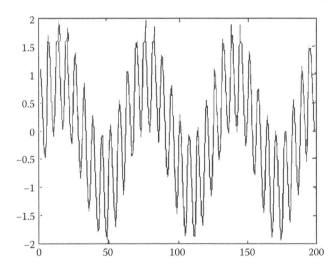

FIGURE 7.23
Plot of the sum of two sinusoids $f(t) = \sin(t) + \sin(10t)$.

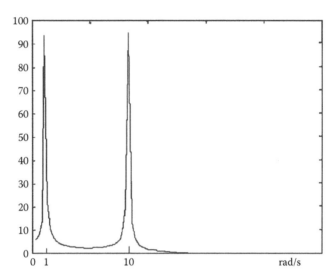

FIGURE 7.24
Fourier transform of the sum of two sinusoids.

We would now like to extract the 1 rad/s component from the composite signal. This can be accomplished using an ideal low-pass filter as defined above, with cutoff frequency above 1 rad/s but below 10 rad/s. If the 10 rad/s component is completely blocked by the filter, the result will be the function $f(t) = \sin(t)$, whose spectrum is shown in Figure 7.25.

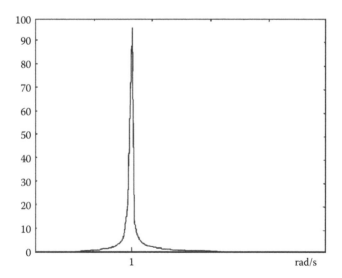

FIGURE 7.25
Fourier spectrum after filtering.

The filtering operation illustrated above relies on the fact that the spectral density contains components that can be individually filtered. In the above example, there were two discrete components in the spectrum. For continuous spectra, the same considerations apply. We will consider this case in the following.

2. *Square wave*

As we discussed in Section 7.1, a square wave, as shown in Figure 7.4a has a Fourier spectral density shown in Figure 7.26. It is made up of components at the fundamental frequency ω_0 and its odd harmonics. By performing a filtering operation on this square wave, we can change its shape in a predictable way. For example, if we apply a low-pass filter with cutoff frequency at $2\omega_0$ rad/s, the output will contain one component at the fundamental frequency. In the time domain, this waveform will be a sine wave. We have therefore converted a square wave into a sine wave through low-pass filtering.

If we apply an ideal band-pass filter to the square wave with cutoff frequencies at $2\omega_0$ rad/s and at $4\omega_0$ rad/s, we are left with one component at $3\omega_0$ that again results in a sine wave at the output of the filter.

3. *Electrocardiogram*

Figure 7.26 is an example of a normal ECG. The production of the normal ECG was discussed in Section 6.3. Let us hypothesize that we would like to design a signal processing algorithm to detect the position of the peak of the R wave (the highest amplitude peak in the ECG). Such a problem is actually very practical and will be considered in great detail later on. Since there are other peaks present in the wave besides the R peak, such as the P and T waves and possible peaks due to noise, we would like to enhance the R wave relative to the other peaks before applying the detection algorithm. Examination of the frequency spectrum of the ECG shown in Figure 7.27, it can

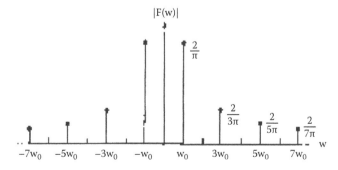

FIGURE 7.26
Fourier spectrum of a square wave.

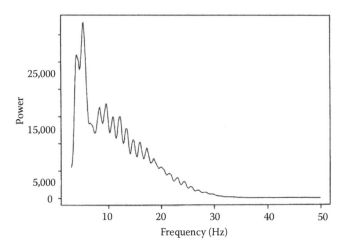

FIGURE 7.27
Fourier spectrum of an electrocardiogram.

be determined that most of the energy in the QRS complex is in the frequency band from 10 to 20 Hz and the energy in the P and T waves are in a band of frequencies below 10 Hz. Therefore, to enhance the R wave, we can band-pass filter the signal between 10 and 20 Hz before R wave detection.

4. *Electroencephalogram*

The electroencephalogram (EEG) is produced by the neurons in the brain and recorded from the surface of the scalp. The production of the normal EEG was briefly discussed in Section 6.3. Figure 7.28

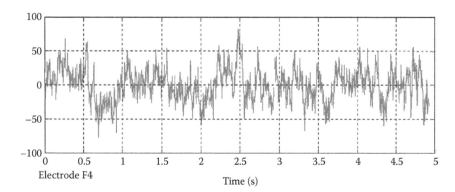

FIGURE 7.28
Normal electroencephalogram.

shows a typical EEG waveform. Many studies have shown that if the frequency spectrum of the EEG is examined, energy will be found in specific frequency bands depending on the mental state of the subject. For example, if the subject is awake, relaxed, eyes closed, there is an increase of the energy in the 8–12 Hz frequency band, called the alpha band. On the other hand, if the subject is awake and active with eyes open, the energy shifts to the 13–30 Hz frequency band, called the beta band. Therefore, band-pass filters can be used to separate the frequency bands and provide information on which band has the greatest energy.

Examine Figure 6.13 for an example of these waves in the time domain.

5. *60 Hz rejection*
Many biological signals contain a 60 Hz component due to inter-ference from the power lines surrounding the experimental setup. This was seen when we examined the ECG found in Figure 6.20a through c. There are two methods of removing the unwanted 60 Hz component from the desired biological signal based on filtering. First, if the energy in the desired signal lies in frequency bands which are all below 60 Hz, we can separate the desired from the undesired components by using a low-pass filter with cutoff frequency to be above the frequency band of the desired signal but below 60 Hz. Second, if the energy in the desired signal includes 60 Hz, a very nar-row band-reject filter at 60 Hz can be used to remove the unwanted noise but retain as much of the desired signal as possible.

Problems

7.1 Determine the complex or exponential form of the Fourier Series (C_n and φ_n) in terms of sine and cosine coefficients (a_n and b_n).

7.2 Find the Fourier coefficients of the following function

$$X(t) = 4\cos(20\pi t) + 5\sin(40\pi t)$$

7.3 Find the Fourier coefficients of the following function, which is not in one of the standard Fourier series forms.

$$X(t) = 2 + 5\sin 6t + 4\cos 8t.$$

7.4   Is the given function even or odd?

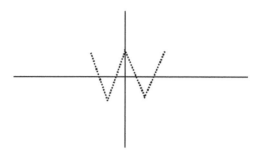

7.5   For an even function the b_n coefficients can be defined as _____.

7.6   Is the given function even or odd?

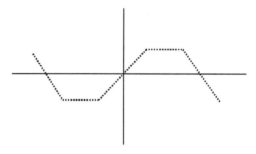

7.7   A 128 point discrete Fourier transform will take 128 time domain samples and produce how many real and imaginary Fourier coefficients?

7.8   Describe the scaling property of Fourier transforms both mathematically and in words.

7.9   Describe the time shifting property of Fourier transforms both mathematically and in words.

7.10 Describe the relationship between a time function $f(t)$ that changes rapidly in time and the resultant Fourier transform of $f(t)$.

7.11 How can the frequency resolution of a discrete Fourier transform be increased?

7.12 Describe the frequency response of an ideal low-pass filter.

References

1. Nilsson, J.W. *Electric Circuits*. Addison-Wesley, Reading, MA, 1993.
2. Ziemer, R., Tranter, W., and Fannin, D. *Signals and Systems: Continuous and Discrete*. Macmillan, New York, 1989.
3. Phillips, C. and Parr, J. *Signals, Systems and Transforms*. Prentice-Hall, Englewood Cliffs, NJ, 1995.

8

DSP System Level Concepts and Digital Filter Implementations

Before we can proceed with digital filters, we first need to review a major DSP system level concept. Previously, we have seen that Fourier analysis breaks down complex signals by looking at them as a sum of sinusoidal waves, which enables us to perform signal analysis in an easier fashion. Similarly, we can simplify complex waveforms in the digital domain by looking at them as a sum of individual pulses. After all, once a signal is digitized, it is merely a series of digital pulses. If we can break down and analyze how a system responds to a single pulse, we can apply that same response when a series of pulses is applied. If a system is linear and time invariant, then once we know how it reacts to one pulse, we can determine how it will respond to multiple pulses, based on the principle of superposition. Thus, the analysis can be broken down and simplified. An unknown system is characterized by how it responds to a simple impulse. To see how the same unknown system responds to a complex sampled signal, we merely consider the sampled signal to be a series of impulses that can be scaled and added.

8.1 Superposition

So what constitutes a linear system and therefore enables us to take this approach? A system that is linear adheres to the principle of superposition. This implies that the system must be both homogeneous and additive. If a system is homogeneous, it implies that when an input is scaled into the system, the result will be an output that is scaled by the same amount. The other is the principle of additivity, which states that if a system responds to an input with a set response and another input with a second response, then, when both signals are input into the system at the same time, the response will be the summation of the two individual output responses. The underlying concept here is that the two inputs do not interfere with one another. Let us define both terms with a few equations. Assume X1 and X2 are two different inputs into a system. Then, the system is homogeneous if the following equations hold good:

Assume input X1 produces output Y1 and input X2 produces output Y2, or

$$X1[k] \rightarrow Y1[k] \quad \text{and} \quad X2[k] \rightarrow Y2[k] \tag{8.1}$$

Then, we must be able to scale both the inputs and get out the same two outputs, only scaled.

$$\alpha X1[k] \rightarrow \alpha Y1[k] \quad \text{and} \quad \alpha X2[k] \rightarrow \alpha Y2[k] \tag{8.2}$$

This defines a homogeneous system. Similarly, we can review the principle of additivity from a mathematical perspective. This states that if we apply the same two inputs X1 and X2 at the same time, then the output is the summation of the two outputs when the signals were applied separately. Basically, the two inputs do not interfere with one another. If

$$X1[k] \rightarrow Y1[k] \tag{8.3}$$

and

$$X2[k] \rightarrow Y2[k] \tag{8.4}$$

then

$$X1[k] + X2[k] \rightarrow Y1[k] + Y2[k] \tag{8.5}$$

A system is time invariant only if a time shift in the input signal produces an identical shift in the output signal. Mathematically, if

$$X[k] \rightarrow Y[k] \tag{8.6}$$

then for a time-invariant system

$$X[k - k0] \rightarrow Y[k - k0] \tag{8.7}$$

Before we get too far with this, let us take a simple example to see why these two principles are so important. Suppose we know that a linear time-invariant system when stimulated with an impulse function responds with an output of $2\delta[n - 2]$. How will this same system respond when it is stimulated with a signal that is $\delta[n] + 2\delta[n - 1]$? In order to solve this problem, we merely need to understand how the first input differs from the second input, then apply that same mathematical relationship to the known output and we will have the unknown response.

Since X1 = $\delta[n]$ and X2 = $\delta[n] + 2\delta[n - 1]$, to make X1 look like X2, we merely need to take X1 and add it to 2X1 delayed by one interval. So now apply this same mathematical relationship to the known output Y1 and we will determine the output to the unknown stimulus, Y2. So take Y1 and add it to 2Y1 delayed by one, or

Y1	$2\delta[n-2]$
{Y1 × 2}Delay 1	$4\delta[n-3]$
Summation	$2\delta[n-2] + 4\delta[n-3]$

Thus, the unknown response is $2\delta[n-2] + 4\delta[n-3]$.

We can now take this a step further. Any signal can be broken down into a set of impulses, each one being a scaled and time-shifted version of the delta function. Therefore, if this signal (an impulse) is input into a linear time-invariant system, the output will be a summation of scaled and shifted impulse responses. Thus, the output can be determined by adding a series of scaled and shifted impulse responses. More importantly, we can calculate the output of any linear time-invariant system when it is stimulated with a signal, once we know how the system responds when it is stimulated by an impulse, this being the system's impulse response. The process by which we sum these responses is convolution and this mathematical operation is highly related to digital filtering. So, now that we understand the driving reason behind this approach, let us examine the details behind digital filtering and convolution.

8.2 Digital Filter and Convolution

One of the simplest and most intuitive filters to understand is the running average filter. In this filter, adjacent samples are averaged together. The result is a smoothing of the input signal. This type of filter does not provide good frequency selectivity; however, the time-domain performance is quite good.

To get an understanding of how this filter works, let us start with an example. We will take a running average filter that averages four consecutive samples. We shall use a triangular input signal and examine the filter's output. Let us look at the individual steps involved.

The input signal is defined in Figure 8.1. The next series of equations show how data are sequentially advanced through the filter.

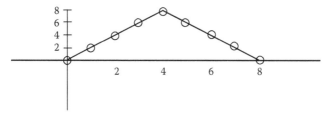

FIGURE 8.1
Input signal into the average-four filter.

The filter must average four consecutive samples. Therefore, the output equation for the filter is defined as follows:

$$Y(n) = \frac{1}{4}\{X[n] + X[n-1] + X[n-2] + X[n-3]\} \qquad (8.8)$$

where
 X is the input signal
 n refers to the input sample number
 $X(n)$ represents the current sample
 $X(n-1)$ represents one sample older in time
 $X(n-2)$ represents a sample that is two samples older than the current sample
 $X(n-3)$ represents a sample that is three samples older than the current sample

As defined, the output $Y(n)$ is always a function of the current sample and three older samples. We can now apply this equation to determine the filter output shown in Figure 8.2.

$$Y(0) = \frac{1}{4}\{X[0] + X[0-1] + X[0-2] + X[0-3]\}$$

$$= \frac{1}{4}\{X[0] + X[-1] + X[-2] + X[-3]\}$$

$$= \frac{1}{4}\{0 + 0 + 0 + 0\} = 0$$

$$Y(1) = \frac{1}{4}\{X[1] + X[1-1] + X[1-2] + X[1-3]\}$$

$$= \frac{1}{4}\{X[1] + X[0] + X[-1] + X[-2]\}$$

$$= \frac{1}{4}\{2 + 0 + 0 + 0\} = \frac{1}{2}$$

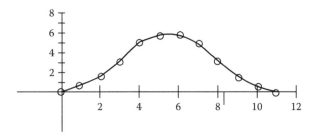

FIGURE 8.2
Digital filter output.

$$Y(2) = \frac{1}{4}\{X[2] + X[2-1] + X[2-2] + X[2-3]\}$$

$$= \frac{1}{4}\{X[2] + X[1] + X[0] + X[-1]\}$$

$$= \frac{1}{4}\{4 + 2 + 0 + 0\} = \frac{3}{2}$$

$$Y(3) = \frac{1}{4}\{X[3] + X[3-1] + X[3-2] + X[3-3]\}$$

$$= \frac{1}{4}\{X[3] + X[2] + X[1] + X[0]\}$$

$$= \frac{1}{4}\{6 + 4 + 2 + 0\} = 3$$

$$Y(4) = \frac{1}{4}\{X[4] + X[4-1] + X[4-2] + X[4-3]\}$$

$$= \frac{1}{4}\{X[4] + X[3] + X[2] + X[1]\}$$

$$= \frac{1}{4}\{8 + 6 + 4 + 2\} = 5$$

$$Y(5) = \frac{1}{4}\{X[5] + X[5-1] + X[5-2] + X[5-3]\}$$

$$= \frac{1}{4}\{X[5] + X[4] + X[3] + X[2]\}$$

$$= \frac{1}{4}\{6 + 8 + 6 + 4\} = 6$$

$$Y(n) = \left\{0, \frac{1}{2}, \frac{3}{2}, 3, 5, 6, 6, 5, 3, \frac{3}{2}, \frac{1}{2}, 0\right\}$$

In this simple running average filter, the filter coefficients are all the same, each having a weight of 1/4.

The filtering operation is defined in Figure 8.3, which demonstrates how the filter operation is performed for a filter 4-taps deep, such as the one defined above. In this figure, {b0, b1, b2, b3} represent the filter coefficients, X[n] the input, and Y[n] the resultant output of the filter.

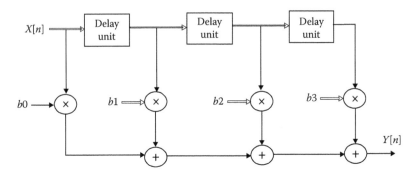

FIGURE 8.3
Digital filter block diagram.

The general formula for performing filtering is defined as follows:

$$Y(n) = \sum_{k=0}^{M} B_k [n - k] \tag{8.9}$$

where
  M is the order of the filter, which equals the number of tap delays
  B_k is the filter coefficient

NOTE: Filter length $= L = M + 1$

Looking merely at time-domain data, we can see that the effects of this filter are threefold. First, it clearly smoothed the input signal. Second, it stretched the input signal. Finally, it delayed the input signal. Whereas previously, the peak of the input signal occurred at the fourth sample, after filtering, the peak was delayed to somewhere in between the fifth and sixth samples, say 5.5.

Reviewing the filtering formula presented in Equation 8.9, the $X[n - k]$ term essentially flips the input, enabling the filter to work with samples that are progressively older in time. This incidentally exactly matches the definition of convolution. Convolution can be a hard concept to grasp, however, once a graphical convolution example is understood, the concept becomes much clearer.

Examine the graphical convolution of the two rectangles shown in Figure 8.4, which are used to demonstrate graphical convolution.

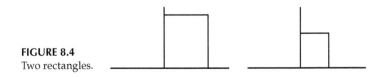

FIGURE 8.4
Two rectangles.

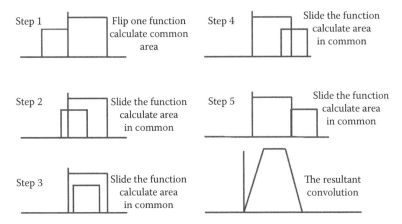

FIGURE 8.5
Graphical convolution.

Figure 8.5 demonstrates how graphical convolution is performed. First, one of the rectangles is flipped and slid past the other, while the common area is integrated. This is exactly what has been done in the filtering example previously provided. Thus, digital filtering is convolution.

So far, we have only discussed the basic sliding window filter. This filter has the unusual quality in that all of the filter coefficients are identical. While the sliding window filter coefficients are fixed, most digital filters have coefficients that are different from one another. Generally, it is the performance requirements of the filter that determine the filter coefficients. Filter coefficients are derived based on the required frequency response of the filter and the number of filter taps available. The process for deriving filter coefficients is mathematically intense and typically requires a reiterative process. Therefore, there are software packages designed to provide this function. The designer typically specifies the desired frequency response and the maximum number of filter coefficients. A software program then takes this information and proceeds through a reiterative process until a set of filter coefficients that achieves the desired performance has been achieved. A flowchart for this process appears in Figure 8.6. This is essentially a flowchart for the Parks–McClellan approach, which is one of the most popular techniques used to derive filter coefficients.

Notice in the aforementioned algorithm that the coefficients need to be windowed. Windowing is a major function performed in DSP applications. The need for windowing arises anytime we look to shorten a function in time. Many of the time–frequency domain transforms rely upon integrating out to ±infinity. This, unfortunately, is impossible to do on a digital computer. Therefore, we need to apply windowing techniques that are designed to help diminish this problem.

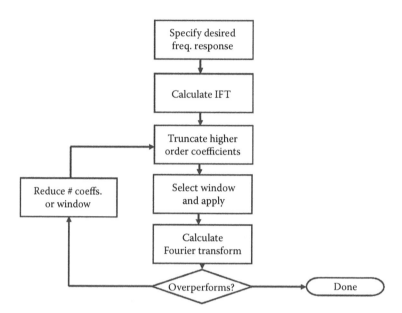

FIGURE 8.6
Parks–McClellan flowchart for deriving filter coefficients.

8.3 Windowing

In this section, we will explore what happens when we attempt to process signals that are present for a finite period of time. In order to clearly understand the problem, consider a sine wave that is present for all time. Let us represent the sine wave as shown in Equation 8.10. We will simplify our story by making the amplitude of the sine wave equal to 1 and the phase equal to 0.

$$f(t) = \sin \omega_0 t, \quad -\infty < t < \infty \tag{8.10}$$

If we take the Fourier transform of $f(t)$, we obtain the resulting magnitude as shown in Equation 8.11.

$$|F(\omega)| = \pi \left[\delta(\omega - \omega_0) + \delta(\omega + \omega_0) \right] \tag{8.11}$$

Notice that we have energy only at one frequency, the frequency of the sine wave.

Now, let us consider the more practical scenario where the sine wave is present only for a finite time period from time $t = -T/2$ to time $t = T/2$. We will derive the resulting Fourier transform by two different methods, the direct method and the window method.

The direct method involves using the definition of the Fourier transform. Applying the definition, we obtain

$$F(\omega) = \int_{-\frac{T}{2}}^{\frac{T}{2}} \sin \omega_0 t \, e^{-j\omega t} \, dt \tag{8.12}$$

If we substitute

$$\sin \omega_0 t = \frac{e^{j\omega_0 t} - e^{-j\omega_0 t}}{2j} \tag{8.13}$$

into Equation 8.12 and integrate, we obtain

$$F(\omega) = \frac{jT}{2}\left[\frac{\sin([\omega+\omega_0]/2)T}{([\omega+\omega_0]/2)T} - \frac{\sin([\omega+\omega_0]/2)T}{([\omega+\omega_0]/2)T} \right] \tag{8.14}$$

To use the window method, we imagine that we will construct a sine wave present only within the interval $t = -T/2$ to $T/2$ by taking a sine wave present for all time and placing it in a window that will only allow the wave to be visible in the interval $t = -T/2$ to $T/2$. The window is shown in Figure 8.7 and is called a *rectangular* or *boxcar* window because of its shape.

Mathematically, the windowing operation is equivalent to multiplying the infinitely long sine wave by the equation of the window. Since multiplication in the time domain is equivalent to convolution in the frequency domain, we will evaluate the Fourier transform of the resulting windowed sine wave by convolving the Fourier transform of the infinitely long sine wave with the

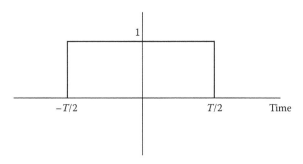

FIGURE 8.7
Rectangular window.

Fourier transform of the window. From Equation 7.74, the Fourier transform of an infinitely long sine wave is

$$F_1(\omega) = j\pi\left[\delta(\omega+\omega_0)-\delta(\omega-\omega_0)\right] \qquad (8.15)$$

The rectangular window $w(t)$ in Figure 8.7 can be expressed mathematically by Equation 8.16:

$$w(t) = \begin{cases} 1 & -\dfrac{T}{2} < t < \dfrac{T}{2} \\ 0 & \text{elsewhere} \end{cases} \qquad (8.16)$$

The Fourier transform of $w(t)$ becomes

$$W(\omega) = T\left(\frac{\sin \omega T/2}{\omega T/2}\right) \qquad (8.17)$$

The convolution operation in the frequency domain is defined as follows:

$$F_1(\omega)W(\omega) = \frac{1}{2\pi}\int_{-\infty}^{\infty} F_1(\omega-\omega')W(\omega')\,d\omega' \qquad (8.18)$$

Substituting Equations 8.15 and 8.17 and performing the integration, we obtain $F(\omega)$ as in Equation 8.14. A plot of $F(\omega)$ is shown in Figure 8.8. Comparing the windowed result in Figure 8.8 with the Fourier transform of an infinitely long sine wave, we notice several differences. In Figure 8.9, all the energy in the signal is concentrated at the frequency ω_0, while the windowed signal has energy smeared in a band around ω_0. As we will see later, the degree of smearing is related to the length of the window.

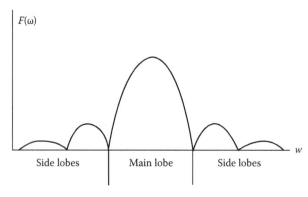

FIGURE 8.8
Spectrum of a windowed sinusoid.

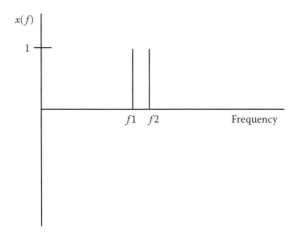

FIGURE 8.9
Spectrum of two infinitely long sinusoids with closely spaced frequencies.

Notice also that the transform of the windowed sine wave has two major components: a *main lobe* and *side lobes*. Both the main lobe width and the existence of side lobes contribute to the distortion in the spectrum due to the windowing operation.

A natural question that arises at this point is whether the shape of the window can be changed to produce an *optimum window*, that is, a window that will produce an impulse function for the spectrum of a finite length sine wave. The answer can be obtained by returning to Equation 8.18 in which W represents the Fourier transform of the window function and F_1 the Fourier transform of the infinitely long sine wave. In order for both the input and output sine waves to have similar spectra (both spectra equal to impulse functions at the frequency of the sine wave), the Fourier transform of the window W must also be an impulse, which implies that the optimum window is a constant for all time. This result is impractical since it requires the signal to exist for all time, thus negating the effect of the window.

A second question to be considered is whether the existence of main lobes and side lobes causes any major problems in interpretation or whether we can still obtain a correct result in calculation. To point out problems in interpretation due to the main lobe and side lobes, consider a signal composed of two sine waves whose frequencies are close together. Figure 8.9 displays the spectrum of this signal when the signal is infinitely long. Notice that the spectrum consists of two impulses at the frequencies of the sine waves and that the existence of two distinct sine waves in this signal is easily discernible. Now, consider what happens when the signal is windowed with a finite length rectangular window. Each sine wave produces a main lobe and side lobes, which interact if the sine waves are close together in frequency. The resulting spectrum for one case is shown in Figure 8.10.

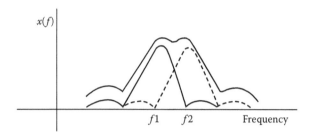

FIGURE 8.10
Spectrum of two windowed sinusoids with closely spaced frequencies.

Notice that it is impossible to decompose the resulting spectrum accurately into the spectra of two sine waves. It is also possible to interpret the spectrum incorrectly and conclude that the spectrum is of some unknown signal with multiple sinusoidal components. We therefore seek suboptimum windows that are finite in length and have narrow main lobes and small side lobes.

We will now consider four common windows and discuss their properties including their equations, shapes in the time domain, main lobe widths, and side lobe heights. We will show that there is no window that simultaneously satisfies all the properties we desire in a window and that we must compromise and choose the most important properties to satisfy.

1. Rectangular window: We have already discussed the equation and time-domain shape of this window. Comparison with other windows of the main lobe and the side lobe properties will be discussed in the following.

2. Triangular (Bartlett) window: The time-domain expression for this window is given in Equation 8.19 and its Fourier transform is given in Equation 8.20. Figure 8.11 shows the shape of the window.

$$w_{tr}(t) = \begin{cases} 1 - \dfrac{|t|}{T} & |t| \leq T \\ 0 & \text{elsewhere} \end{cases} \qquad (8.19)$$

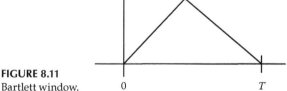

FIGURE 8.11
Bartlett window.

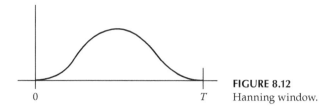

FIGURE 8.12
Hanning window.

$$W_{tr}(\omega) = T\left[\frac{\sin(\omega T/2)}{\omega T/2}\right]^2 \tag{8.20}$$

3. Hanning window: Equation 8.21 gives the time-domain expression and Equation 8.22 the Fourier transform for this window. The shape of the window is shown in Figure 8.12.

$$w_{ha}(t) = \begin{cases} \frac{1}{2}\left(1+\cos\frac{\pi t}{T}\right) & |t| \le T \\ 0 & \text{elsewhere} \end{cases} \tag{8.21}$$

$$W_{ha}(\omega) = \left(\frac{\sin\omega T}{\omega}\right)\left(\frac{\pi^2}{\pi^2 - \omega^2 T^2}\right) \tag{8.22}$$

4. Hamming window: Equation 8.23 gives the time-domain expression and Equation 8.24 the Fourier transform of the window. The shape of the window is shown in Figure 8.13.

$$w_H(t) = \begin{cases} 0.54 + 0.46\cos\frac{\pi t}{T} & |t| \le T \\ 0 & \text{elsewhere} \end{cases} \tag{8.23}$$

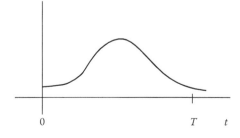

FIGURE 8.13
Hamming window.

$$W_H(\omega) = \frac{2\sin\omega T}{\omega}\left[\frac{0.54\pi^2 - 0.08\omega^2 T^2}{\pi^2 - \omega^2 T^2}\right] \tag{8.24}$$

The discrete forms of the rectangular, Hanning, and Hamming windows are given by Equations 8.25 through 8.27 respectively. In these equations, N represents the total number of points in the window.

$$w_R(n) = \begin{cases} 1 & 0 \le n \le N-1 \\ 0 & \text{elsewhere} \end{cases} \tag{8.25}$$

$$w_{ha}(n) = \begin{cases} \frac{1}{2}\left[1 - \cos\left(\frac{2\pi n}{N-1}\right)\right] & 0 \le n \le N-1 \\ 0 & \text{elsewhere} \end{cases} \tag{8.26}$$

$$w_H(n) = \begin{cases} 0.54 - 0.46\cos\left(\frac{2\pi n}{N-1}\right) & 0 \le n \le N-1 \\ 0 & \text{elsewhere} \end{cases} \tag{8.27}$$

We will now examine the main lobe widths and side lobe heights of the four windows discussed above. Figure 8.14 shows this comparison. Notice that the main lobe widths of the triangular, Hanning, and Hamming windows are all about the same but are about twice as wide as the width of the rectangular window. On the other hand, the side lobe height of the rectangular window is much greater than that of the other three. To see this comparison in another way, consider the contents of Table 8.1. In this table, N is the number of points in the window.

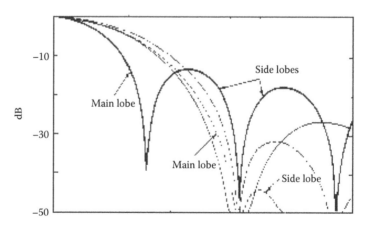

FIGURE 8.14
Comparison of four windows.

TABLE 8.1

Properties of Four Window Functions

Window Function	Main Lobe Width	Side Lobe Height (dB)
Rectangular	$2/N$	−14
Bartlett	$4/N$	−25
Hanning	$4/N$	−32
Hamming	$4/N$	−44

Notice that the widths of the main lobe of the Bartlett, Hanning, and Hamming windows are twice the width of the rectangular window. Notice also that the side lobe height decreases from the rectangular window through the Bartlett, Hanning, and Hamming windows. Therefore, in choosing a window, a decision must be made as to which is the more important consideration, reducing the main lobe width or reducing the side lobe height. The decision most often made is to reduce the side lobe height as much as possible. Therefore, the Hanning or Hamming windows are most often used. It should also be noted that there are many other windows available with other relationships between main lobe width and side lobe height. A discussion of these windows can be found in [1].

Problems

8.1 A linear time-invariant system is stimulated with a unit step function $U[n]$. The output $Y1[n] = 2\delta[n] - 3\delta[n - 1] + 2\delta[n - 2]$ occurs when $X1[n] = U[n]$. If the same system is stimulated by $X2[n] = 2U[n] + 4U[n - 3]$, find the resultant output $Y2[n]$. Use the principles of linearity and time invariance to solve.

8.2 Determine the equation that represents the following filter block diagram (Figure P8.2).

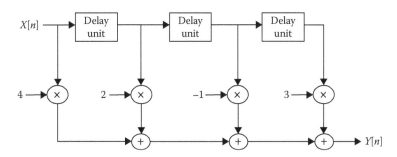

FIGURE P8.2
Filter block diagram.

8.3 A linear time-invariant system is described by the following equation:

$$Y[n] = 5X[n] - 2X[n-1] + 3X[n-2]$$

Draw the block diagram of a system that can perform this operation.

8.4 A linear time-invariant system is described by the following equation:

$$Y[n] = 4X[n] + X[n-1] - 2X[n-2]$$

When the input to this system is as follows:

$X[n] = 0$ for $n < 0$
$\qquad N + 1$ for $n = 0, 1$
$\qquad\quad 1$ for $n \geq 2$

a. Compute the values of $Y[n]$ for $0 <= n <= 6$.
b. Draw a block diagram of this system.
c. Determine the response of this system to a unit impulse function $X[n] = \delta[n]$.

8.5 The given input signal $X[n]$ is input into a filter with coefficients equal to {1, 2, 1}. Determine the filter output.

$X[n] = 0$ for $n < 0$
$\qquad 1$ for $n = 0$
$\qquad 2$ for $n = 1$
$\qquad 1$ for $n = 2$
$\qquad 0$ for $n = 3$

8.6 A sliding window average-3 filter with coefficients of {1/3, 1/3, 1/3} is used to filter the following input signal, X1.

$$X1[n] = 3U[n-1] + 3U[n-2] - 3U[n-3] - 3U[n-4]$$

$$Y1[n] = \frac{1}{3}X[n] + \frac{1}{3}X[n-1] + \frac{1}{3}X[n-2]$$

Determine the output of the filter Y1[n]. What has the filter done to the input signal?

8.7 What are the two characteristics that a linear system must possess in order to meet the principle of superposition?

8.8 Why are time functions windowed?

Reference

1. Oppenheim, A.V., and Willsky, A.S. *Signals and Systems*. Prentice-Hall, Englewood Cliffs, NJ, 1997.

Further Reading

McClellan, J.H., Schafer, R.W., and Yoder, M.A. *DSP First: A Multimedia Approach*. Prentice-Hall, Upper Saddle River, NJ, 1999.

9

Techniques and Examples for Physiological Signal Processing

In this chapter, we will investigate, discuss, and compare several signal processing techniques. We have already discussed in detail, the Fourier series and the Fourier transform techniques and have discussed the advantages of the technique. There are, however, situations where the standard Fourier technique either does not adequately yield the desired information or yields incorrect information. Under those situations, other special techniques are needed. We will discuss several of these special techniques including autoregressive (AR) modeling, time–frequency analysis, and wavelet analysis.

9.1 AR Modeling

The methods most commonly used for spectral analysis are based on the discrete Fourier transform (DFT) (implemented as the fast Fourier transform [FFT]) and AR modeling. The spectrum derived from the FFT is obtained from all the data present in the recorded signal, regardless of whether its frequency components appear as specific spectral peaks or as nonpeaked broadband powers. In contrast, the AR procedure uses the raw data to identify a best-fitting model from which the final spectrum, consisting of a DC component and a variable number of peaks, is derived.

The major advantage of the AR technique is its ability to limit the number of spectral peaks in the model depending on the order of the model. For example, as we will see in the next chapter, the analysis of heart rate (HR) variability is usually concerned with investigation of two spectral peaks. Therefore, a second-order AR model is used, which will produce exactly two peaks. If the data can be assumed to be derived from a series of fixed-rate oscillators, the AR methods are suitable because of their ability to identify the central frequency of the oscillation in an analytic way. Further, the AR approach is particularly useful when the number of samples available for the analysis is low because the frequency resolution of the AR derived spectrum is not as dependent as the FFT method on the length of the recording. On the other hand, when the analysis is focused on broadband powers, the AR method is accurate only when a high model order is used.

We will now consider some of the mathematics of the AR process. This will not be an exhaustive study, and the reader is advised to consult references [1,2] for a more detailed treatment. The reader is also assumed to have a basic background in the theory of probability and random variables. Readers lacking this background can consult references [3,4].

Consider a time series x_t that satisfies

$$x_t = a_1 x_{t-1} + a_2 x_{t-2} + \cdots + a_p x_{t-p} + \varepsilon_t \tag{9.1}$$

where ε_t is white noise with zero mean and variance σ_t^2. This is an AR process of order p with zero mean. The AR coefficients $a_1 \ldots a_p$ satisfy the Yule–Walker equations, which are linear equations and will be derived in the following.

Burg's algorithm is an alternative to the Yule–Walker equations for fitting AR models. The Levinson–Durbin recursion allows the determination of the coefficients for a model of order k, given the coefficients for a model of order $k - 1$. This allows the calculation of the model for many orders in a simple manner so that an optimum order may be found.

Once the optimum AR model is found, the spectrum $s(f)$ of the AR process with coefficients $a_1 \ldots a_p$ is given by

$$s(f) = \frac{\sigma_\varepsilon^2}{\left| 1 - a_1 e^{-2\pi i f} - \cdots - a_p e^{-2\pi i p f} \right|^2} \tag{9.2}$$

where σ_ε^2 is the variance of the process ε_t.

The derivation of the Yule–Walker equations is based on the theory for linear mean square estimation of random variables. To begin, consider the two random variables y and x, each with zero mean. The problem is to determine the value of a constant a, such that y is the best estimate of ax. This will be true if the mean square error $E[(y - ax)^2]$ is minimized.

The solution of this problem is given by the orthogonality principle, which states that the mean square error will be minimized if $E[(y - ax)x] = 0$, that is, if the error $y - ax$ is orthogonal to the known quantity x. On expanding the expected value, we obtain

$$E(yx) - aE(x^2) = 0$$

which can be solved for a yielding

$$a = \frac{E(yx)}{E(x^2)} \tag{9.3}$$

Expanding this to many random variables we can restate the problem as follows:

Given random variables $s_0, s_1, \ldots, s_n$ all with zero means, define $R_{ij} = E(s_i s_j)$. We want to estimate s_0 by a linear combination $a_1 s_1 + \cdots + a_n s_n$, such that the mean square error $E[(s_0 - (a_1 s_1 + \cdots + a_n s_n))]^2$ is minimum. By the orthogonality principle,

$$E\left[\left\{s_0 - \left(a_1 s_1 + \cdots + a_n s_n\right)\right\} s_i\right] = 0 \quad i = 1 \ldots n \tag{9.4}$$

On expanding Equation 9.4, we obtain

$$R_{0i} = a_1 R_{1i} + a_2 R_{2i} + \cdots + a_n R_{ni} \quad i = 1 \ldots n \tag{9.5}$$

The constants a that satisfy Equation 9.5 are the solutions to the equations and the coefficients of the AR model. On writing Equation 9.5 in matrix forms, we obtain the familiar Yule–Walker formulation:

$$\begin{bmatrix} R_{01} \\ \vdots \\ R_{0n} \end{bmatrix} = \begin{bmatrix} R_{11} & R_{21} & \cdots & R_{n1} \\ R_{12} & R_{22} & \cdots & R_{n2} \\ \vdots & \vdots & \vdots & \vdots \\ R_{1n} & R_{2n} & \cdots & R_{nn} \end{bmatrix} \begin{bmatrix} a_1 \\ a_2 \\ \vdots \\ a_n \end{bmatrix} \tag{9.6}$$

The last step involves assuming that the random variable s in our example actually represents points in a time series as shown in Equation 9.1.

9.2 Time–Frequency Analysis

Both FFT analysis and AR spectral analysis assume that the signal to be analyzed is stationary over the entire period of the analysis. Loosely speaking, a stationary process is one whose mean and autocorrelation do not change with time. (This is actually called wide-sense stationarity.) This means that if we were to split a given sequence of measurements into two equal segments and plot a windowed Fourier spectrum of each segment, the two plots are expected to be similar. If we were to estimate AR parameters for each segment, the two sets of parameters are expected to be similar.

However, in many cases, the signal is not stationary and, in fact, the non-stationary nature of the signal is what we are interested in examining. For example, if we record and digitize two speech segments and repeat the above procedure, the results for the two segments are likely to be different. We will explore some examples of this later. Therefore, we will examine a method for exhibiting the spectrum for nonstationary signals without distorting the resulting spectrum. This method is called time–frequency analysis, and it allows for a display of the signal spectral characteristics over time. We are therefore producing a video or a dynamic display of the spectral characteristics instead of a static display. With this dynamic display, we can answer questions such as "what happened before," "what happens next," "what changed," "how fast did it change," and "how much did it change."

FIGURE 9.1
Sine wave of varying frequency.

Before we delve into some of the mathematics of time–frequency analysis, let us examine some simple examples illustrating the advantages of the method. Consider the waveform shown in Figure 9.1. Notice that we have a sine wave whose frequency is changing with time. If we take the Fourier transform of a segment of the signal in which only one frequency is present, the signal is stationary and the spectrum will clearly show the predominant frequency. However, if we take the Fourier transform of the section that is outlined in the figure (a nonstationary section), we will get a result that can be interpreted in many different ways. We will get two spectral peaks indicating that there are two frequencies present in the interval, but we cannot tell when the components were present, for how much time the components

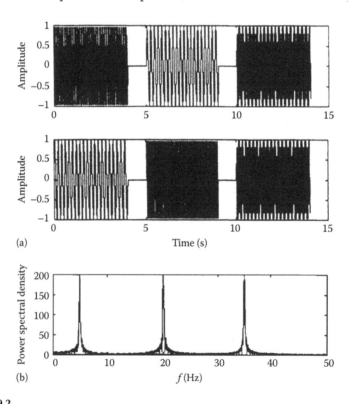

FIGURE 9.2
(a) Two plots illustrating sine waves of three different frequencies present for different time intervals. (b) Spectrum for both plots of Figure 9.2a.

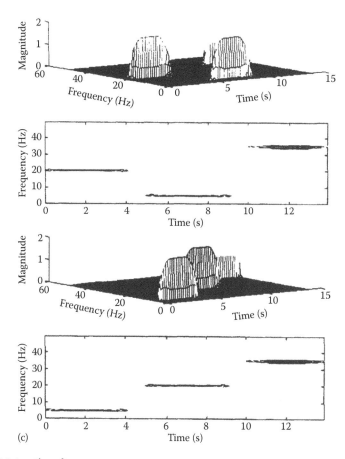

FIGURE 9.2 (continued)
(c) Time–frequency plots for the sine waves in Figure 9.2a.

were present, or whether they were simultaneously present or individually present. Several of these possibilities are shown in Figure 9.2. Notice that the waves shown in the two plots of Figure 9.2a contain the same frequencies (5, 20, and 35 Hz) but occur at different times. Figure 9.2b shows the FFT for both waves illustrating that both waveforms have the same FFT. Figure 9.2c shows the corresponding time–frequency plots for the plots of Figure 9.2a. Note that although the FFT plots are identical, the time–frequency plots are quite different, illustrating how time–frequency analysis can be used to uniquely describe the signals in both time and frequency.

A display that clearly shows the information obtainable from time–frequency analysis is shown in Figure 9.3. The figure shows the time–frequency distribution of a whale sound. The time waveform is on the left running up the page. The Fourier transform is shown below the main figure. Notice that both the time waveform and the Fourier transform do not show a clear picture of what is happening in time and in frequency. However, the time–frequency

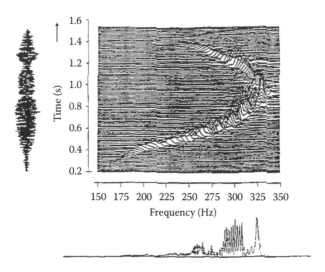

FIGURE 9.3
Time–frequency plot of a whale sound.

distribution shows that at a time of about 0.3 s the major frequency component is about 175 Hz. As time progresses to about 1 s, the frequency increases
to a maximum of about 340 Hz, at which time the frequency decreases to
about 250 Hz at about 1.4 s.

Let us now explore some of the mathematics involved in the time–frequency
analysis. We must first discuss the uncertainty principle which is, in general,
a problem with spectral analysis but not with time–frequency analysis. If we
consider the Fourier transform (refer to Section 7.3 to aid in this discussion), we
recall that if we consider a sine wave that goes on forever, the Fourier transform
is an impulse (spike) at the frequency of the sine wave. On the other hand, if
we window the sine wave (cut it to a finite time), the spectrum widens in the
frequency domain. Therefore, there exists an inverse relationship between the
length of a signal in the time domain and the width of the spectrum in the frequency domain. The narrower the signal is in time (time localization), the wider
(less localized) it is in frequency and vice versa. It is not possible using Fourier
transform techniques to localize a signal in both time and frequency. This is
known as the uncertainty principle and can be expressed mathematically as

$$\sigma_t \sigma_\omega \geq \frac{1}{2} \tag{9.7}$$

where t is in seconds and ω is in Hz. An explanation of Equation 9.7 is as
follows:

Given a time signal $s(t)$. The energy of the signal $s(t)$ is given by

$$E = \int |s(t)|^2 \, \mathrm{d}t \tag{9.8}$$

Therefore, $|s(t)|^2$ can be considered as the energy per unit time at time t or the energy density. The average time can therefore be defined as

$$\langle t \rangle = \int t \, |s(t)|^2 \, dt \qquad (9.9)$$

and the variance can be written as

$$\sigma_t^2 = \int \left(t - \langle t \rangle \right)^2 |s(t)|^2 \, dt \qquad (9.10)$$

Similarly, we can define $|S(\omega)|^2$ as the energy per unit frequency and the variance in frequency written as

$$\sigma_\omega^2 = \int \left(\omega - \langle \omega \rangle \right)^2 |S(\omega)|^2 \, d\omega \qquad (9.11)$$

The uncertainty principle can then be written as

$$\sigma_t \sigma_\omega \geq \frac{1}{2} \qquad (9.12)$$

This limitation does not allow a localization in both time and frequency using Fourier transform techniques but can be overcome using time–frequency analysis. Of course, other problems arise with time–frequency analysis that will be discussed in the following.

We now want to derive the function $P(t, \omega)$ as the density in time and frequency. Such a function P should have the following properties to be useful as a density function:

1. $P(t, \omega)$ is a joint density. Therefore, $P(t, \omega)\Delta t \Delta \omega$ is the energy in time interval Δt and frequency interval $\Delta \omega$ at time t and frequency ω.
2. $P(t, \omega) \geq 0$ for all t and all ω.
3. $\iint P(t, \omega)dt\, d\omega = \int |s(t)|^2 \, dt = \int |S(\omega)|^2 \, d\omega$ equals total energy in the signal.
4. Marginals:

$$\int P(t, \omega)\, dt = P_2(\omega) = |S(\omega)|^2$$

$$\int P(t, \omega)\, d\omega = P_1(t) = |s(t)|^2$$

$P_1(t)$ and $P_2(\omega)$ are called the marginals of the distribution and must be equal as shown above.

5. Time and frequency shifts:

$$\text{If } s(t) \rightarrow s(t+t_0), \quad \text{then } P(t,\omega) \rightarrow P(t+t_0,\omega)$$

and

$$\text{if } S(\omega) \rightarrow S(\omega+\omega_0), \quad \text{then } P(t,\omega) \rightarrow P(t,\omega+\omega_0)$$

9.2.1 Short-Time Fourier Transform

If we attempt to achieve time localization by taking the Fourier transform of a signal in a short time window, we have derived the short time Fourier transform (STFT). As mentioned above, the STFT must satisfy the uncertainty principle and therefore does not have proper frequency localization (the spectrum smears in the frequency domain). In addition, the STFT suffers from the following:

It does not satisfy the marginals.

The analysis includes the properties of the window used in the transform.

The result is not necessarily zero when the signal is zero. This last problem (lack of time support) is quite serious because it does not adequately follow transients in the signal.

9.2.2 Quadratic Distributions

The STFT can be considered as a linear distribution. In an attempt to overcome some of the problems mentioned above, a new class of distributions called quadratic distributions was developed. As an example, the square magnitude of the STFT (called the spectrogram) is a quadratic distribution. This started with the work of Wigner in 1932 and now includes over 21 distributions [5]. Many of these can be derived from a single member and are known as the Cohen class. We will start with a discussion of the Wigner–Ville distribution (WVD) and then move on to other examples.

9.2.2.1 The Wigner Distribution

The Wigner distribution is defined as

$$W(t,\omega) = \frac{1}{2\pi}\int s^*\left(t-\frac{1}{2}\tau\right)s\left(t+\frac{1}{2}\tau\right)e^{-j\omega\tau}\,d\tau \tag{9.13}$$

where
$s(t)$ is the input signal
τ represents a time shift
$W(t,\omega)$ is the two-dimensional (2D) time–frequency distribution

As an example of the Wigner distribution, consider the following signal, which is known as a chirp or linear frequency modulation (FM):

$$y(t) = \begin{cases} e^{j\left(\omega_0 t + \mu t^2\right)} & -T/2 \le t \le T/2 \\ 0 & \text{otherwise} \end{cases} \tag{9.14}$$

The WVD for the chirp is given by

$$W(t,\omega) = 2(T - 2|t|)\frac{\sin\left[(\omega - \omega_0 - 2\mu t)(T - 2|t|)\right]}{(\omega - \omega_0 - 2\mu t)(T - 2|t|)} \tag{9.15}$$

In particular, when T tends to infinity the WVD approaches $2\pi\delta_D(\omega - \omega_0 - 2\mu t)$, so it becomes concentrated on the line $\omega(t) = \omega_0 + 2\mu t$.

The Wigner distribution has the following properties:

1. It is always real, that is, $W^*(t,\omega) = W(t,\omega)$.
2. It is uniquely related to the signal $s(t)$ up to a constant phase factor.
3. It satisfies the marginals exactly.
4. It takes on negative values for all signals except a chirp. This is a problem since $W(t, \omega)$ is a density. However, ignoring the negative regions usually gives good results.
5. It gives the instantaneous frequency exactly.
6. For a finite duration signal, it is zero before the signal starts and after the signal ends (finite support).
7. If a signal is band limited, the Wigner distribution is zero outside the band.

The above can be considered advantages of the Wigner distribution. However, the Wigner distribution has two problems or disadvantages. The first is its behavior in noise. Noise at one time will appear in the Wigner distribution and at other times nonlocal. In other words, the noise will spread over larger time durations than in the original signal. The second concerns the generation of cross terms. If two signals are added, that is, $s(t) = s_1(t) + s_2(t)$, then the corresponding Wigner distribution becomes

$$W(t,\omega) = W_{11}(t,\omega) + W_{22}(t,\omega) + W_{12}(t,\omega) + W_{21}(t,\omega)$$

where
$W_{11}(t,\omega)$ is the Wigner distribution of $s_1(t)$
$W_{22}(t,\omega)$ is the Wigner distribution of $s_2(t)$
$W_{12}(t,\omega)$ and $W_{21}(t,\omega)$ are cross terms (noise)

The cross terms add distortion to the result and pose a problem, which will be discussed more fully in the following. To summarize, the major strengths of the Wigner distribution are as follows:

It gives a good picture of the time–frequency structure of a signal.
It can calculate the instantaneous frequency exactly.
It reveals components in a multicomponent signal.

The major weaknesses of the Wigner distribution are as follows:

It can become negative and negative regions cannot be interpreted.
It produces cross terms or interference terms.

A modification of the Wigner distribution, which results in a reduction in the cross terms is the pseudo-Wigner distribution (PWD). This is equivalent to the windowing operation for the Fourier transform (see Sections 7.3 and 8.3). It serves the triple purpose of reducing the cross terms, attenuating the Fourier transform side lobes (when truncating the signal to a finite duration), and reducing computational complexity in discrete-time processing. The PWD is defined by

$$PW(t,\omega) = \int_{-\infty}^{\infty} s\left(t + \frac{1}{2}\tau\right) s^*\left(t - \frac{1}{2}\tau\right) h\left(\frac{1}{2}\tau\right) h^*\left(-\frac{1}{2}\tau\right) e^{-j\omega\tau} \, d\tau \qquad (9.16)$$

where $h(t)$ is a user defined window function. The effect of the window is to localize the integration to the vicinity of $\tau = 0$. Therefore, the distribution at time t is mainly affected by signal values in the vicinity of t.

The ability of the PWD to eliminate cross terms is limited and can be improved by an additional windowing operation in the frequency domain. This produces the smoothed pseudo-Wigner distribution (SPWD) defined by

$$SPW(t,\omega) = \frac{1}{2\pi} \int_{-\infty}^{\infty} \int_{-\infty}^{\infty} W_h(0,v) g(\theta) W(t - \theta, \omega - v) \, dv \, d\theta \qquad (9.17)$$

where
$g(t)$ is another window function (which may be equal to $h(t)$ or different)
W is the Wigner distribution of $s(t)$
W_h is the Fourier transform of the window function $h(t)$

9.2.2.2 General Class of Distributions

Before discussing the general class of time–frequency distributions, it is useful to define the ambiguity function (AF) of the signal $y(t)$:

$$A(v,\tau) = \int_{-\infty}^{\infty} y(t+0.5\tau)y*(t-0.5\tau)e^{jvt}\,dt \qquad (9.18)$$

It can now be shown that the AF is, up to a constant factor, the 2D inverse Fourier transform of the WVD. Therefore, the properties of the AF are the duals of the properties of the WVD.

Both the STFT and the Wigner distribution belong to a general class of distributions (Cohen's class), which can all be described by one general equation, which includes a kernel that determines the particular distribution. The general form is as follows:

$$C(t,\omega) = \frac{1}{2\pi} \iiint e^{j(vu - vt - \omega\tau)}\phi(v,t)s*\left(u - \frac{1}{2}\tau\right)s\left(u + \frac{1}{2}\tau\right)du\,dv\,d\tau \qquad (9.19)$$

where $\phi(\theta,t)$ is the kernel. Comparing with Equation 9.18, we see that Equation 9.19 can be written in terms of the AF as follows:

$$C(t,\omega) = \frac{1}{2\pi} \iint \phi(v,\tau)A(v,\tau)e^{-j(vt+\omega\tau)}\,dv\,d\tau \qquad (9.20)$$

Therefore, the time–frequency distribution C is obtained by multiplying the AF of the signal by the kernel and taking the 2D Fourier transform of the result. An equivalent procedure is to compute the WVD of the signal, take its 2D inverse Fourier transform, multiply by the kernel, and take the 2D Fourier transform of the result.

As an example, for the Wigner distribution, the kernel is equal to 1. Therefore, once a kernel is chosen, the distribution is generated. The properties of the distribution can be determined by examining the properties of the kernel. For example, in order to have a real distribution, it can be shown that the kernel must satisfy $\phi(v,\tau) = \phi*(-v,-\tau)$. Similarly, other properties can be related as follows: If the kernel is independent of t, then the distribution is a translation invariant in time. Also, if the kernel is independent of frequency, then the distribution is a translation invariant in frequency.

The distribution satisfies the time-marginal property if and only if $\phi(v,0) = 1$ for all v, and satisfies the frequency marginal property if and only if $\phi(0,\tau) = 1$ for all τ.

The observation of the Cohen class of distributions has shown that it is possible to achieve cross-term reduction through proper choice of the kernel.

The crucial observation for this reduction is that while the auto terms tend to be concentrated around the origin of the AF, the cross terms tend to occupy regions far from the origin. An example of a kernel that achieves reduced interference is the Born–Jordan distribution, which has a kernel given by

$$\phi(v, \tau) = \frac{\sin 0.5 v \tau}{0.5 v \tau} \tag{9.21}$$

Another example is the Choi–Williams distribution, which has a kernel given by

$$\phi(v, \tau) = e^{-v^2 \tau^2 / \sigma} \tag{9.22}$$

where the parameter σ controls the width of the kernel. This kernel, when multiplying the AF, will attenuate the away-from-origin regions, thus attenuating the cross terms. The parameter σ should ideally be chosen to preserve the auto terms of the distribution as much as possible.

The following procedure can be used to construct a kernel with cross term reduction properties. The basic idea is to choose a real-valued window function $h(t)$ with the following properties:

$$\int_{-\infty}^{\infty} h(t) \, dt = 1$$

$$h(t) = h(-t)$$

$$h(t) = 0 \quad \text{for } |t| > 0.5$$

$H(\omega)$ the Fourier transform of $h(t)$, has a low-pass characteristic. The kernel is then $\phi(v, \tau) = H(v\tau)$. The family of distributions thus obtained is called the reduced interference distributions (RID). The Born–Jordan kernel is obtained by taking $h(t)$ as the rectangular window function.

9.3 Examples of Physiological Signals Processing

We will now use our knowledge of biomedical simulation and signal processing to examine several physiological systems. We will examine the autonomic nervous system by means of HR variability. We will also examine the analysis of physiological signals such as the electrocardiogram (ECG), the electroencephalogram (EEG), and the electromyogram (EMG). In addition, we will discuss the topic of circadian rhythms and then attempt to connect all of these signals and systems in our discussion of physiological stress. The reader is urged to consult a text on human physiology to supplement this discussion [6].

9.3.1 Physiology of the Autonomic Nervous System and Heart Rate Variability

One of the important applications that we will consider in this chapter is HR variability, which is a noninvasive technique that gives us a window into the functioning of the autonomic nervous system. The autonomic nervous system is generally responsible for control of the "unconscious" systems of the body and is, in turn, controlled by the hypothalamus. The autonomic nervous system consists of two branches, the sympathetic and parasympathetic branches. Table 9.1 lists a short group of physiological systems and what happens to each system during sympathetic and parasympathetic nervous system activity. In general, the two branches act in a reciprocal manner, with sympathetic stimulation occurring during "stressful" periods ("fight or flight") and parasympathetic stimulation occurring during "relaxed" periods ("rest and recuperation"). The reader is urged to consult a book on human physiology for a more detailed description of the autonomic nervous system [6].

In 1981, there was a landmark paper [7] that began the field of HR variability as a tool for examining the functioning of the autonomic nervous system. The paper outlined the signal processing necessary to achieve this goal. The basic measurement is the time interval between heart beats as calculated from the ECG. Figure 9.4 shows a normal ECG. If the time interval between beats is measured, it is noticed that the interval is not constant but varies in a cyclical manner. Most measurements of HR variability involve the determination of the R-wave positions as markers of each beat position. The R wave is used because of its distinctive shape and prominent amplitude making it the easiest part of the beat for a computer to detect.

Basic knowledge about HR variability includes the following [8]:

1. HR variability can be measured by the R–R intervals in the ECG.
2. It depends on the activity of both the sympathetic and parasympathetic branches of the autonomic nervous system.

TABLE 9.1

Physiological Systems

System	Sympathetic Effect	Parasympathetic Effect
Pupil of eye	Dilates	Contracts
Heart	Increase rate	Decrease rate
Gastrointestinal (GI) tract	Decrease motility	Increase motility
Fat cells	Increases fat breakdown	—
Liver	Releases glucose	—

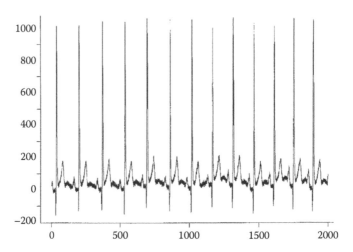

FIGURE 9.4
Normal ECG.

3. It is affected mainly by respiration (respiratory sinus arrhythmia) and by blood pressure (baroreceptor control system).

4. Respiration affects HR through the parasympathetic nervous system only.

Previous studies on HR variability have shown the following:

1. HR variability has been shown to be a predictor of the occurrence of cardiac symptoms [9]. In particular, Bigger has shown in a landmark study that in a group of patients who had post-myocardial infarction, the incidence of a second myocardial infarction was inversely correlated to the degree of parasympathetic activity, as measured by HR variability.

2. It decreases with age [5]. Therefore, any study involving HR variability must use age-matched controls.

3. It decreases in smokers and in sedentary people. Conversely, HR variability increases in people who habitually exercise [10].

Figure 9.5 shows the steps involved in constructing an interbeat interval (IBI) signal, which will be used to examine the autonomic nervous system. In Figure 9.5a, we see a normal ECG. We will assume that the ECG is windowed with a Hanning window and sampled at 200 Hz. Each R wave is detected and a pulse is produced at the position of each R wave. This pulse train is shown in Figure 9.5b. In Figure 9.5c, the height of each pulse is adjusted to be the length of the previous R–R interval. For example, in the figure, two successive beats occur at time 2 s, $T(m-1)$, followed by a beat at time 2.9 s, $T(m)$. Therefore, the interval of 0.9 s becomes the height of the pulse

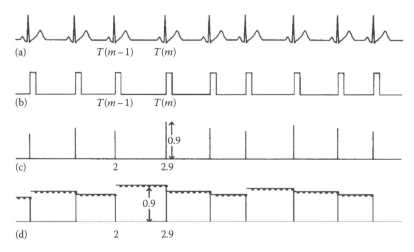

FIGURE 9.5
Steps in the construction of an IBI signal.

at time $T(m)$. In Figure 9.5d, the pulse wave of Figure 9.5c is interpolated to produce a wave with equally spaced samples. This type of interpolation is called backward step interpolation, where the height of the wave in a time interval is kept constant at the value of the length of the time interval. The wave shown in Figure 9.5d is called an IBI signal and will become the basis from which information on HR variability will be obtained.

In order to obtain the pertinent HR variability information, the IBI signal of Figure 9.5d is first demeaned (the mean or DC value of the signal is removed) thus removing the steady state HR information. The resulting signal is then transformed using the FFT to produce a frequency spectrum. A typical result is shown in Figure 9.6. Three frequency regions can be identified in the frequency spectrum. The area under the spectrum in the band of frequencies between 0.15 and 0.4 Hz, called the *high-frequency band,* has been shown to be an indicator of parasympathetic nervous system activity. The major activity in this band is due to respiration and a predominant peak usually occurs at the respiration frequency. The area under the spectrum in the frequency band between 0.05 and 0.15 Hz, called the *low-frequency band,* has been shown to be influenced by both sympathetic and parasympathetic nervous system activity and plays a large part due to the variability in the blood pressure control system. The band of frequencies between 0.02 and 0.05 Hz is called the *very low-frequency band* and is due mainly to the activity of the temperature control system.

To illustrate the information obtained from HR variability spectra, Figure 9.7a shows two plots taken under two different conditions. It shows the spectrum of 1 min of data taken under resting conditions. The solid line shows the HR variability spectrum, while the dotted line shows the spectrum of the respiration signal. Notice that the predominant frequency of respiration is at about 0.2 Hz (12 breaths per min) and that the predominant peak in the

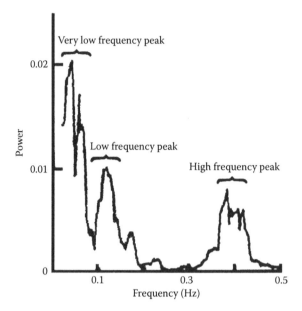

FIGURE 9.6

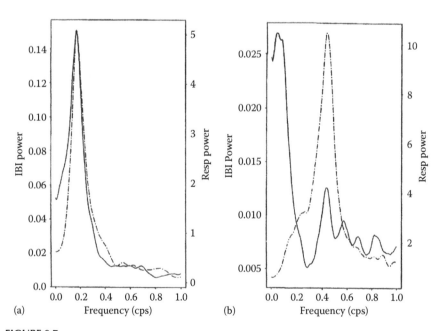

FIGURE 9.7
Frequency spectrum of an IBI signal. (a) Spectra of IBI (solid line) and respiration (dotted line) during rest. (b) Spectra of IBI (solid line) and respiration (dotted line) during exercise.

high-frequency region occurs at the respiration frequency. Also notice the absence of a peak in the low-frequency region. This is due to the high parasympathetic activity in this subject at rest.

Figure 9.7b shows the spectrum of 1 min of data taken during running on a treadmill at 4 mph. Notice the differences between the plots of Figure 9.7a and b. First, the frequency of respiration has shifted to 0.4 Hz (24 breaths per min) showing the increase of breathing rate due to exercise. Second, a low-frequency peak is now evident and is in fact higher than the high-frequency peak. This shows the decrease in parasympathetic activity and the increase in sympathetic activity, which occurs during exercise.

Signal processing concepts play a large role in the analysis of HR variability. As was pointed out above, a great deal of information can be obtained from the frequency spectrum of the IBI signal. This spectrum can be obtained by using either the FFT, as described in Chapter 7, or by using the AR model, as described in Section 9.1. The FFT approach is routinely used by Mathias et al. [11] in their work, while the AR approach is routinely used by Malliani et al. [12] in their work. As was pointed out in Section 9.1, each approach has its advantages and disadvantages and both are approximations to the real data. For example, the AR approach allows the choice of model order, which results in a controlled number of frequency peaks. The FFT approach, on the other hand, produces more accurate results for broadband signals.

Note that the frequency band of interest for HR variability studies ranges from 0.02 to 0.4 Hz. A sinusoid of 0.02 Hz frequency has a time period of 50 s. Therefore, in order to have enough data to accurately determine the frequency components, it is necessary to have data sets of several minutes duration. Typically, in our research, we do not analyze data sets of less than 2 min duration. This presents a problem since in 2 min, the characteristics of the autonomic nervous system can change, producing a nonstationary data set. Both AR spectra and Fourier spectra require that the data set is stationary. In order to overcome this, time–frequency analysis is used. The basics of time–frequency analysis was described in Section 9.2. In addition to small nonstationarities in the data as mentioned above, time–frequency analysis also allows the following of rapid changes in HR variability due to onset or recovery from "stressful" events. This will be discussed in more detail in the following.

As an illustration of the use of time–frequency analysis, Figure 9.8 shows the high-frequency activity plotted as a function of time during an exercise test on a normal subject. The protocol for this subject was as follows. The subject was told to rest for 2 min, then was given 2 min on an exercise bicycle, to begin pedaling and to get his HR up to a specified level. At that point, the subject was instructed to keep pedaling at a constant rate for 4 min. Finally, at the 8 min mark, the subject was instructed to stop pedaling and sit quietly during recovery. From Figure 9.8, we can see the high-frequency activity (parasympathetic activity) during each phase of the procedure as well as the change in activity during the transitions from one stage to the next. Notice that the activity is high during rest, drops quickly at the beginning

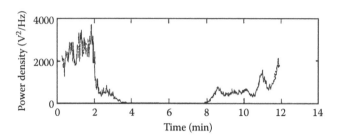

FIGURE 9.8
High-frequency activity as a function of time during exercise.

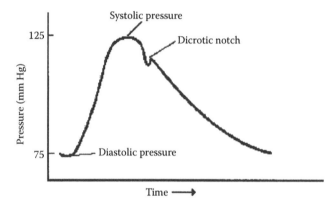

FIGURE 9.9
Arterial blood pressure wave.

of exercise, remains at a very low value during the exercise, and recovers toward its initial value after the exercise.

The blood pressure also carries the information that is useful in the study of HR variability. Figure 9.9 shows a typical arterial blood pressure wave, which can be recorded from a major artery such as the aorta. In Figure 9.10, the systolic peaks are detected and a systolic blood pressure variability wave is constructed by interpolating between each systolic peak in the same way as described above for HR variability. If the Fourier transform of the blood pressure variability wave is calculated, a spectrum similar to that of Figure 9.6 is obtained, including a high-frequency peak at the respiratory frequency and a low-frequency peak at about 0.1 Hz as was true for the HR variability spectrum.

The question that arises at this point is whether the blood pressure variability spectrum contains any additional information not present in the HR variability spectrum. There are two approaches to the answer to this question. First, it has been stated in the literature [13] that the low frequency area in the blood pressure variability spectrum is related purely to sympathetic activity, rather than to a mixture of sympathetic and parasympathetic activity, as was the case for the HR variability spectrum. Experiments must be

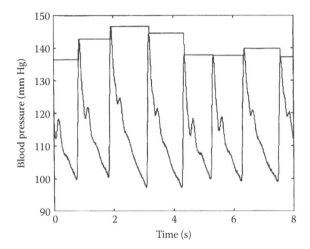

FIGURE 9.10
Production of a systolic blood pressure variability wave.

performed to verify this. Second, it is instructive to examine the coherence between the HR variability spectrum, blood pressure variability spectrum, and respiration spectrum. Coherence was discussed in Section 7.9 and relates in this context to whether there is a linear relation between HR variability, blood pressure variability, and respiration at any frequencies within the high- or low-frequency bands. Again, experiments must be performed to verify this and to note the difference in coherence between normal and pathological subjects. It is felt that the coherence measure will add significant information to allow researchers to better understand the difference in the physiology between normal and pathological subjects.

9.4 Measurement of Physiological Stress

In this section, we will discuss the general concept of physiological stress and how we might, through signal processing techniques, measure such a variable. Stress can be defined as the body's reaction to a perceived mental, physical, or emotional distress. Although this definition may appear to be circular, there are important concepts contained in it. First, the important result of stress on the body is in terms of what is perceived, not necessarily the actual stress that is present. Second, the stress may be mental, emotional, or physical. This implies that exercise (physical stress) evokes the same physical reaction as taking an examination (mental stress). Implied in this definition is the fact that the opposite is also possible, that is, relaxation induces the reverse physiological responses to stress.

Because of the changes to the autonomic nervous system due to stress, as discussed above, the HR variability spectrum and systolic blood pressure variability spectrum, combined with time–frequency analysis, can be used to observe the time course of stress in the body. However, many other physiological variables have also been shown to be related to the level of stress and relaxation. Respiration, both rate and volume, gives an indication of metabolic level, which is certainly correlated to stress level. Blood levels of substances such as cortisol and catecholamines have been shown to be related to stress level.

Other variables of interest include EEG effects and effects on the peripheral circulation. The EEG is the integrated voltage observed on the surface of the scalp due to the activity of neurons in the brain. There are two measures of stress and relaxation, which can be observed from the EEG. From the EEG spectrum, activity can be observed in the 8–12 Hz (alpha) frequency band, where increased activity is related to increased relaxation (decreased stress). A more sophisticated measure of stress and relaxation is the phase coherence, which shows the phase synchrony, as a function of frequency, from two different spatial locations on the surface of the scalp. It has been shown that during meditation, there is a marked increase in coherence, especially in the alpha-frequency band, and especially between frontal EEG sites [14].

The last measure to be discussed is the change in peripheral blood flow occurring during stress or relaxation. It has been shown [15] that blood volume in the periphery (fingers and toes) decreases during periods of stress and increases during periods of relaxation. In a study by Nketia and Reisman [16], finger temperature and blood volume were measured during stress and relaxation in normal subjects. Figure 9.11 shows a typical result where the

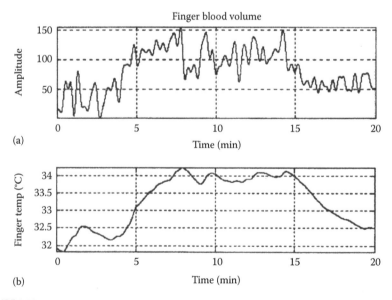

FIGURE 9.11
Finger blood volume and finger temperature during (a) stress and (b) relaxation protocol.

subject sat with eyes open for 3 min, sat with eyes closed (relaxed) for 12 min, and was stressed by a loud noise for 5 min. Note that both the finger blood volume and the finger temperature increase during relaxation and decrease during stress. The temperature, however, responds more slowly due to the heat capacity of the finger.

9.5 Circadian Rhythms

The subject of circadian rhythms illustrates the connection between biomedical simulation, signal processing, and instrumentation. We will consider all aspects in this discussion. Circadian rhythms are oscillations within the body, which occur with a period of about 24 h. Experiments performed on both humans and animals show that these rhythms are inherent in the body and are not caused by the 24 h light-dark cycle. These experiments include putting the animal or human into an environment, where there are no cues from the outside. They can eat when they want and sleep when they want. Under these conditions, the circadian rhythms continue to oscillate but with a period of not exactly but close to 24 h. However, as soon as they are put into the outside environment, the period *entrains* to 24 h. This phenomenon of entrainment will be discussed further in the following.

Variables that exhibit circadian rhythms, which are easily measurable, include body temperature, sleep-wake activity, feeding, and physical activity. Other circadian rhythms are exhibited by some blood hormone levels.

Many questions still remain concerning the origin and purpose of circadian rhythms. For example:

1. What is the mechanism causing the circadian rhythms?
2. Is there one master oscillator and a group of slaves causing all the rhythms or are there separate but coupled oscillators?
3. Are we seeing the oscillator itself or just an external manifestation of the oscillator?

To begin the discussion on modeling circadian rhythms, let us consider Equation 9.23, which is the differential equation for a *linear sinusoidal oscillator*.

$$\frac{d^2y}{dt^2} + \omega^2 y = 0 \qquad (9.23)$$

where ω is the frequency of oscillation. The solution of Equation 9.23 is

$$y(t) = A\cos(\omega t + \phi) \qquad (9.24)$$

Now consider what happens when an external forcing function is added, which is at a frequency different from the natural frequency of the oscillator. This case is shown in Equation 9.25.

$$\frac{d^2y}{dt^2} + \omega_1^2 y = E \cos \omega_2 t \tag{9.25}$$

The solution for this system is

$$y(t) = C_1 \cos(\omega_1 t + \phi_1) + C_2 \cos(\omega_2 t + \phi_2) \tag{9.26}$$

It can be seen that, because the system is linear, both frequencies ω_1 and ω_2 are present in the solution. Therefore, entrainment, where the oscillator frequency shifts to the external frequency, cannot occur and the modeling of circadian rhythms must be done with nonlinear oscillators.

An example of a simple nonlinear oscillator that exhibits the property of entrainment is the Van der Pol oscillator. It is given by Equation 9.27:

$$\frac{d^2y}{dt^2} - \mu(1-y^2)\frac{dy}{dt} + \omega^2 y = 0 \tag{9.27}$$

In this system, the degree of nonlinearity is determined by the value of the parameter μ. If the parameter is set equal to zero, the system becomes linear and the output $y(t)$ is sinusoidal. As the parameter μ is increased in value, the shape of the output $y(t)$ changes from sinusoidal and becomes more square. If an external sinusoidal stimulus is applied to the system, Equation 9.28 results:

$$\frac{d^2y}{dt^2} - \mu(1-y^2)\frac{dy}{dt} + \omega_1^2 y = E \cos \omega_2 t \tag{9.28}$$

The degree of entrainment of this system will depend on the value of E (the larger the value, the greater the entrainment), the value of $|\omega_2 - \omega_1|$ (the larger the value, the lesser the entrainment), and the value of μ (the larger the value, the greater the entrainment).

In order to verify the theory presented above, circadian rhythms were monitored on a group of rhesus monkeys who were placed in a controlled environment, where the number of hours of light and dark could be controlled. The core temperature of the monkeys was monitored and sampled every 10 min to observe the circadian rhythms. The result for a monkey under "free running conditions" is shown in Figure 9.12. Free running occurs when the lights on the monkey's environment are left on continuously so that the natural frequency of the circadian oscillator can be observed. It can be determined that the fundamental period of the oscillator is slightly less than 24 h.

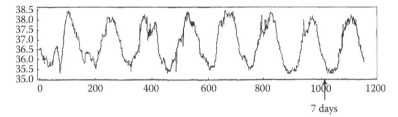

FIGURE 9.12
Circadian temperature rhythm in a monkey under free running conditions. Data are sampled at one point every 10 min.

Notice also that the oscillation is not purely sinusoidal and that each cycle is not the same as the previous cycle. The nonsinusoidal nature of the rhythm is due to the nonlinearity of the oscillator. However, the nonperiodic nature of the signal is due to the presence of another oscillator, oscillating at a different natural frequency, and coupled to the first oscillator. We therefore investigated the modeling of coupled oscillators to attempt to match the experimental data with the output of the model.

Figure 9.13 schematically illustrates the coupling of two oscillators. The figure shows coupling in each direction although the degree of coupling in each direction may be different. The natural frequencies of each oscillator can be determined by calculating the Fourier spectrum of the temperature wave of Figure 9.12. Two examples of time series and their corresponding spectra are shown in Figure 9.14. Frequencies of 26 and 12.5 h were determined to most closely match the experimental data. Note that these frequencies are not harmonically related, which explains how different cycles can look. It was also determined by comparing experimental and model results that velocity coupling produced the closest correspondence between the two results. The equations for velocity coupling of two oscillators x and y are shown in Equation 9.29.

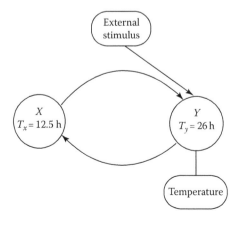

FIGURE 9.13
Schematic illustration of the coupling of two oscillators.

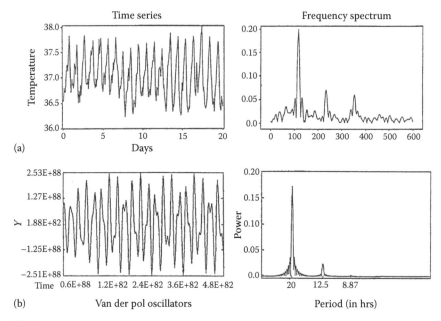

FIGURE 9.14
Two examples of time series and their corresponding spectra.

$$\frac{d^2x}{dt^2} - \mu_x(1-x^2)\frac{dx}{dt} + \omega_x^w x + F_{yx}\frac{dy}{dt} = 0$$

$$\frac{d^2y}{dt^2} - \mu_y(1-y^2)\frac{dy}{dt} + \omega_y^2 y + F_{xy}\frac{dx}{dt} = 0$$

(9.29)

9.6 EEG, Spectral Analysis, and Coherence

In this section, we will consider another physiological signal that is commonly acquired from the surface of the body, namely, the EEG. We will then discuss two types of signal processing applications for the EEG, spectral analysis, and coherence. We will begin with some basic physiology and measurement techniques related to the EEG.

The EEG measures the electrical activity of the brain as seen from electrodes placed on the scalp. The amplitude of the acquired EEG is generally in the range of 50–100 μV and 0.05–100 Hz. The EEG has been shown to reflect the state of activity of the brain and can be viewed in either the time domain or in the frequency domain. The EEG is usually characterized by four frequency bands, named alpha, beta, theta, and delta. The time domain EEG

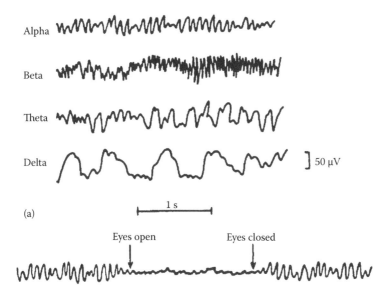

FIGURE 9.15
Frequency bands in the EEG.

characteristic of each of the four bands is shown in Figure 9.15. Activity in the beta band, 14–50 Hz is generally observed during intense mental activity and with eyes open. Activity in the alpha band, 8–13 Hz is generally observed during a quiet resting state with eyes closed. Activity in the theta band, 4–7 Hz is generally observed during disappointment or frustration in adults or during meditation. Activity in the delta band, below 4 Hz is generally observed during sleep. At the bottom of the figure is illustrated how the EEG changes when a subject changes from eyes open to eyes closed and vice versa. Notice that during eyes closed, the EEG is predominantly alpha (low frequency, high amplitude) whereas during eyes open, the EEG is predominantly beta (low amplitude, high frequency).

In addition to the spectrum of the EEG, the coherence is another measure that provides information about the state of the brain. As was discussed in Section 7.9, the coherence is a measure of the linear association of each frequency of one signal on the same frequency of another signal. The coherence was defined as the ratio of the cross spectral density of the two signals to the product of the individual spectral densities. In the case of the EEG, it has been shown [17] that during deep meditation, the coherence increases between symmetrical areas on both sides of the head. Therefore, the coherence can be used as a measure of the depth of meditation or of relaxation of the subject. As an example consider Figure 9.16 that shows coherence as a function of time and frequency in one subject during eyes open and eyes closed periods. Notice how the coherence increases in the alpha-frequency range, when the subject sits with eyes closed.

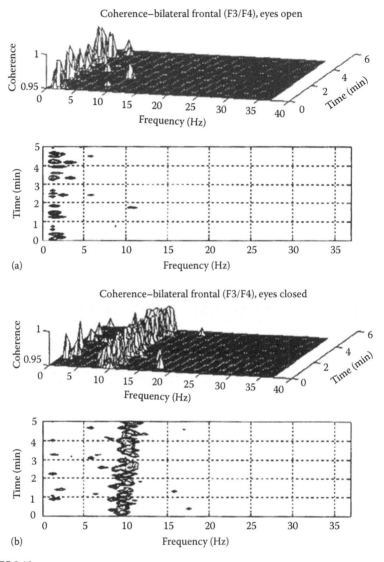

FIGURE 9.16
Coherence as a function of time and frequency during (a) eyes open and (b) eyes closed periods in one subject.

9.7 EMG, Spectral Analysis, and Mean Power Frequency

The last biological signal that we will consider in this chapter is the EMG. The EMG is an electrical signal that is produced when muscles contract and can be measured either on the body surface near the muscle of interest or directly from the muscle by penetrating the skin with needle or fine

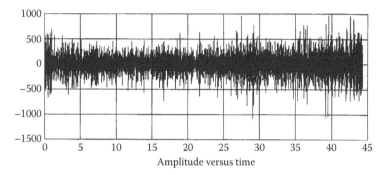

FIGURE 9.17
Normal EMG recorded with surface electrodes.

wire electrodes. The measured EMG is usually a summation of the individual action potentials from the fibers forming the muscle being examined. Typical EMG characteristics include a voltage range of $50\,\mu V$–$5\,mV$ and a frequency range of 10–3000 Hz. Figure 9.17 shows a typical EMG signal recorded with surface electrodes.

Spectral analysis of the EMG can be used to investigate the amount of fatigue present in the muscle under test. Fatigue is characterized as a progressive increase in discomfort arising from the active muscle as prolonged constant force contractions at moderate load levels are maintained. This discomfort is accompanied by specific changes in the EMG power spectrum of the muscle. Note that fatigue is now considered to be a continuous time dependent process throughout the entire contraction. This process is associated with a spectral shift of the EMG toward lower frequencies. Figure 9.18 shows the spectrum of an EMG both before and after fatigue. Mathematical measures

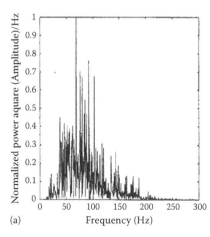

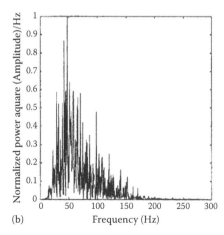

FIGURE 9.18
Spectrum of EMG (a) before and (b) after fatigue.

for fatigue can be obtained by calculating the median frequency and the mean power frequency. The median frequency is defined by Equation 9.30.

$$\frac{\sum_{f=a}^{f_{med}} P(f)}{\sum_{f=a}^{f=b} P(f)} = \frac{1}{2}$$

(9.30)

where
  $P(f)$ is the power spectral density at the frequencies of interest ranging from frequency (a) to frequency (b)
  f_{med} is the median frequency

The mean power frequency is defined by Equation 9.31.

$$f_{mean} = \frac{\sum_{f=a}^{f=b} f \times P(f)}{\sum_{f=a}^{f=b} P(f)}$$

(9.31)

It has been shown [18] that if median frequency is plotted as a function of time during fatigue, a linear relation is obtained with high correlation. Figure 9.19 shows an example of such a plot for a normal subject. Note a correlation coefficient of 0.933 for this subject. Therefore, the slope of the linear regression line is a measure of fatigue and can be compared with subjects with muscle disease.

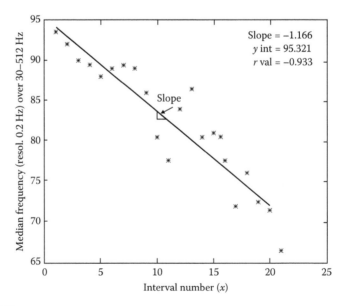

FIGURE 9.19
Plot of median frequency as a function of time during fatigue study.

Problems

9.1 What is the major advantage of the AR technique?

9.2 Explain the uncertainty principle.

9.3 List three weaknesses of the STFT.

9.4 List the two major weaknesses of the Wigner distribution.

9.5 The frequency spectrum of the IBI signal identifies three frequency regions. The area under the high-frequency band is an indicator of which nervous system activity.

9.6 Define stress.

References

1. Thevien, C. (1992). *Discrete Random Signals and Statistical Signal Processing.* Prentice Hall, Englewood Cliffs, NJ.
2. Porat, B. (1994). *Digital Processing of Random Signals.* Prentice Hall, Englewood Cliffs, NJ.
3. Leon-Garcia, A. (1994). *Probability and Random Processes for Electrical Engineering.* Addison-Wesley, Reading, MA.
4. Viniotis, Y. (1998). *Probability and Random Processes.* McGraw-Hill, Boston, MA.
5. DeMeersman, R. (1993). Aging as a modulator of respiratory sinus arrhythmia. *J. Gerontol.* 48: B74–B78.
6. Vander, A., Sherman, J., and Luciano, D. (1994). *Human Physiology.* McGraw-Hill, New York.
7. Axelrod, S., Gordon, D., Ubel, F., Shannon, D., Barger, A., and Cohen, R. (July 1981). Power spectrum analysis of heart rate fluctuations: A quantitative probe of beat-to-beat cardiovascular control. *Science* 213: 220–222.
8. Kamath, M. and Fallen, E. (1993). Power spectral analysis of heart rate variability: A noninvasive signature of cardiac autonomic function. *Crit. Rev. Biomed. Eng.* 21(3): 245–311.
9. Bigger, T., Fleiss, J., Steinman, R., Rolnitsky, L., Klieger, R., and Rottman, J. (1992). Frequency domain measures of heart power variability and mortality after myocardial infarction. *Circulation* 85: 164–171.
10. DeMeersman, R. (1993). Aerobic fitness and heart rate variability. *Am. Heart J.* 125: 726–731.
11. Mathias, J., Muller, T., Perrott, M., and Cohen, R. (November/December 1993). Heart rate variability: Principles and measurement. *ACC Curr. J. Rev.* 10–12.
12. Malliani, A., Pagani, M., Lombardi, F., and Cerutti, S. (August 1991). Cardiovascular neural regulation explored in the frequency domain. *Circulation* 84(2): 482–492.
13. Malliani, A., Pagani, M., and Lombardi, F. (1994). Physiology and clinical interpretation of variability of cardiovascular parameters with focus on heart rate and blood pressure. *Am. J. Physiol.* 73: 3C–9C.

14. Newandee, D. and Reisman, S. (March 1996). Measurement of the electroen-cephalogram (EEG) coherence in group meditation. *Proceedings of the IEEE 22nd Northeast Bioengineering Conference*, New Brunswick, NJ, pp. 95–96.

15. Halperin, J., Cohen, R., and Coffman, J. (1983). Digital vasodilation during stress in patients with Raynaud's disease. *Cardiovasc. Res.* 17: 671–677.

16. Nketia, P. and Reisman, S. (May 1997). Relationship between thermoregulation and hemodynamic response of the skin to relaxation and stress. *Proceedings of the IEEE 23rd Northeast Bioengineering Conference*, Durham, NC, pp. 27–28.

17. Orme-Johnson, D., Dillbeck, M., and Wallace, R. (1982). Intersubject EEG coherence: Is consciousness a field? *Int. J. Neurosci.* 16: 203–209.

18. Maniar, R., Reisman, S., Krivikas, L., and Taylor, A. (May 1995). Normal values for median frequency in upper extremity muscles during isometric contraction at 50% of maximal voluntary contraction. *Proceedings of the IEEE 21st Northeast Bioengineering Conference*, Bar Harbor, ME, pp. 38–40.

Part III

Biomechanics

10

Biomechanics: Bone

10.1 Introduction

This chapter is designed to familiarize the reader with the basic concepts and functional anatomy associated with the study of biomechanics. The application of mechanical principles to the study of the human musculoskeletal system is not new. The origins of these applications date back many hundreds or perhaps thousands of years to the time of the ancient Egyptians and Da Vinci. Today, the field of biomechanics is often a subspecialty of general programs in biomedical engineering. Officially, the field has been present for approximately 40 years if one considers the genesis of official societies, departments, and conferences as the initial starting point of a discipline.

10.1.1 A History of Biomechanics

Perhaps one of the most famous personalities in science was Galileo Galilei (1564–1642). Though trained as a physician, he moved to the fields of mathematics and physics and eventually founded the telescope for which he was more recognized. However, what is less known is that Galileo was one of the first individuals to develop the theories associated with cantilevered beams. In fact, he began to examine bone structures and observed that the cross-sectional area of long bones such as the femur increases faster than length in order to support the weight of large animals. Today, this concept is recognized as an increase in the moment of inertia, with a corresponding stiffness that is proportional to the fourth power of the radius.

The next individual to impact the history of biomechanics was Giovanni Borelli (1608–1679). He was trained as a mathematician and physicist and subsequently studied physiology. His works included a book entitled *On the Movement of Animals*, which was published a year following his death so as not to face the wrath of the Catholic Church. The book displayed, for the first time, detailed images and concepts associated with the common tasks encountered during activities of daily living and included actions such as lifting, jumping, gait, and the equilibrium of human joints. Moving forward from Galileo, Borelli applied the principles of levers in an attempt to understand the function and mechanisms of muscles as they applied to motion of the joints. Another vital piece of information brought to light by Borelli was the concept and determination of the center of gravity (COG) as it applied to the human body. His theory indicated that forward motion occurs only if the COG is propelled beyond the area of skeletal support. In addition, he found that the swinging of limbs was a natural function that humans intuitively perform in an attempt to maintain balance and stability of the COG.

Moving forward approximately 100 years we arrive at Thomas Young (1773–1829), an English physician and physicist. His notable achievements in fields other than mechanics included demonstration of the wave theory of light and initial deciphering of Egyptian hieroglyphics. Working with von Helmholtz, he developed the theory of color. He also discovered astigmatism. In the field of mechanics, he is credited with the definition of the modulus of elasticity, commonly referred to as Young's modulus, which is computed as the ratio of the stress to the strain in the elastic region of a material.

Perhaps one of the most commonly applied principles in musculoskeletal mechanics was hypothesized by German anatomist Julius Wolff (1836–1902). Wolff hypothesized that bone is optimized to provide maximum strength while maintaining the mass to a minimum.

10.1.2 Wolff's Law

> Every change in the form and the function of a bone or of their function alone is followed by certain definite changes in their internal architecture, and equally definite secondary alterations in their external confirmation in accordance with mathematical laws.

> **(Wolff, 1986)**

In a more succinct verbalization, one can state that bone will adapt its internal architecture in response to external constraints and loads. This is a crucial and vital phenomenon that is to be considered when applying biomechanical principles to the design of stabilizing devices in cases of trauma. In addition, the design of arthroplasty or replacement devices such as total hip and knee prostheses must take into account that the adjacent bone is no longer under a true physiological loading transfer condition due to the presence of a nonbiological interface. As we will see in the proceeding sections, Wolff's law will come into play whenever implantation of a manufactured device is introduced into the body. The concept of biologic adaption can also be extended to include a biological reaction due to the presence of a foreign material within the body. In short, the biology of the human body will always react so as to counteract the impact of the implanted device. It is up to the biomechanicians and designers of implant devices to introduce arthroplasty solutions that restore functionality without adversely inducing mechano-biological reactions by the body. Thus far, the collaboration of these two engineering fields has led to successful designs in hip, knee, and shoulder arthroplasty or replacement devices. Designs have also emerged for spinal disc and nucleus replacements. Traditional designs for fixation devices such as screws, plates, rods, and spacers are continuously being improved to reduce their deleterious effects upon the body with respect to mechanical reactions under the premise of Wolff's law. As well the perpetual research in materials science is undertaking the challenge to minimize the biological interactions that occur with introduction of implant devices into the human body.

10.2 Mechanical Concepts

The human body is an efficient and highly specialized biological machine. As such, it obeys the laws of motion and, with that, the laws of Newtonian mechanics. In order to begin the study of biomechanics, a short review of both mechanics and materials is appropriate. The majority of the materials within this section will deal primarily with two-dimensional (2-D) mechanics. While it is recognized that the true motion and loading within the body is in fact three-dimensional (3-D), to include this additional dimension within the computations would require additional mathematical developments and would go beyond the concept of an introductory chapter in biomechanics. In actuality, the 2-D treatment of biomechanical musculoskeletal problems does provide results that are useful and applicable in many circumstances.

To lay some groundwork for definitions commonly utilized, one can start with the terms of dynamics and statics. Dynamics involves the study of forces applied to an object that produce a change in the motion of the object. That is, an unopposed force acting on an object will accelerate the object and result in a nonequilibrium condition. Statics involves the study of the forces applied so as to ensure stability of the object so as to result in no change in the motion of the object. In contrast to many problems associated with the study of classical statics and dynamics, changes in size and shape may occur in the body under the influences of forces. For the purpose of this chapter, we shall define biomechanics as the study of forces and moments upon the human body in the healthy, pathologic, and reconstructed state. Biomechanics can be studied from the joint level (macroscopic) to the cellular level (microscopic). The resulting forces and moments within the body are dependent upon:

- The structure of the materials that comprise the tissue
- The state of health associated with the tissue
- External and internal loading

10.2.1 Structure of Materials

Tissues are optimally designed for specific mechanical functions. The structure of tissues is dependent upon the role that the tissue must play within the body. For example, external loading is borne by bone within the skeleton, smooth motion within the joints is facilitated by articular cartilage, and internal loading via moments is facilitated by the specific orientation and actions of muscles, tendons, and ligaments and is the basis of skeletal locomotion.

10.2.2 Tissue Health

Unlike nonbiological materials, the health of the material is not often a consideration in traditional mechanics. That is, the material properties of an "engineering" material are not affected by the health of the material. One may argue that chemical processes such as corrosion play a role but, in general, if one were to obtain the material properties of two samples of steel fabricated 20 years apart under the same process, the results would be identical barring any external processes. Biological tissues display inherent maturation, healing, and degeneration processes that are unique to each individual and hence, can lead to significant variations in material properties for identical tissues from individuals of identical age and sex. On some level, all biological tissues display the property of viscoelasticity with normal healthy tissues manifesting greater viscoelasticity than that displayed by pathologic tissue of the same kind. Further, biological tissues can be conditioned/optimized (bone, muscle, tendon, and ligament). You may commonly refer to these two concepts as "warming up" prior to commencing any physical activity. It is difficult to image the application of this preconditioning to a material such as steel or concrete, though in the elastomer and plastics field this would certainly be considered prior to material evaluation.

10.2.3 Internal and External Loading

The resulting forces at a joint are termed the joint reaction forces (JRF) and can often be many times greater than body weight. This loading is the result of internal soft tissue restraints and constraints that will overload the joints in order to balance any one specific joint and thus result in overall body balance. That is, these internal soft tissue restraints and constraints precompress the joints prior to the application of external loading such as walking or strain climbing. It is a common misconception that forces and bones act so as to move our respective joints. In reality, musculoskeletal motion is better described as the process by which limbs are rotated about joints as a result of the movements produced by the skeletal flexor/extensor muscles that generate moments. In other words, the musculoskeletal system is a collection of linked lever systems, where the lever arms of muscles and tendons are small and the moment arms of the extremities are large.

10.2.3.1 Musculoskeletal Motion

As was stated, the human body, specifically the musculoskeletal system, is designed to be an efficient machine incorporating mechanical principles for motion and stability. Common terms employed in relation to joint movement are mechanical advantage or disadvantage. To illustrate this point consider Figure 10.1a, where the force is being applied near the hinges of the door (L_1) in an attempt to open the door. In order to swing the door open, a considerably large force will be required. In contrast, the application of a relatively small force at a greater distance from the hinges (L_2) will cause the door to swing easily. This example is analogous to the creation of a moment by muscles in order to move a joint such as the knee. If the quadriceps (or thigh) muscle was located directly upon the tibia (or shin), it would require considerably more force to move the knee than if it were connected to the patella (or knee cap) where a moment arm is created.

For the purpose of establishing the groundwork for further discussions, a brief review of statics is appropriate. One of the primary features in the study of statics is forces, all of which obey Newton's laws. It is important to recognize that all forces are vector quantities and display both magnitude (value) and orientation. The latter property is crucial when one considers the computations involving JRF. When a force is generated at a distance offset from the axis corresponding to a point of articulation, a lever arm results. Such a condition yields a moment or torque about the rotation axis of the joint and generated motion of the joint.

It follows that a brief discussion of acceleration is also warranted. There are two types of acceleration: linear and angular. Linear acceleration is produced by a force occurring along a straight line. In contrast, angular acceleration is produced by a torque about an axis of rotation. These are illustrated in Figure 10.2.

In dealing with the subject of acceleration one often encounters the term "deceleration." This is a misconception and should be avoided. A deceleration by default implies that a "deforce" is being applied. Recall that both acceleration and force are vector quantities. "Deceleration" simply reflects the fact that

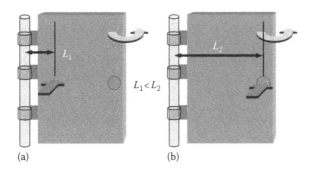

(a) (b)

FIGURE 10.1
(a) A mechanical disadvantage results when generation of large forces is needed to create moments large enough for movement. (b) A mechanical advantage results when small forces applied at a distance (moment arm) are needed to create moments large enough for movement.

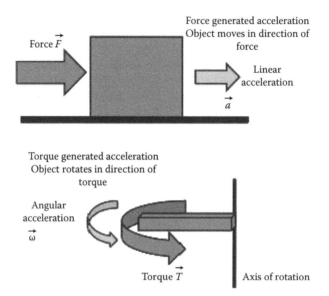

FIGURE 10.2
Linear acceleration results when a force is applied to an object that is sufficient to cause motion. Angular acceleration results when a torque is applied to an object about a rotation axis that is sufficient to cause rotation.

a force has been applied in the direction that opposes the current direction of motion. The opposing forces cancel out via vector addition and the result is a net force in the direction opposite to the motion, which generates the acceleration in that direction. In short, the acceleration has only changed the magnitude and direction (Figure 10.3).

By definition, acceleration is equal to the change in speed divided by the time interval during which the change occurs. This is an important point with respect to biomechanics. Due to the viscoelastic nature of biological tissues, it turns out that, if the length of time that a force acts upon a tissue is lengthened, the effects of the force are diminished and result in a reduction with respect to dynamic loading via slower joint motion. A common example of this is the use of soft heels in shoes.

For many of the biomechanical problems presented in this text, static analyses will be utilized. As such, we must define the equilibrium condition under which these computations take place:

- The sum of all forces and moments acting on an object is zero.
- No linear or angular acceleration is present.
- The system of objects is at rest or at a constant velocity.

With these conditions, it is possible to solve many biomechanical joint conditions using the physical principles acquired through traditional mechanics lectures (Figure 10.4).

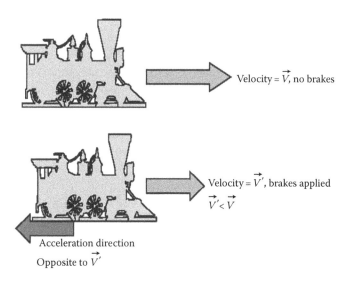

FIGURE 10.3

A reaction force is acting to accelerate (in the opposite direction of travel) to eventually STOP the motion of the train.

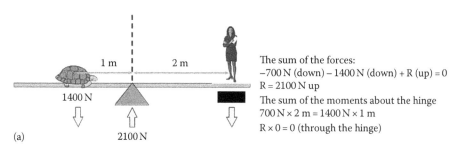

The sum of the forces:
$-700\,N\,(down) - 1400\,N\,(down) + R\,(up) = 0$
$R = 2100\,N\,up$
The sum of the moments about the hinge
$700\,N \times 2\,m = 1400\,N \times 1\,m$
$R \times 0 = 0$ (through the hinge)

(a)

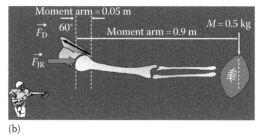

(b)

FIGURE 10.4

Traditional statics techniques can be used to solve biomechanical problems under equilibrium conditions.

10.3 Material Concepts

One of the most useful and simple methods for the mechanical characterization of a material is the conduction of a compressive (or tensile) test. Under the testing conditions, a specimen of the material consisting of a known geometry (typically a cylinder or cube) is subjected to a loading regimen (either compressive or tensile). The particulars of loading rate, clamping, and explicit geometry are generally specified by a unique testing standard developed for the field in which the material is to be utilized. To briefly review the nomenclature, we will use the picture illustrated in Figure 10.5, which depicts a material in the shape of a cylinder under a compressive load.

Consider the cylinder possessing an initial length L_o and cross-sectional area A when a force of magnitude F is applied along the longitudinal axis of the cylinder. Under this compressive force, the cylinder becomes deformed and now possesses a shorter length L.

To eliminate geometric effects, force is divided by the cross-sectional area and deformation (or elongation) is divided by the original length which produces geometrically normalized measures of force and deformation, namely, stress σ and strain ε.

In a more familiar setting, the difference between the initial preloaded length and the final loaded length is represented by $\Delta = L - L_o$. We define the applied stress as the ratio of the force per unit area and is given algebraically as $\sigma = F/A$. The resulting strain is defined as the change in deformation relative to the original length and is stated as $\varepsilon = \Delta/L_o$.

If one were to plot a graph of the applied stress versus the resulting strain, the resulting profile of the curve would be comparable to that seen in Figure 10.6. Such a plot describes the structural behavior of a material under loading.

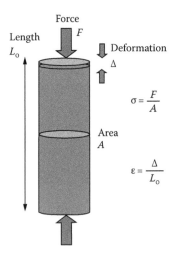

FIGURE 10.5
A compression test for a material using a cylindrical geometry.

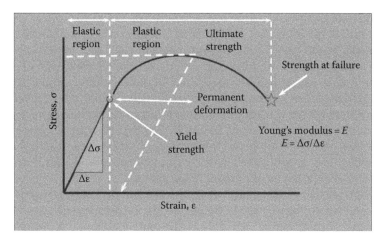

FIGURE 10.6
A typical stress versus strain curve. This curve is characteristic of most materials.

The stress versus strain plot is independent of geometry and therefore describes only the material behavior of the specimen.

The slope of the stress versus strain plot determines the structural stiffness and is directly proportional to both the cross-sectional area A and is inversely proportional to the length L. The slope of the curve in this region is linear and is defined by the ratio of stress to strain. This quantity is termed Young's modulus and is expressed as $E = \sigma/\varepsilon$.

All stress versus strain curves will display a linear or elastic region followed by a nonlinear or plastic region. The point at which the linear and plastic regions coincide is termed the yield point and represents the point at which further application of increasing stress will result in permanent deflection of the material via the initiation of microstructural damage. Plastic deflection represents permanent deflection, that is, there is no recovery or no return to the original dimension.

When the load is released in the elastic portion of the curve, the curve will return to zero deflection. Any deflection beyond the yield point is permanent by definition. When the load is released in the plastic portion of the curve, a line drawn from the point of load release parallel to the slope of the elastic portion of the curve will intersect the x-axis (deflection). The postyield region will terminate with a maximum deflection and stress and is known at the point of ultimate strength.

While these concepts are appropriate for material property determination, when one considers the evaluation of mechanical properties of the elements that comprise the musculoskeletal system, prefabricated samples from tissues such as bone or ligament are often difficult to obtain. Tissues are generally irregular in shape, possess multiple tissue types within a given sample, and display anisotropic properties. As a consequence, many investigations involving biomechanical testing are conducted on entire structures such as a femur (thigh bone) or tendon. Under these testing configurations, the use of a stress versus strain curve is less appropriate. For example, whole bone specimens of equal material modulus but with dissimilar geometries will display a unique and distinct structural stiffness. Bones from different anatomical regions will possess varying cross-sectional areas and lengths, thus making the use of stress versus strain plots of whole bones difficult in determination of the mechanical properties of the bone. Though the stress versus strain curve is less useful in whole or structural applications, the use of the force versus deflection curve under these conditions can provide data useful for mechanical evaluation as illustrated in Figure 10.7.

The characteristics associated with a force versus deflection curve are similar and analogous to those displayed in a stress versus strain curve. Rather than ultimate stress, one defines ultimate load and in a similar manner one would define the modulus as $E = \sigma/\varepsilon$; the stiffness K is defined as $K = \Delta F/\Delta L$ and is the slope of the force versus deflection curve. One characteristic of this curve that is in contrast to the stress versus strain curve is the presence of a "toe" region. This feature is located immediately prior to the elastic region and is representative of specimen "settling." Recall that in the evaluation of material properties, the material is uniformly subjected to loading. However, in the case of these structural tests, the force application is not always uniformly distributed across a specimen, especially in the case of biological tissues.

Unlike engineering materials such as steel, biological tissues all display viscoelastic properties to some degree. While it is easy to visualize such a phenomenon for tissues such as tendons, ligaments, and muscles, the manifestation of viscoelastic effects for harder tissues such as bone has also

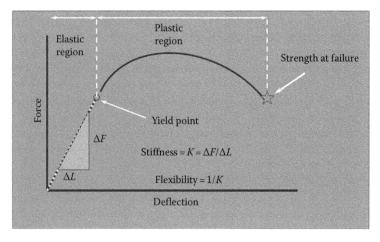

FIGURE 10.7
A typical force versus deflection curve. This curve is characteristic of most materials.

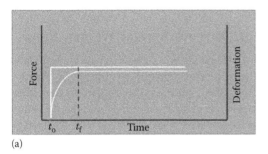

(a)

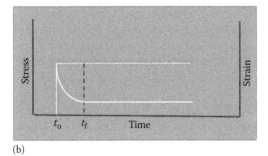

(b)

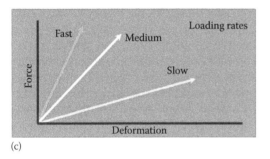

(c)

FIGURE 10.8
Viscoelastic properties displayed by tissues. (a) Creep, (b) stress relaxation, (c) loading rates can influence the rate and magnitude of deformation.

been shown. Viscoelastic parameters include viscosity, creep, and stress relaxation (Figure 10.8). Viscosity is a time-dependent property, which resists shear and flow. The observed deflection depends not only on the applied force but also on the rate of force application. That is the rate of deformation varies directly with the magnitude of the force and there is no elastic return when the force is removed. All human tissues exhibit viscous properties to some degree.

In response to a constant and continuous load, there is an initial deflection due to an elastic response followed by a viscous component of flow called creep or cold flow. Stress relaxation is manifested by a decrease in stress with time in a viscoelastic specimen under constant strain. Viscoelastic materials such as tendons, ligaments, and muscles such as tissues exhibit higher loads to fracture with less elongation when stretched faster.

10.4 Structure and Function of Musculoskeletal Tissues: Bone

Bone is a well-organized tissue, from the modulation of hydroxyapatite crystals at the molecular level to strain patterns of the trabecular framework (Figure 10.9). The synergy of the molecular, cellular, and tissue arrangement provides a tensile strength nearly that of cast iron, with surprisingly low weight. Bone is the primary structural element of the human body. It serves to protect vital organs and provides a framework that permits skeletal motions. Bone differs from engineering structural materials in that it is self repairing and can alter its properties and geometry in response to changes in mechanical demand.

Bone density reductions are known to occur with aging, disuse, and certain metabolic conditions. Increased bone density occurs with heavy exercise and after treatment with certain therapeutic agents. Changes in bone geometry are observed with fracture healing, aging, exercise, and after certain operative procedures.

In the Unites States, more than 250,000 hip and 500,000 vertebral fractures occur each year in people older than 45 years of age, of which 60% of those aged more than 70 years who sustain a hip fracture will die.

Bone consists of two forms: woven and lamellar (Figure 10.10). Woven bone is considered immature or primitive bone (found in the embryo, newborns, fracture callus, and in metaphyseal region of growing bone). This type of bone is also found in tumors, osteogenesis imperfecta, and pagetic bone. Woven, or primary, bone is coarse fibered and contains no uniform orientation of the collagen fibers. It has more cells per unit volume than lamellar bone, its mineral content varies, and its cells are randomly arranged. The relatively disoriented collagen fibers of woven bone give it isotropic mechanical characteristics when tested. The mechanical behavior of woven bone is similar regardless of the orientation of the applied forces.

Lamellar bone begins to form 1 month after birth. By 1 year of age, it is actively replacing woven bone, as the latter is resorbed. By age 4, most

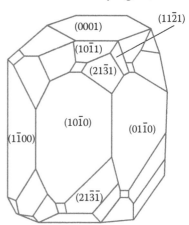

FIGURE 10.9
Structure of the hydroxyapatite crystal.

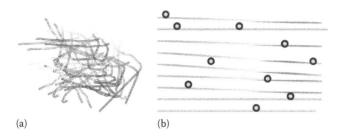

FIGURE 10.10
Bone may be divided into two types: (a) woven and (b) lamellar.

normal bone is lamellar bone. Lamellar bone is a more mature bone that results from the remodeling of woven or previously existing bone. Lamellar bone is found in several structural and functional systems like trabecular lamellae, outer and inner circumferential lamellae, interstitial lamellae, and osteons with concentric lamellae (Figure 10.11).

The highly organized, stress-oriented collagen of lamellar bone gives it anisotropic properties; that is, the mechanical behavior of lamellar bone differs depending on the orientation of the applied forces, with its greatest strength parallel to the longitudinal axis of the collagen fibers. Woven

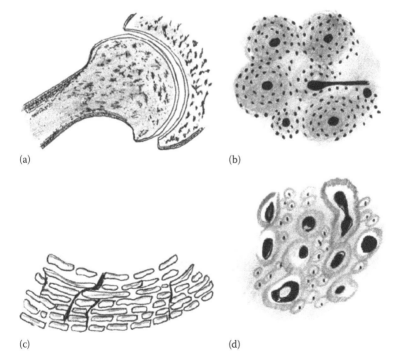

FIGURE 10.11
Mature bone is classified as (a) cancellous, (b) compact (or cortical), (c) plexiform (mostly in animals), or (d) haversian. (www.floridastaugustine.com)

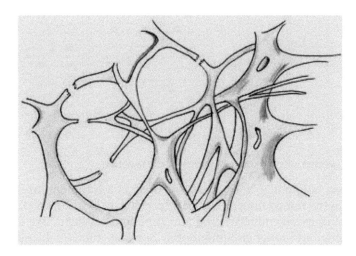

FIGURE 10.12
Branch network of trabecular bone. (www.archive.nyu.edu)

and lamellar bone is structurally organized into cancellous (spongy) bone
and cortical (compact) bone.

Cortical bone has four times the mass of trabecular (cancellous) bone,
although the metabolic turnover of trabecular bone is eight times greater
than that of cortical bone. Bone turnover is a surface event.

Trabecular bone surface area is greater than that of cortical bone. Trabecular
bone is found at metaphysis and epiphysis of long bones and in cuboid bones
(vertebrae). The internal beams of trabecular bone form (3-D) branching lat-
tice aligned along areas of stress. Trabecular bone is subjected to a complex
set of stresses and strains, although compression seems to predominate
(Figure 10.12).

Cortical bone is found as the "envelope" in cuboid bones and as the
"diaphysis" in long bones. Cortical bone is subject to bending, torsional, and
compressive forces. In small animals, there is no special arrangement of the
vascular network in cortical bone; it consists simply of layers of lamellar bone.

In larger animals that experience rapid growth, cortical bone is made up of
layers of lamellar bone and woven bone with the vascular channels located
mainly in the woven bone (plexiform bone). Such an arrangement of vascu-
larity and bone allows rapid growth and the accumulation of large amounts
of bone over a short time.

Haversian bone is the most complex type of cortical bone and is composed
of vascular channels circumferentially surrounded by lamellar bone. This
complex arrangement of bone around the vascular channel in haversian bone
is called an osteon. The osteon is an irregular branching cylinder composed
of a more or less centrally placed neurovascular canal surrounded by cell-
permeated layers of bone matrix. Osteons are usually oriented in the long axis
of the bone and are the major structural units of cortical bone (Figure 10.13).

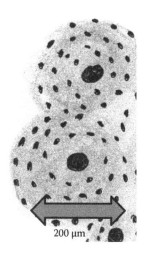

FIGURE 10.13
Osteon in cortical bone. (www.eatingforenergy.ca)

200 μm

Cortical bone is, therefore, a complex of many adjacent osteons and their interstitial and circumferential lamellae. The major types of bone cells are the osteoblasts, osteocytes, and osteoclasts.

The bone-forming cells are the osteoblasts and osteocytes; the principal difference between these cells is location. Osteoblasts line the surface of bone; osteocytes are osteoblasts embedded in a mineralized matrix. All of these cells are derived from the same osteoprogenitor cell line. An osteoblast is defined as a cell that produces type I collagen, is responsive to parathyroid hormone (PTH), and produces osteocalcin (Figure 10.14).

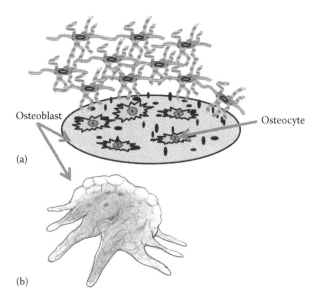

Osteoblast

Osteocyte

(a)

(b)

FIGURE 10.14
(a) Osteocyte and osteoblast cells in bone matrix. (b) Osteoblasts make bone.

Once an osteoblast becomes surrounded by bone matrix, which then becomes mineralized, the cell is characterized by a higher nucleus to cytoplasm ratio. Such a cell is the osteocyte of bone. Osteocytes are arranged concentrically around the central lumen of an osteon and between lamellae. They are uniformly oriented with respect to the longitudinal and radial axes lamellae.

Osteoblasts and osteocytes have extensive cell processes that project and establish contact and "communication" between adjacent osteocytes and the central canal osteons. Osteocytes can metabolically manipulate their environment independent of surface resorption and accretion and is important in the cellular regulation of calcium exchange.

Bone crystals are extremely small and have a surface area of approximately $100 \, m^2/g$ or a total of 100 acres of surface area in the adult human body.

Most of these bone crystals, buried away from the endosteal and periosteal bone surfaces, appear to be unavailable to affect the necessary exchange with extracellular fluid, making it difficult to explain the immediate exchange of bone mineral with the extracellular fluid.

There is, however, a vast surface area on the haversian canal and lacunar walls and an even larger area on the canalicular walls, which in the adult totals 3 acres, where bone mineral exchange with extracellular fluid can take place. Osteoclasts are the major resorptive cells of bone and are characterized by their large size (20–100 μm in diameter) and their multiple nuclei. Osteoclasts are derived from cells of the bone marrow, which are the precursors that also give rise to monocytes and macrophages. Osteoclasts are distinguished from macrophages by virtue of their ability to resorb bone and express certain cell surface markers as well as by their acid phosphatase activity.

It is presumed that at some point during mononuclear cell development, the cell becomes committed to form either a macrophage or an osteoclast. Osteoclasts lie in regions of bone resorption in pits called Howship's lacuna. Osteoclasts appearing some distance from the surface of bone do not have ruffled borders and are called "inactive" or "resting" osteoclasts. Observations show that the ruffled border of the osteoclasts sweeps across the surface of bone. The infolding of the ruffled border appears only over disrupted bone surfaces and increases the surface area of the plasma membrane. The infolds of the ruffled border end in channels and vesicles in the cell cytoplasm, within these lie mineral crystals. Osteoclasts bind to the bone surface through attachment proteins called integrins and resorb bone by isolating an area of bone under the attachment site. The osteoclasts then lower the pH of the local environment by production of hydrogen ions through the carbonic anhydrase system. The lowered pH increases the solubility of the hydroxyapatite crystals and the organic components of the matrix are removed by acidic proteolytic digestion (Figure 10.15).

One may ask as to why osteoclasts are required. Recall that Wolff's law states that bone remodels according to the loading conditions. If loading is not sufficient to warrant the current volume of bone, the osteoclasts will remove the excess bone material, thereby conserving energy and nutrients.

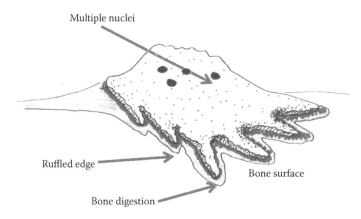

FIGURE 10.15
Osteoclasts remove bone via acidic proteolytic digestion.

Such a condition occurs in astronauts who are subjected to low gravity and hence do not experience loading upon the skeleton.

10.4.1 Material Composition of Bone

Bone is a composite structure consisting of a ductile organic polymer called collagen and a harder brittle mineral component called hydroxyapatite. The combination of these two elements yields a structure that is resistant to loading (strong) and yet is deformable (flexible). The hard hydroxyapatite mineral gives bone excellent resistance to compression and is similar to the properties displayed by common chalk; brittle in bending yet can sustain compression loading. The collagen displays good tensile strength, though the resistance of collagen to compression is minimal due to the inherent low modulus (Figure 10.16).

This two-phase structure allows for resistance to fracture by microcracking in the mineral portion and energy dissipated in the low modulus collagen, thereby permitting resistance to tension, compression, and shear though not in equal magnitudes. In order, bone is strongest in compression (compressive

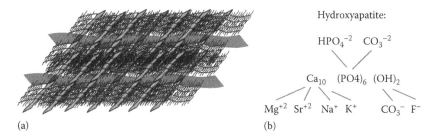

(a)   (b)

FIGURE 10.16
The ductile collagen component (a) of bone allows for flexibility, while the hard hydroxyapatite mineral (b) provides strength.

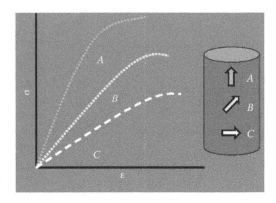

FIGURE 10.17
The strength of bone is dependent upon
the loading direction (anisotropic).

strength of 30 kpsi), displays intermediate strength in tension (tensile strength
of 20 kpsi), and is weakest in shear (Figure 10.17).

In the mineral phase of bone under plastic deformation (i.e., post-yield loading), the osteons in the lamellar structure display both ductility and toughness (area under the stress versus strain curve). However, under severe loads,
the osteons undergo differential slipping, which eventually results in pullout
and separation from the bone matrix and is displayed as micro-fracturing.

Material behavior of bone describes how bone tissue behaves mechanically,
regardless of where the tissue is located in any whole bone. As with any evaluation of fundamental material behavior, mechanical tests are performed on
standardized specimens under controlled mechanical and environmental
conditions. These tests are designed to eliminate any behavior associated
with the specimen geometry. The requirement for the validity of bone data
obtained is that it should be used only for bone with the same microstructure
and the environment as the test specimens.

10.4.2 Cortical Bone: Elastic Behavior

The elastic properties of isotropic materials do not depend on the orientation
of the material with respect to the loading direction and are characterized by
a single modulus (Young's modulus). Most conventional engineering materials, such as 316L stainless steel, are isotropic. In general, bone is anisotropic
and hence displays different mechanical and material properties depending
on the direction of applied loading.

Besides modulus, the other parameter necessary to fully characterize the
elastic behavior of an isotropic material is Poisson's ratio. Poisson's ratio is
a measure of how much a material bulges when compressed (or contracts
when stretched). It is defined as the negative of the strain perpendicular to
the loading direction divided by the strain along the loading direction
(≈0.3 for metals). The elastic properties of anisotropic materials depend on
their orientation with respect to the loading direction. This is true for bone;
however, the elastic properties of human cortical bone display a certain
degree of symmetry, which reflects the bone's osteonal microstructure.

10.4.3 Cortical Bone

The elastic properties of human cortical bone for loading in the plane transverse to the longitudinal axis are approximately isotropic and are substantially different from those for loading in the longitudinal direction, which is parallel to the axis of the osteons (along the longitudinal axis of the bone) (Figure 10.18).

Therefore, human cortical bone usually is considered to be a transversely isotropic material. Young's modulus and Poisson's ratio are used to describe the elastic properties of isotropic material. Five parameters needed to describe a transversely isotropic material are the longitudinal, transverse, and shear moduli and two Poisson's ratios. The modulus of cortical bone in the longitudinal direction is approximately 1.5 times the transverse modulus and greater than five times the shear modulus. Cortical bone bulges more than metals when subjected to uniaxial compression (Poisson's ratio >0.6) (Figure 10.19).

The strength of cortical bone also depends on whether it is loaded in tension, compression, or torsion and thus, represents asymmetry with respect to the strength properties. Consequently, it is not precise to specify the strength of cortical bone with a single number.

Cortical bone is stronger in compression than in tension. For example, the tensile strength in longitudinal loading is approximately 130 MPa, while the corresponding compressive strength is 190 MPa. For transverse loading, the tensile strength is very low (50 MPa), while the compressive strength (130 MPa) is comparable to the tensile strength in longitudinal loading.

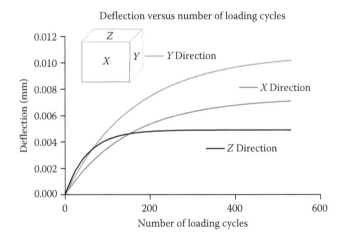

FIGURE 10.18
Cyclic loading of a bone cube in the orthogonal X, Y, and Z directions. Z corresponds to axial loading. Not only does this direction yield the smallest deformation, it also displays the fastest settling.

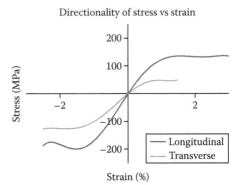

FIGURE 10.19
Stress–strain curves for monotonic tension and compression loading of cortical bone, both in the longitudinal and transverse directions.

These data suggest that cortical bone has adapted to a situation in which compressive loading is greater than tensile loading. This situation is consistent with the combined bending and axial compressive loads that are on the femoral diaphysis during walking. Under these loading conditions, the maximum compressive stresses exceed the maximum tensile stresses.

Because the tensile and compressive yield strengths of cortical bone are approximately equal to the respective ultimate strengths, the maximum stresses that bone can sustain are close to its yield strength. Thus, when cortical bone is loaded close to its yield point, it is also close to fracture. Furthermore, bone loaded by stresses that are just above its yield strength will deform by a relatively large amount compared to its elastic behavior. Therefore, cortical bone undergoes relatively large deformations just prior to fracture.

Materials that absorb substantial energy before failure are classified as tough materials. Biomechanically, toughness is important in traumatic events in which bone is forced to absorb energy, such as automobile accidents or falls. If the energy delivered to the bone is greater than the energy the bone can absorb, fracture will result. Thus, for longitudinal loading, cortical bone is a tough material because it can absorb substantial energy before fracture.

Furthermore, because the ultimate strain for longitudinal loading is greater than the yield strain for transverse loading, cortical bone can be classified as a relatively ductile material under longitudinal loading. Bone is tougher under compressive loads than under tensile loads. Consequently, bone has the lowest resistance to loading regimes that cause tensile stresses in the transverse direction. For example, those stresses that can arise as "hoop" stresses when a cementless hip prosthesis is press fit into the canal of the femur (thigh bone).

Cortical bone displays viscoelastic behavior because its mechanical properties are sensitive to both the strain rate and the duration of the applied loads (Figure 10.20).

The in vivo strain rate for bone can vary by more than an order of magnitude in the course of daily activities such as slow walking (strain rate—$0.001\,s^{-1}$) and brisk walking (strain rate—$0.01\,s^{-1}$).

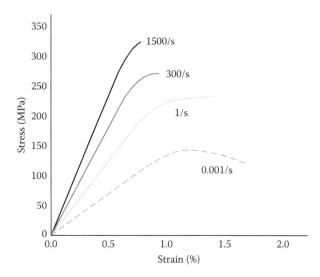

FIGURE 10.20
Due to the viscoelastic properties of bone, a strain rate dependency is observed.

The increase in the initial slope of the stress–strain plot as the strain rate increases indicates that cortical bone has a higher modulus at higher strain rates. Cortical bone exhibits a ductile to brittle transition as the strain rate increases. However, for the range of strain rates typical of more normal activity (less than $0.1\,\mathrm{s^{-1}}$), ductility increases (ultimate strain increases) as the strain rate increases.

If bone tissue is subjected to a constant stress for an extended period of time, it will continue to deform (creep). Cortical bone exhibits the same three characteristic stages of creep behavior as do many conventional engineering materials. In the primary stage, specimen strain continues after loading and the creep (increase in strain) rate gradually decreases. In the secondary stage, there is a lower, usually constant creep rate. Finally, in the tertiary stage, there is a marked increase in the creep rate just before creep fracture (Figure 10.21).

The modulus and strength properties of cortical bone progressively deteriorate with aging for both men and women. The slope of the stress–strain curve after yielding increases by 8% per decade. The decrease in energy absorption by approximately 7% per decade results mainly from reductions in the ultimate strain. Taken together, these data indicate that the cortical bone material in the human femur becomes less stiff, less strong, and more brittle with aging.

In vivo cortical bone is exposed primarily to repetitive, low intensity loading that produces lower stress levels than those required to fracture bone during monotonic loading. This cyclic loading of bone can result in damage at the microstructural level. Not all loads result in immediate damage to bone, but, if damage does occur and accumulates over time, bone strength is reduced.

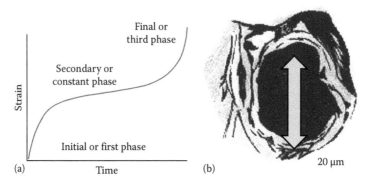

FIGURE 10.21
(a) Creep curve. (b) Osteon has been pulled out of the matrix. (www.doitpoms.ac.uk)

Fatigue properties of cortical bone may be the stimulus for bone remodeling. Physiologic levels of loading can induce fatigue damage and failure in vitro; therefore, bone remodeling, which occurs continuously in vivo, may repair the damage caused by relatively low intensity loading, such as occurs in walking. It has been hypothesized that bone remodeling occurs to repair microcracks that form in bone as a result of the repetitive loading of daily activity. The fatigue life of bone is better correlated with strain than with stress. Fatigue properties are temperature dependent and decrease as temperature increases.

10.4.4 Trabecular Bone

There is a large variation in density for trabecular bone. Both spatial and temporal variations in trabecular bone density can occur as a result of changes in anatomic location and age. For example, trabecular bone material properties within the proximal tibia can vary by up to two orders of magnitude because of changes in density alone. To further complicate matters, material properties of trabecular bone also depend on its architecture, which like its density, depends on the anatomic site and, to a lesser extent, on age.

While cortical bone is essentially a low porosity solid, trabecular bone is best described as an open-celled porous foam (Figure 10.22). Depending on the type and orientation of these basic cellular structures, the mechanical properties can vary by at least an order of magnitude. In general, the modulus of trabecular bone can vary from 10 to 2000 MPa, depending on the anatomic site and age.

Trabecular bone is less stiff than cortical bone (modulus 17,000 MPa). It has been demonstrated (Carter and Hayes) that the modulus E of trabecular bone in any loading direction is related to its apparent density ρ_a through

$$E = 3790 \left(\frac{d\varepsilon}{dt} \right) e^{0.06} \left(\rho_a \right)^3 \left(\text{bovine trabecular bone} \right)$$

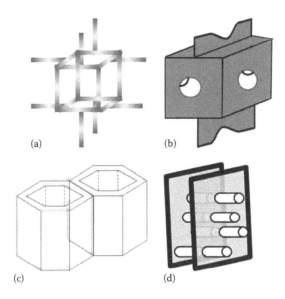

(a)
(b)
(c)
(d)

FIGURE 10.22
While there are several foam models that have been used to represent trabecular bone it is best described as an open-celled porous foam. (Redrawn from Gibson, L.J. and Ashby, M.F. *Cellular Solids: Structure and Properties*, 2nd Edition, Cambridge University Press, Cambridge, 1997.)

$$E \propto \left(\frac{d\varepsilon}{dt} \right) e^{0.06} \left(\rho_a \right)^2 \text{ (human trabecular bone)}$$

$$\sigma_{ult} = 68 \left(\frac{d\varepsilon}{dt} \right) e^{0.06} \left(\rho_a \right)^2 \text{ (bovine/human trabecular bone)}$$

Consequently, a 25% reduction in density, as has been observed in elderly cadaveric vertebrae, results in a 56% decrease in modulus. The main mechanism of deformation in these open-celled foams is bending of individual trabeculae, even though the bulk specimen is compressed without bending. There are large variations in trabecular bone architecture over different anatomic sites.

The shape of the basic cellular structure can also affect the Poisson's ratio of trabecular bone. However, theoretical values of Poisson's ratio for open-celled foam materials such as trabecular bone may be negative (the material contracts when compressed) or the value becomes greater than 1 (metals are less than 0.5).

Mean values of Poisson's ratio for trabecular bone is in the range 0.06 ± 0.03 to 0.95 ± 1.29. The orientation of individual trabeculae is controlled mainly by the direction of the forces applied to the skeleton, according to the widely cited, but still qualitative Wolff's law. Modulus may vary as much as a factor of 10 depending on the direction and apparent density. In order to arrive at a measure of the true modulus for trabecular bone, mechanical data is to be correlated with stereologic description of the architecture. As a result, the individual results are not generally applicable.

10.4.5 Trabecular Bone: Viscoelastic Behavior

Because very few experiments have been performed in which constant loads have been held on trabecular bone for extended periods of time, its creep behavior is poorly understood. However, both modulus and strength of trabecular bone have been shown to be weakly dependent on strain rate (power law). Bone marrow may also play a role in trabecular bone's load-carrying capacity but only at very high strain rates (above $10 s^{-1}$, as in gunshot wounds). Restricting marrow flow through the intertrabecular spaces, in very high strain rates, can enhance the mechanical properties.

For tensile loading, failure occurs by fracture of the individual trabeculae. As more trabeculae fracture, the specimen can take less and less load, until finally complete fracture occurs (similar to reinforced concrete), which is typical of engineering materials designed to resist primarily compressive forces. At physiologic rates, the marrow plays a negligible role in the viscous behavior of trabecular bone.

Comparison of the compressive and tensile behaviors of trabecular bone indicates that the postyield load-carrying capacity of trabecular bone is high for compression and almost negligible for tension. Thus, trabecular bone loaded beyond the ultimate strength in compression can still carry substantial load; it will not significantly overload surrounding trabecular bone and failure will not spread to the surrounding tissue. In this case, the surrounding tissue must carry the full load and thus may be overloaded. Although local failure of trabecular bone in compression is not likely to lead to failure of a whole bone, local failure in tension could have catastrophic consequences (Figure 10.23).

The reductions in density depend on a number of factors, including gender and anatomic site. In the lumbar spine, direct measurements have shown a decrease in trabecular bone density of approximately 50% from ages 20 to 80 years. These studies have demonstrated that, with decreasing density, the number and thickness of the trabeculae decrease, while the size of the intertrabecular spaces increases. Regardless of density, the

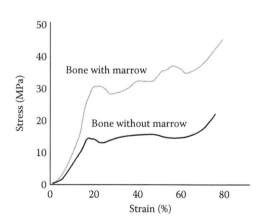

FIGURE 10.23
Effects of marrow. The viscoelastic effects effectively increase the strength of trabecular bone.

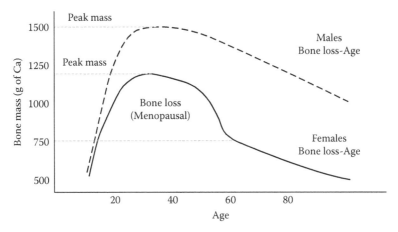

FIGURE 10.24
Effects of age on bone mass.

number of horizontal trabeculae in the lumbar spine is less than the number of vertical trabeculae, and the number of vertical trabeculae decreases at twice the rate of the horizontal trabeculae (Figure 10.24).

This loss of trabeculae may be more damaging to the structural integrity of trabecular bone than mere thinning because lamellar new bone can be formed only on existing surfaces, making complete loss of a single trabecula irreversible. Reduced resistance to failure caused by buckling of individual trabeculae has been referred to as a "triple jeopardy," because of three independent factors:

- Reduction in number
- Decrease in thickness
- Increase in length

10.4.6 Fracture

It has been suggested that these cracks are caused by fatigue loading and that such cracks may play a role in bone remodeling, age-related fractures, subchondral collapse after aseptic necrosis of the femoral head, degenerative joint disease, and other bone disorders. The only fatigue data available for trabecular bone are for bovine bone. Preliminary data suggest that the uniaxial compressive strength of trabecular bone can be reduced by an order of magnitude by 1,000,000 cycles of loading. The mechanism for fatigue cracking in trabecular bone involves fracture and buckling of individual trabeculae and, as such, differs for cortical bone where cracks accumulate within the bone matrix.

In the following example, we will consider the fast fracture of cortical bone [3]. Imagine an inflated balloon pricked with a pin. If viewed in slow motion one will observe that the rubber of the balloon will rupture and result in a "fast fracture." Failure in the rubber occurs despite the fact that the stress within the balloon was lower than the yield stress of the rubber. Obviously, the pin creates a defect in the balloon wall. Rupture is determined by the propagation of the defect. That is, if the defect grows, then a rupture and subsequent "pop" will occur. If the defect does not propagate, then rupture will not occur. Consider what would happen if the balloon was only slightly inflated. Contact with a pin would not cause the balloon to break. The determination as to the progression of the defect is best addressed through the use of an energy balance at the surface.

In the case of the balloon, the air pressure in the balloon provides the energy by doing work on the balloon wall. This work can be used to change the elastic energy stored in the rubber or it can be used to create a tear.

Define [3]:

∂U^{el}, the increase in the elastic energy of the rubber comprising the balloon.

∂W, the incremental work done on the rubber by external loads, that is, the pressure in the balloon.

∂U^{cr}, the energy required to increase the defect size, that is, the energy needed to propagate the defect.

By definition, a defect will only propagate if the increase in work exceed or is equal to the sum of the elastic energy increase in the rubber and the energy required to propagate the defect. If this condition is not fulfilled, there is insufficient energy to grow the defect. Algebraically, this can be stated as

$$\partial W \geq \partial U^{el} + \partial U^{cr}$$

Suppose one has a rectangular block of thickness, t, with a preexisting crack of length, a (Figure 10.25) [3]. You may have this preexisting crack because you

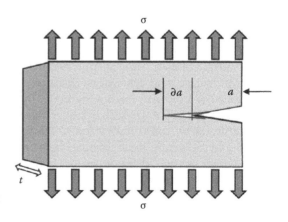

FIGURE 10.25
Rectangular block with preexisting crack.

went running last night and decided to run the extra mile that you would not normally run. As well, let us assume that the block is subjected to a constant tensile stress, σ.

One can define the fracture toughness, G_c, to be the energy absorbed by the material per unit area of the *new* crack surface created. It is important to realize that the crack surface is NOT the visible area enclosed by the crack, but is the area of the actual crack interface. In order to enlarge the crack from length a to a length ∂a will therefore require $G_c t\,\partial a$ units of energy or ∂U^{cr}.

∂U^{el} and ∂W depend on the specifics of the problem as they are unrelated to the material properties. We shall assume that the edges of the block are clamped so as to remain stationary. Although this clamping results in externally generated stresses, since the edges are immovable, this results in zero work being done ($\partial W = 0$) [3], and therefore the energy equation for defect propagation may be written as

$$\partial U^{el} + \left(G_c t\,\partial a\right) \le 0.$$

In the case of $\partial U^{el} < 0$, the crack propagation will release elastic energy. Under uniaxial tension, the energy stored per unit volume is given by the quantity of $\sigma^2/2E$, where σ is the resulting stress due to the applied tension and E is Young's modulus.

The crack propagation causes a local stress relaxation in the material around the crack. The stress decays continuously from the crack interface to its relieved value far from the crack. For simplicity, we will say that the stress is equal to zero at some finite region surrounding the crack and is relieved outside this finite region (Figure 10.26).

For a defect of length a, the stress relief region will have a surface proportional to $1/2\pi a^2$ and results in a sample volume $V = 1/2\pi a^2 t$. The stress energy relief is then given by [3],

$$U^{el} = -\left(\frac{\sigma^2}{2E}\right)\left(\frac{1}{2}\pi a^2 t\right) = \frac{-\sigma^2 \pi a^2 t}{4E}$$

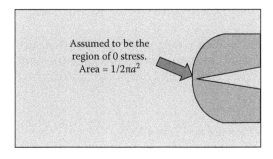

Assumed to be the region of 0 stress. Area = $1/2\pi a^2$

FIGURE 10.26
Crack propagation.

As the defect grows by an amount ∂a, then U^{el} will change by

$$\partial U^{el} = \frac{\partial}{\partial a}\left(\frac{-\sigma^2\pi a^2 t}{4E}\right) = \left(\frac{-\sigma^2\pi a t}{2E}\right)\partial a$$

and the defect propagation equation becomes

$$\frac{1}{2}\sigma^2\pi\alpha \geq G_c E$$

This oversimplification with respect to how the stress is actually relieved around a crack returns values that are approximately 50% of those physically recorded [3]. A significantly more complex derivation shows that the critical condition for crack propagation is actually given by

$$\sigma^2\pi a \geq G_c E$$

Regardless, of the 1/2 factor, it should be noted that the quantity $\sigma^2\pi a$ contains only material properties and is therefore by definition a material property and may be recognized as the fracture toughness designated as $K_c = \sqrt{EG_c}$. One can also define a stress intensity factor, $K = \sigma\sqrt{\pi a}$. The defect propagation can then be written as $K \geq K_c$. Notice that the intensity factor, K, depends on both the imposed stress σ and the crack size a. If the combination is too large then fracture occurs.

10.4.6.1 Fatigue Fracture

In general, crack propagation in bone is generally observed at stress levels considerably less than those that cause catastrophic or fast fracture. Under normal activities of daily living, we apply stress under a repeated or cyclic loading and unloading pattern. Suppose that the applied stress σ varies periodically between σ_{max} and σ_{min} rather than being a constant value as in the previous derivation. The stress intensity factor, $K = \sigma\sqrt{\pi a}$ will also vary cyclically with the stress instantaneous values of σ. In addition, each time the stress completes a single cycle, the crack length a will increase by a small amount, ∂a (Figure 10.27) [3].

For most materials, after a rapid initiation phase, cracks grow at a steady state until suddenly a fracture occurs. That is, the crack grows at a constant rate until it reaches a critical size. Steady state crack growth can be expressed by Paris' Law

$$\text{Rate of defect size increase} = \frac{da}{dN} = C(\Delta K)^m$$

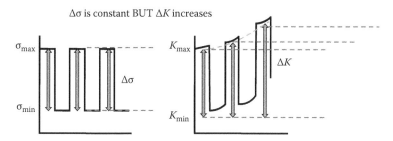

FIGURE 10.27
Rate of fracture size increase.

where
    N is equal to the number of cycles
    the constants C and m are empirically determined values

Consider an initial crack of length equal to a_o [3]. One can integrate Paris' law to determine how many cycles (N) it would take to grow the crack to a final size equal to a_f and hence lead to fracture (Figure 10.28).

$$\frac{da}{dN} = C(\Delta K)^m$$

$$\frac{da}{C(\Delta K)^m} = dN$$

$$N_f = \int_0^{N_f} dN = \int_{a_o}^{a_f} \frac{da}{C(\Delta K)^m}$$

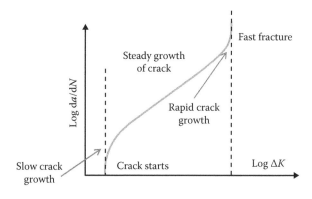

FIGURE 10.28
Crack propagation. (Redrawn from Eithier, C.R. and Simmons, C.A. *Introductory Biomechanics. From Cells to Organisms,* Cambridge University Press, Cambridge, 2007.)

Substituting $\Delta K = \Delta \sigma \sqrt{\pi a}$ yields

$$N_f = \left[\left[\frac{1}{C(\Delta\sigma)^m \pi^{1/2m}} \right] \frac{a^{1-m/2}}{1-(m/2)} \right]_{a_o}^{a_f}$$

That is, the number of cycles to failure is related to the defect length, the changes in stress, and several constants.

References

1. Wolff, J. *The Law of Bone Remodelling*, Springer, Belin Heidelberg, New York, 1986 (translation from the German 1892 edition).
2. Gibson, L.J. and Ashby, M.F. *Cellular Solids: Structure and Properties*, 2nd Edition, Cambridge University Press, Cambridge, 1997.
3. Eithier, C.R. and Simmons, C.A. *Introductory Biomechanics. From Cells to Organisms*, Cambridge University Press, Cambridge, 2007.

11

Structure and Function of Musculoskeletal Tissues, Connective Tissue, and Spine

Connective tissue consists of ligaments, tendons and muscle, and cartilage. These tissues are essential for joint motion and consist primarily of an extracellular matrix (ECM) comprising collagen and proteoglycans distributed within water and cells. As seen in Table 11.1, the type of connective tissue is defined by the relative quantities of collagen, proteoglycans, and water.

11.1 Collagen

There are many types of collagen that have been discovered and while many types exist there are only a few that are common in the musculoskeletal system. Type I is most commonly found in ligaments, tendons, and bone, while Type II is most common in cartilage. Regardless of collagen type, all possess the same basic structure. Basic unit of cartilage is the tropocollagen or the collagen triple helix that contains three α chains about 1.5 nm in diameter and 280–300 nm in length and held together by hydrogen bonds. The helix is bonded covalently to other helix molecules at 1/4 lengths by covalent cross-link bonds. When five tropocollagens are assembled together, they form a microfibril with 50–500 nm microfibril assemblies called fibrils. Figure 11.1 shows the basic collagen structure.

11.2 Ligaments

Ligaments are connective tissues that span from one bone to another bone. They participate in, and to some extent actually guide, the movement of the joints that they span. One way in which this joint movement is controlled is via maintenance of joint congruency. That is, the tensile forces within the ligaments generate a net compressive force across the joint and thus result in the joint following the path of rotation defined by the joint geometry.

Ligaments are composed predominantly of Type I and some Type III collagen as well as elastin fibers and proteoglycans. The relative percentage of these components is dependent upon the required function of the ligament.

TABLE 11.1

The Relative Quantities of Collagen, Proteoglycans, and Water Define the Type of Connective Tissue

Tissue Type	Collagen	% Dry Weight Elastin	Proteoglycan	Hydrated Sample Weight % H$_2$O
Tendon	Mean 80 (Range 75–85)	<3	Mean 1.5 (Range 1–2)	Mean 67.5 (Range 65–70)
Ligament (extremities)	Mean 77.5 (Range 75–80)	<5	Mean 2 (Range 1–3)	Mean 60 (Range 55–65)
Articular cartilage	Mean 62.5 (Range 50–75)	Minimal	Mean 25 (Range 20–30)	Mean 70 (Range 60–80)
Fibro-cartilage	Mean 70 (Range 65–75)	Minimal	Mean 2 (Range 1–3)	Mean 65 (Range 60–70)

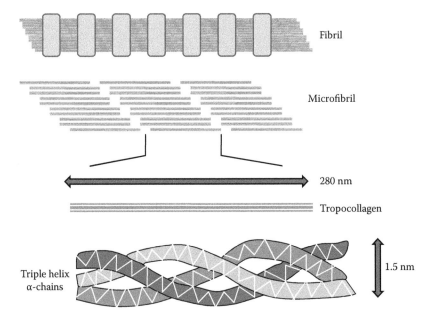

FIGURE 11.1
Structure of collagen molecules. (Redrawn from Nordin, M. and Frankel, V.H. *Basic Biomechanics of the Musculoskeletal System,* Lea and Febiger, Malvern, PA, 1989.)

Those ligaments that are required to stretch a significant amount will contain more elastin while those required to be stiffer will contain a considerable amount of collagen as in the anterior cruciate ligament (ACL).

The collagen within the central portion is formed of fibers that aggregate to form fascicles (Figure 11.2). In order to provide tensile resistance, the collagen fibers are for the most part oriented along the long axis of the ligament. It should be recognized that approximately 2/3 of ligament weight is water.

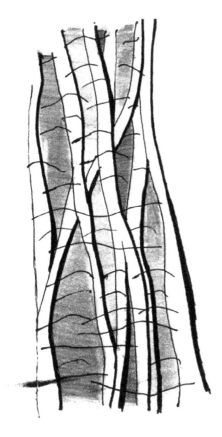

FIGURE 11.2
Structure of collagen in a ligament. (Redrawn from Nigg, B.M. and Herzog, W. *Biomechanics of the Musculo-Skeletal System*, John Wiley, Chichester, UK, 1989.)

As a matter of cellular identification, ligament cells are referred to as fibroblasts. While they orient themselves along the ligament axis, they do in fact display variations in morphology, size, and density of dispersion.

11.3 Tendon

Tendons connect muscles to bone and provide the connection from the power generator (muscle) to the joint for locomotion (recall that ligaments only guide motion, they do not generate or transfer the power to move the joint). The structure of tendons is similar to that of ligaments with respect to overall architecture. Some tendons are surrounded by a sheath that is filled with synovial fluid in order to aid in gliding as in the finger joints. Where tendon connects the muscle is termed the myotendinous junction and, when the insertion of the tendon is into the bone, the site is called the osteotendinous junction. The mid-regions of the tendon display substantially more fibrocartilages that become more calcified as it nears the point of bone insertion

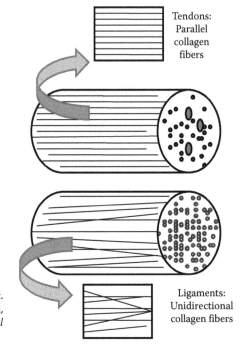

FIGURE 11.3
Fiber orientations in tendon and ligament.
(Redrawn from Nordin, M. and Frankel,
V.H. *Basic Biomechanics of the Musculoskeletal
System*, Lea and Febiger, Malvern, PA, 1989.)

(Figure 11.3). At both insertion points the collagen fibers within the tendon
become more aligned with the loading direction (Figure 11.4).

11.4 Cartilage

Cartilage is divided into three main types: hyaline cartilage, elastic cartilage,
and fibrocartilage. These types are differentiated based upon the proportions
of proteoglycans and elastin. Hyaline cartilage is most common within the
musculoskeletal system as it is found in joints where motion is demonstrated.
Examples of these joints include the hip, knee, and shoulder. Due to the presence of this type of cartilage in the articulating joints, this cartilage is often
termed articular cartilage. Articular cartilage cells are termed chondrocytes.

Within the joints containing this articular cartilage, a tissue capsule surrounds the entire joint and is filled with a lubricating fluid called synovial
fluid. This fluid surrounds and resides within the moving cartilage layers
effectively reducing the coefficient of friction between cartilage layers to
0.005. To place this in perspective, the coefficient of friction for ice on ice is
approximately 0.05. Mature cartilage does not contain nerves, blood vessels,
or lymphatics. Cartilage is both nourished and drained only by diffusion or
convection processes via the interaction with the surrounding synovial fluid.

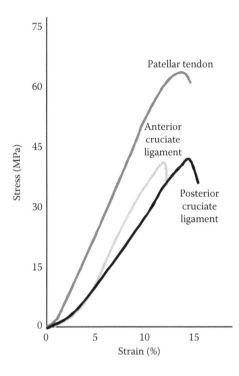

FIGURE 11.4
Strength of tendons and ligaments is dependent upon the desired function, collagen, water and proteoglycan content. (Redrawn from Butler, D.L., Kay, M.D., and Stouffer, D.C. *J. Biomech.* 29, 1131, 1996.)

Elastic cartilage is found in locations such as the external ear and larynx. Due to its substantial elastin content, it is quite flexible when compared to articular cartilage. Fibrocartilage contains thick dense layers of collagen fibers between layers of hyaline cartilage and is found in the meniscus, discs of the spine, and in the transition regions between bone and tendon insertions.

11.5 Muscle

There are several types of muscles present in the body (Figure 11.5). Muscles can be classified as voluntary and involuntary. Involuntary muscles include smooth muscles such as those that surround the lumen of tubes and generate peristaltic waves as in the gastrointestinal tract. Another type of involuntary muscle is cardiac muscle that is exclusive to the heart. In this section, we will address the voluntary muscles that comprise the skeletal muscles.

In normal healthy individuals, skeletal muscle constitutes between 40% and 45% of the body weight. As pointed out in the previous section, all skeletal muscle is connected to bone that comprises a joint via tendon insertions. Skeletal muscle is the power generator that is responsible for locomotion. It must be recognized that with all of the connective tissues discussed, the only mechanical force that can be generated within these tissues is tension.

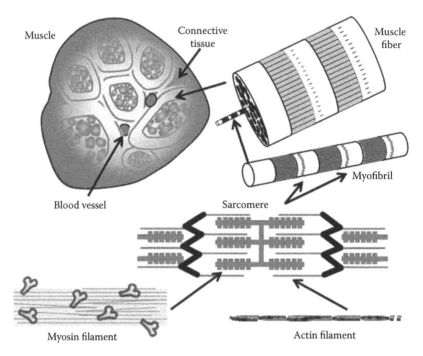

FIGURE 11.5
Structure of muscle. (Redrawn from Vamder, A.J., Sherman, J.H., and Luciano, D.S. *Human Physiology: The Mechanisms of Body Function*, 4th edition, McGraw-Hill, New York, 1985; Desaki, J. and Uehara, Y. *J. Neurocytology* 10, 101, 1981.)

The tension or strength within skeletal muscles is achieved via contraction of the muscle. These contractions are controlled by nerves located within the muscle fibers.

The strength of muscle contraction is dependent upon several factors. These include health of the tissue as well as the cross-sectional area of the muscle. For any muscle, there is an ideal resting length. In addition, for a fixed length, there exists an ideal length for maximum tension that may be generated within the muscle. The complete tension versus length profile for skeletal muscle consists of two components. There exists a passive component that is founded upon the native resistance of the tissue that comprises the muscle. The second component termed the active component is related to the muscle length and is controllable. Graphically, these two components are illustrated in Figure 11.6 with the resulting muscle tension versus length profile.

The efficiency of a muscle is dependent upon several elements that involve the orientation of the individual muscle fibers, the actual desired function and the tendon insertion locations with respect to the joint. To illustrate this point, in humans, the insertion of the biceps muscle and tendon is placed relatively close to the elbow joint. Such a configuration provides a considerable amount of mobility and speed (just ask a baseball pitcher). In contrast,

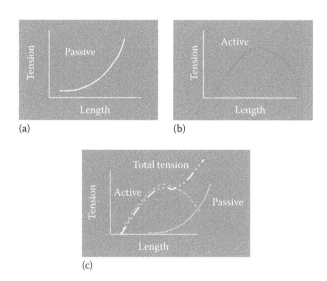

(a)

(b)

(c)

FIGURE 11.6
Muscle displays (a) a passive tensile resistance and (b) an active tensile resistance to loading. (c) The overall muscle tension performance of the muscle is the net combination of the passive and active tension profiles.

consider a primate such as a gorilla or chimpanzee. The location of the biceps insertion is located at a significantly greater distance from the elbow joint. We have all observed primate hanging from tree branches for what seems an extensively long period of time as compared to humans. The increased moment arm due to the increased distance of the insertion results in a large increase in power (Figure 11.7). However, under this condition the speed and mobility are sacrificed.

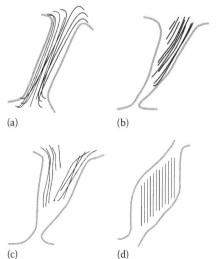

(a)

(b)

(c)

(d)

FIGURE 11.7
Muscles are optimized for function based on fiber orientation and tendon insertion locations.

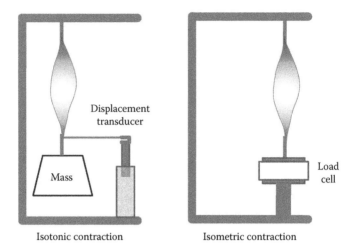

FIGURE 11.8
Muscle contraction: (a) isotonic (constant load) and (b) isometric (constant length). (Redrawn from Eithier, C.R. and Simmons, C.A. *Introductory Biomechanics. From Cells to Organisms.* Cambridge University Press, Cambridge, 2007.)

Muscle contraction can be evaluated under two scenarios. Isotonic contraction evaluation is performed by application of a prescribed known tension and recording the resultant muscle length. An alternative method is to apply isometric contraction. Under this setup, the muscle is maintained at a constant length while the force generated during contraction is recorded (Figure 11.8).

11.6 Muscle (Connective Tissue) Response

Although this example is being applied to muscle tissues, adjustment of some of the specific parameters to reflect properties of other tissues such as ligaments and tendons can be employed. Consider the following model (Figure 11.9) for our discussion of muscle response to loading under isometric contraction. In this model, we shall define several elements. We shall include a dashpot that represents a damping or absorbing component and possesses a damping coefficient of η. Included in the model is another element represented by a spring and having a spring constant designated by K. Finally a force generator that applies a constant force is included. The resulting tension in the muscle is designed by the variable T. We can identify three locations 1, 2, and 3 from which two muscle lengths can be applied. From location 1 to location 2 shall be designated

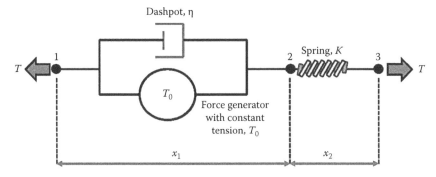

FIGURE 11.9
Model for muscle contraction.

as muscle length x_1 and from location 2 to location 3 shall be assigned muscle length x_2.

In this computation, we shall assume that the mass of the muscle is negligible and that the tension T is initially equal to zero until it instantaneously reaches a value of T_0 and lasts for a duration of C (Figure 11.10).

The force produced by the dashpot is given by $F_d = \eta(dx_1/dt)$. The total muscle tensile force is represented as a function of the contact tension, the time of activation, the spring constant and the damping coefficient. Algebraically the muscle tension can be represented as

$$T = f\left(T_0, C, K, \eta\right)$$

If the initial tension prior to activation is zero, one can ask what the muscle tension will be at a time beyond the onset of isometric contraction. Recall

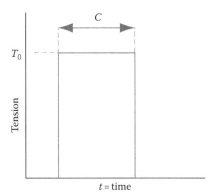

FIGURE 11.10
Definition of applied force, T, in Figure 11.9.

that under isometric contraction where the muscle length is constant, therefore, one can state that

$$x_1 + x_2 = \text{constant}$$

and therefore

$$\left(\frac{dx_1}{dt}\right) + \left(\frac{dx_2}{dt}\right) = 0 \tag{11.1}$$

Looking at location 2, one can state that

$$T - T_{\text{spring}} = m_{\text{spring}} a_{\text{spring}}$$

but we have assumed that the muscle mass is negligible and thus $m_{\text{spring}} = 0$ and therefore

$$T = T_{\text{spring}} = K\left(x_2 - x_{2\text{initial}}\right) \tag{11.2}$$

where $x_{2\text{initial}}$ is the spring's original length. If we look at the dashpot and the force generator when the muscle is active (i.e., $0 \le t \le C$), the muscle tension is given by

$$T = T_0 + F_d = T_0 + \eta\left(\frac{dx_1}{dt}\right) \tag{11.3}$$

If one rearranges Equations 11.1 and 11.3 to isolate the time derivative of x_1 and computed the time derivative of the tension in Equation 11.2, one is left with three equations and three unknowns and the result is a first-order differential equation given by

$$\frac{1}{K}\frac{dT}{dt} + \frac{T - T_0}{\eta} = 0 \tag{11.4}$$

If one assumes that the initial tension T_0 in the muscle is zero, the solution to the above equation is given by [6]

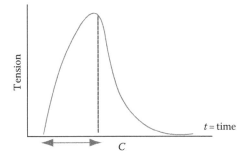

FIGURE 11.11

Graphical depiction of solution to Equation 11.4.

$$T(t) = T_0\left[1 - e^{-\kappa t/\eta}\right] \quad \text{for } 0 \leq t \leq C$$

$$T(t) = T_0\left[1 - e^{-\kappa t/\eta}\right]e^{-\kappa(t-C)/\eta} \quad \text{for } t > C$$

Graphically, this is seen in Figure 11.11.

11.7 Extremities

The previous sections have established the function and structure of some of the elements comprising the musculoskeletal system. We can now move from the tissue level to the structure or joint level. This section will look at the extremities of the human body. More specifically, the knee, hip and shoulder joints will be examined. The characteristics of each joint are unique with the individual configuration of bone geometry and connective tissue contributing to the overall biomechanical function of the joint.

11.7.1 Knee

Often, we consider the knee as a hinge due to the overall gross motion of rotation in the plane of motion. In reality, the knee does in fact display all six degrees of freedom. That is, the true motion of the knee is such that it manifests three rotations and three translations (Figure 11.12). The rotations associated with the knee are defined as flexion/extension (about the X axis), varus/valgus (about the Y axis), and internal/external rotation (about the

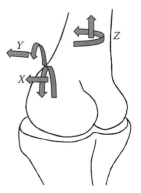

FIGURE 11.12
The knee displays six degrees of freedom—three translations
and three rotations. (scielo.br)

Z axis). With respect to the translations, these are termed anterior/posterior
(along the X axis), compression/distraction (along the Z axis), and medial/
lateral (along the Y axis).

Motion	Value
Flexion/extension	$-15°$ to $-140°$
Varus/valgus	$6°$–$8°$
Internal/external rotation	$25°$–$30°$
Anterior/posterior	5–10 mm
Compression/distraction	2–5 mm
Medial/lateral	12 mm

The knee is the largest synovial joint (a joint containing a capsule containing synovial fluid) in the body and is composed of three bones and three
joints (Figure 11.13). The three bones of the knee are comprised of the femur
(thigh bone), tibia (shin bone), patella (knee cap), and, to a lesser degree, the
fibula.

The knee joint itself is made up of the tibiofemoral joint, which involves
articulation of the femur upon the tibia and is comprised of a medial compartment and a lateral compartment. The second joint is called the patellofemoral joint and involves the articulation of the patella upon the femur.
The final joint comprising the knee is the tibiofibular joint. While it does
play a role in knee function, it is not often included as a component of the
"true" knee joint.

11.7.1.1 *Femur*

The femur is both the longest and strongest bone in the body with a shaft
displaying a nearly cylindrical and relatively uniform cross section. The
femoral shaft can be seen to manifest a bow-like geometry when viewed

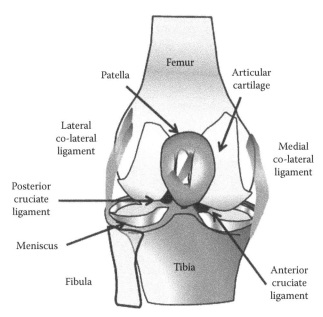

FIGURE 11.13
Functional anatomy of the knee. (www.webmd.com)

anteriorly. As well, the distal aspect of the femur broadens into the medial and lateral condyles that are covered in cartilage and promote articulation with the patella and the tibia. The inferior, posterior oblong portions of the condyles articulate smoothly with the tibial plateau, whereas the central anterior surface between the condyles articulates with the facets of the patella. The larger medial condyle has a more symmetrical curvature. The condyles are separated by a groove called the femoral trochlea. A feature called the sulcus represents the deepest point in the trochlea relative to the mid-plane between the condyles. While the medial condyle may be large in geometry, the lateral condyle has a greater posterior excursion than the medial condyle.

11.7.1.2 Tibia

The proximal tibia flares out laterally and medially to form a somewhat mating congruent surface with the medial and lateral condyles of the femur.

The tibia may be considered as the weight-bearing bone of the leg, whereas the fibula serves the purpose of muscular attachments and completion of the ankle joint (Figure 11.14). In the intact knee, the menisci reside upon the tibia and serve to enlarge the contact area as well as increase conformity between the articulating surfaces of the femur and tibia. The menisci serve two primary functions: load bearing and stability. Each meniscus provides joint lubrication that prevents capsule/synovial impingement and acts as a shock absorber.

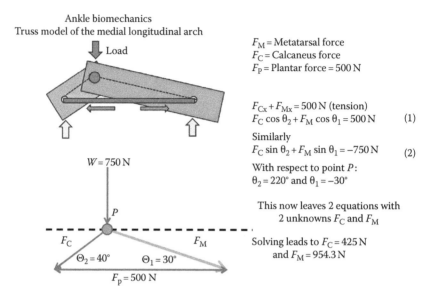

Ankle biomechanics
Truss model of the medial longitudinal arch

Load

F_M = Metatarsal force
F_C = Calcaneus force
F_P = Plantar force = 500 N

$F_{Cx} + F_{Mx} = 500$ N (tension)
$F_C \cos \theta_2 + F_M \cos \theta_1 = 500$ N   (1)

Similarly
$F_C \sin \theta_2 + F_M \sin \theta_1 = -750$ N   (2)

With respect to point P:
$\theta_2 = 220°$ and $\theta_1 = -30°$

This now leaves 2 equations with 2 unknowns F_C and F_M

$W = 750$ N

Solving leads to $F_C = 425$ N
and $F_M = 954.3$ N

P

F_C F_M

$\Theta_2 = 40°$ $\Theta_1 = 30°$

$F_p = 500$ N

FIGURE 11.14
Ankle biomechanics.

11.7.1.3 Patella

The main biomechanical function of the patella is to increase the moment of the quadriceps muscle mechanism. In fact, it can increase the function of the muscle by almost 30% via increasing the moment arm (increases rotational torque). The load across the joint rises as flexion increases but because the contact area also increases, higher forces are dissipated over a larger area. The anterior surface of the patella is convex. The superior border is thick and gives attachment to the tendinous fibers of the rectus femoris and vastus intermedius muscles. The lateral and medial borders are thinner and receive the tendinous fibers of the vastus lateralis and vastus medialis muscles, respectively. These muscles comprise the quadriceps muscle complex. The articular surface of the patella is described as possessing seven regions covered by the thickest (6.5 mm) cartilage in the body.

At 10°–20° of flexion at the distal pole, the patella first contacts the trochlea in a narrow band across the medial and lateral aspect. As flexion increases the contact area moves proximally and laterally. The most extensive contact is made at about 45° where the contact area is an ellipse across the central portion of the medial and lateral region. At 90° the contact area is shifted to the upper part of the medial and lateral region (Figure 11.15). With further flexion, the contact area separates into two distinct medial and lateral patches.

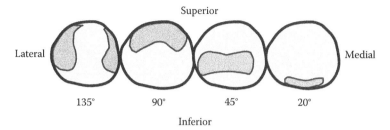

FIGURE 11.15
Contact area of the patella upon the femur under increasing flexion. (www.pt.ntu.edu.tw)

11.7.1.4 Anterior Cruciate Ligament

The ACL is the static restraint to anterior translation of the tibia on the femur. The ligament is taut in extension and slack in flexion. The ACL accounts for approximately 85% of the resistive force observed in anterior translation. The ACL also plays a small role in resisting internal/external rotation. The maximum tensile strength of the ACL is approximately $1725 \pm 270\,\text{N}$.

11.7.1.5 Posterior Cruciate Ligament

This ligament is the primary restraint to posterior displacement of the tibia on the femur and is responsible for over 90% of the resistive force observed in posterior translation. In contrast to the ACL, the posterior cruciate ligament (PCL) is relaxed in extension, taut in flexion. The PCL also plays a role in restraint to varus/valgus forces and internal rotation of tibia on femur. The PCL is considered to be the primary stabilizer of the knee because it is located close to the central axis. Rotation of the joint is almost twice as strong as the ACL. The PCL is maximally taut at full flexion and also becomes tighter with internal rotation. The PCL functions in conjunction with the lateral collateral ligament (LCL) and popliteus tendons to stabilize the knee.

11.7.1.6 Overall Motion

The active movements associated with the motion of the knee joint are described primarily as flexion, extension, medial rotation and lateral rotation. The flexion and extension at this joint differs from those of a true hinge because the axis of rotation with respect to the knee is not fixed (Figure 11.16). Recall, that the ACL and PCL act to restrain translation. While they reduce translation, these ligaments do not eliminate it completely and hence, under flexion and extension, the center of rotation is in fact mobile and not fixed as would be the case in a hinge.

FIGURE 11.16
The knee does not display a fixed center of rotation (COR) as would be the case in a hinge. The COR in the knee displays translation under flexion–extension.

The knee displays limited stability from the bony architecture and relies on the ligamentous and muscular structures to remain stable under loading and motion. At the beginning of flexion, the knee "unlocks" with an external rotation of the femur on the tibia via the interplay of the meniscal, articular, and ligamentous structures, and the contraction of the popliteus muscles. An incremental increase in the forces generated by the quadriceps mechanism and the PCL maintains stability during flexion.

Under extension, parts of both cruciate ligaments, the collateral ligaments and the posterior capsule ligaments, are under tension. There is also passive or active tension in the hamstrings and gastrocnemius muscles. In addition, the anterior parts of the menisci are squeezed between the joint surfaces comprising the femur and tibia.

11.7.2 Hip

Unlike the knee, the hip is comprised of a single joint mechanism and is represented as a ball and socket configuration (Figure 11.17). The bones that comprise the hip joint consist of the femur (the proximal section) and the acetabulum, which is a component of the pelvis. The business end of the femur that is involved in hip mechanics is called the femoral head, hence the ball and socket classification of the hip joint. The convex component of the ball and socket joint of the hip joint can be visualized as being approximately 2/3 of a sphere. The articular cartilage is thickest on the medial-central surface and thinnest toward the periphery with most of the load being transmitted through the superior quadrant.

The angle between the femoral neck and the femoral shaft is of significant biomechanical importance. In the frontal plane, most adults display an angle of 125° of anteversion (the projection of the long axis with respect to the

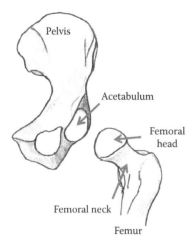

FIGURE 11.17
Functional anatomy of the hip. (www.ortho pediat-rics.com)

femoral condyles). In the transverse plane, the angle is on average 12°. It has been noted clinically that when this angle is greater than 12°, the femoral head becomes uncovered and generates internal rotation during a walking or gait cycle. Conversely when this angle is less than 12°, one observed retro-version and is manifested by external rotation during gait. Thus, the neck/shaft angle facilitates range of motion (ROM). In most individuals hip ROM (Figure 11.18) is given as

Motion	Range
Flexion–extension	140°
Abduction–adduction	75°
Internal–external rotation	90°

During a gait or walking cycle, the dominate motion at the hip in flexion–extension with normal individuals showing between 40° and 60° of range.

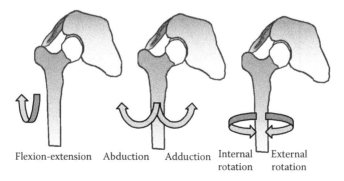

FIGURE 11.18
Range of motion for the hip.

While the hip is considered a ball and socket joint is does generate motion in the other two axes of rotation. During gait, abduction/adduction and internal/external rotation is only a few degrees. While we have treated the knee and hip joint as separate entities, coupled with the ankle joint, they form a kinematic chain (Figures 11.19 and 11.20) that propels the body.

Moment balance

$$(Fy_{ab} \sin 45°) (5\,cm)/\sin 45° = 5/6\,BW\ 15\,cm$$
$$Fy_{ab}\ 5 = 12.5\,BW = 2.5\,BW$$

Force balance

$$Fy_{JR} = Fy_{ab} + 5/6\,BW = 3.3\,BW$$
$$Fx_{JR} = Fx_{ab} = Fy_{ab} = 2.5\,BW$$
$$F_{JR} = [Fx_{JR}{}^2 + Fy_{JR}{}^2]^{1/2} = [2.5\,BW^2 + 3.3\,BW^2]^{1/2} = 4.1\,BW$$

It has been shown that during gait, the loads across the hip can be significantly greater than simple body weight as a result of internal muscle loading and impact loading upon the contact of foot strike with the floor (Figure 11.21).

	Stance Phase	Swing Phase
One leg stance	2.6 BW	
Walking	1.6 BW	0 BW
Running	5.0 BW	3.0 BW

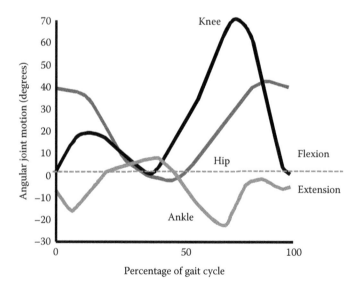

FIGURE 11.19
Propulsion of the human body is achieved via a kinematic chain linking the hip, knee, and ankle.

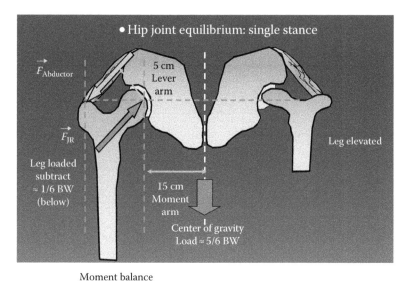

Moment balance

$(Fy_{ab} \sin 45°) * (5 \text{ cm})/\sin 45° = 5/6 \text{ BW} * 15 \text{ cm}$

$Fy_{ab} * 5 = 12.5 \text{ BW} = 2.5 \text{ BW}$

Force balance

$Fy_{JR} = Fy_{ab} + 5/6 \text{ BW} = 3.3 \text{ BW}$

$Fx_{JR} = -Fx_{ab} = Fy_{ab} = 2.5 \text{ BW}$

$F_{JR} = [Fx_{JR}{}^2 + Fy_{JR}{}^2]^{1/2} = [2.5 \text{ BW}^2 + 3.3 \text{ BW}^2]^{1/2} = 4.1 \text{BW}$

FIGURE 11.20
Hip joint equilibrium, single stance.

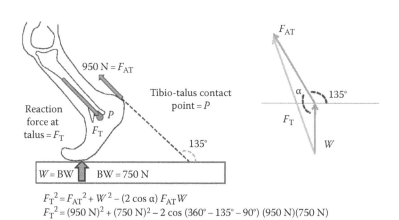

$F_T{}^2 = F_{AT}{}^2 + W^2 - (2 \cos \alpha) F_{AT} W$

$F_T{}^2 = (950 \text{ N})^2 + (750 \text{ N})^2 - 2 \cos (360° - 135° - 90°) (950 \text{ N})(750 \text{ N})$

$F_T = 1572.5 \text{ N}$

FIGURE 11.21
Momentum balance.

11.7.3 Shoulder

The shoulder displays the greatest range of motion of all joints in the body. The shoulder joint is a compilation of several bones that work in unison with the connective tissue so as to achieve this large range of motion (Figure 11.22).

11.7.3.1 Humerus

The humerus is the long bone that contains the biceps muscle (Figure 11.23). Where it becomes part of the shoulder joint, the surface is covered with articular cartilage with a surface area between 33 and 35 mm. The humerus is also positioned at 32° of retroversion (rotated posteriorly).

11.7.3.2 Scapula

The scapula is a thin bone sheet that most of us refer to as the shoulder blade. Where the scapula interfaces with the humerus resides the glenoid fossa, which is approximately 41 mm by 25 mm in area and possesses a pear-shaped, 7.4° posterior retrotilt. The scapula is encompassed by a restraint

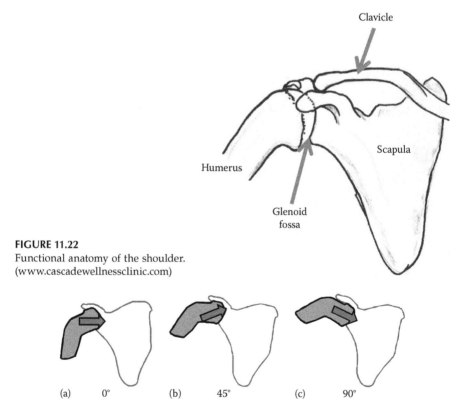

FIGURE 11.22
Functional anatomy of the shoulder.
(www.cascadewellnessclinic.com)

FIGURE 11.23
Direction of the joint reaction force (JRF) during humeral elevation (abduction).

mechanism termed the glenoid labrum consisting of the joint capsule, the glenohumeral ligaments and the head of the biceps tendon.

11.7.3.3 Glenohumeral Joint

The glenohumeral joint is the "true" shoulder joint and, in a manner similar to the hip, is functionally a ball and socket joint. The large ROM displayed by the shoulder may be attributed to several elements. Unlike the hip joint, where approximately coverage was nearly 2/3 of the surface the coverage of the ball/ socket mechanism in the shoulder is considerably less. Add to this shallow coverage, a relatively lax or slack joint capsule and limited ligament support and one is able to achieve a significant range of motion. The shallow socket of the glenoid on the scapula faces slightly anterior. As a consequence a thick rim of fibrocartilage may be found in this region that acts to effectively increase the socket contact area by about 75%. The joint capsule (as with all synovial joints) provides stability via the anterior glenohumeral ligaments. However, in the posterior, location support is only provided by the capsule itself.

Shoulder range of motion (Figure 11.23):
Flexion 180° – extension –60° (sagittal plane)
Abduction 180° – adduction –75° (frontal plane)
Internal rotation 90° – external rotation –90° (transverse plane)

11.8 Spine

In contrast to the joints of the extremities, the spine is a collection of inter-connected joints serving to achieve a range of motion, much as one would envision a bicycle chain. Unique to the spine with regard to articulation are the spinal discs. While other joints are lined with articular cartilage in order to generate smooth kinematics, the discs within the spine deform in order to perform the motions required for the activities of daily living. The spine has four main *functions*: support, mobility, housing/protection, and control.

11.8.1 Spine Function

11.8.1.1 Support

In human beings, the vertebral or spinal column is the principal supporting structure. The spine bears heavy loads, in particular the cervical and lumbar regions. These loads include the weight of the upper body and any loads being lifted, lowered, carried or held.

11.8.1.2 Mobility

The joints of the extremities are attached through a series of joints to the spine. Muscles, ligaments and tendons connect parts of the spine to each other as well as to the other limbs and allow a diverse range of movement. The cervical and lumbar spine regions are the most flexible. Forward (flexion) and backward (extension) bending of the lumbar spine produce the greatest range of movement.

11.8.1.3 Housing/Protection

The spinal column protects the spinal cord and nerves as they pass from the brain to the upper and lower limbs. With the center of rotation under torsional loads located in the spinal canal, the spinal cord is protected from injury.

11.8.1.4 Control

The movement of each segment of the spine is controlled, actively by muscles and passively by ligaments. Without muscular support, the spine is unstable.

11.8.2 Vertebral Column

The vertebral column or spine is made up of a series of bones called vertebrae (Figure 11.24). The spine is a strong and flexible column of bones made up of the cervical (C), thoracic (T), lumbar (L), sacral, and coccygeal regions.

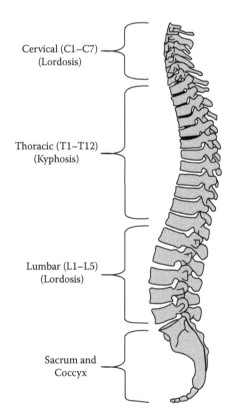

Cervical (C1–C7)
(Lordosis)

Thoracic (T1–T12)
(Kyphosis)

Lumbar (L1–L5)
(Lordosis)

Sacrum and
Coccyx

FIGURE 11.24
Anatomy and regions of the spinal column.

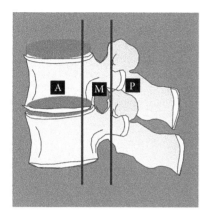

FIGURE 11.25
Three loading columns of the spine: anterior (A), middle (M), and posterior (P) columns.

The cervical or neck region is normally made up of 7 bones, the thoracic or chest region has 12 bones, and the lumbar or low back region is made up of 5 bones. The sacral region consists of 5 fused bones, and the coccygeal region 3–5 tiny bones. The spine is curved and in profile looks like an elongated letter "S" in shape. The cervical and lumbar regions have a forward or lordotic curve and the thoracic region a backward or kyphotic curve, when viewed from the side. It has been suggested that the "S" shape of the spine facilitates energy absorption, protects the spinal structures against impact by increasing its strength, helps maintain balance in the upright position, and absorbs shocks during gait. The vertebral column is divided into three distinct loading regions: the anterior, middle, and posterior columns (Figure 11.25). The ratio of load bearing is 80% in the anterior column and the remaining 20% between the middle and posterior columns.

Motion or articulation of the spine is caused by moments (or torques) generated by muscular contractions and is balanced by ligaments and tendons. Without the adjoining musculature, the vertebral column is unstable and buckles under low compressive loading.

11.8.2.1 *Vertebral Body*

The vertebral body is the weight-bearing part of the vertebrae. The vertebral body is a thick, disc-shaped cylindrical block of bone flattened at the back and with roughened top and bottom surfaces (Figure 11.26). It is made up of spongy (cancellous) bone on the inside, which enables it to resist compression and absorb energy, and a thin outer covering of compact (or cortical) bone.

11.8.2.2 *Facet Joint*

The facet joints are located in the posterior column. They are a "true" joint with respect to articulation. Each of the two joints is covered with a capsule filled with fluid, which serves to lubricate the articular cartilage lining the bony structures. The facet joint serves two functions. The two facet joints

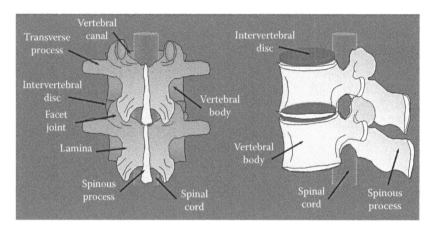

FIGURE 11.26
Anatomy of the vertebral body.

aid in guiding spinal motion by providing a restraint mechanism via the joint capsule, the bony geometry and the adjoining ligaments. The secondary function is to act as a safeguard to loading. As we age, the discs between the vertebral bodies become degenerated and exhibit a reduction in load bearing. The normal 80:20 ratio of anterior to posterior loading is altered and the posterior column is required to bear additional loads. Under these conditions, torsional loading becomes important and the bony structures of the facet joints aid in torsion resistance.

11.8.2.3 Intervertebral Discs

The intervertebral disc is comprised of concentric layers of tissue (collagen) with each layer possessing fibers oriented at +30° or −30° to the horizontal (Figure 11.27).

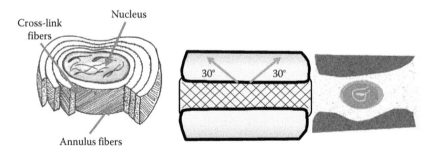

FIGURE 11.27
Collagen fiber orientations in the disc annulus provide flexibility in all loading modes. The nucleus located at the center of the disc provides hydrostatic pressure resistance to compression. (Redrawn from White, A. and Panjabi, M. *Clinical Biomechanics of the Spine.* Lippincott, PA, 1990.)

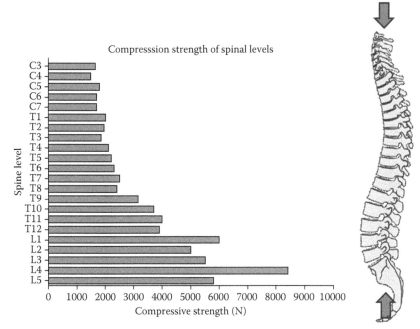

FIGURE 11.28
Load-bearing capacity in increased as one descends from the cervical to the lumbar region. (Redrawn from White, A. and Panjabi, M. *Clinical Biomechanics of the Spine*. Lippincott, PA, 1990.)

At the center of the disc lies the nucleus. The combination of the hydro-static pressures within the nucleus region combined with the annulus fibrosis results in a high load bearing yet flexible construct capable of motion in all directions. Lumbar disc mechanics differ from cervical disc mechanics. The lumbar spine must be able to withstand large "physiological" loads whereas the cervical is required to display increased range of motion (Figure 11.28).

Lumbar	Range of Motion
Forward flexion	90°
Extension	30°
Left lateral flexion	30°
Right lateral flexion	30°
Left lateral rotation	30°
Right lateral rotation	30°

Cervical	Range of Motion
Forward flexion	45°
Extension	45°
Left lateral flexion	45°
Right lateral flexion	45°
Left lateral rotation	80°
Right lateral rotation	80°

The normal activities that we perform on a daily basis give rise to very high loading conditions across the intervertebral discs. It should be kept in mind that unlike many of the other joints discussed when loading can, for the most part, be confined to a single plane resulting in motion within that plane. The spine, however, displays coupled motion (Figure 11.29). That is, motion or loading in one direction can induce motions or reactions to the loading in other directions. The biological configuration of bony geometry, flexible disc and articulating facet joints allows the spine to bear high compressive loading for stability yet, permit mobility in various loading directions. Each of these elements plays a role in overall ROM well as the continuous stability/control of the motion through the ROM. The relative contribution of each element is dependent on the predominant motion under consideration.

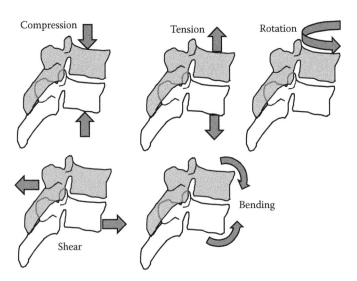

FIGURE 11.29
Spine encompasses all loading modes in a concept called coupled motion.

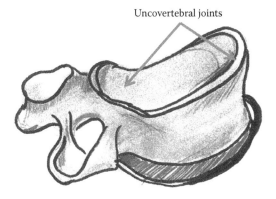

Uncovertebral joints

FIGURE 11.30
Uncovertebral joints in the cervical spine provide resistance to lateral shear.

In lateral bending, the two features known as the uncovertebral joints aid in directing of the cervical vertebrae. The two joints function as a rail, providing resistance to lateral shear (Figure 11.30). The center of rotation (COR) for lateral bending is in the superior body and determined by the radius of the uncovertebral joints. In flexion–extension, the facet joints slide over one another and result in some translation of the vertebral body. During extension, the spinous processes limit extension via contact with each other. The COR is now located in the anterior region of the distal vertebra (Figure 11.31).

Coupled motion is required for axial rotation and results in distraction of intervertebral space once rotation exceeds ≈2°–3°. While this motion is observed at all levels, it is not a constant value and ranges from 2° of coupled axial value

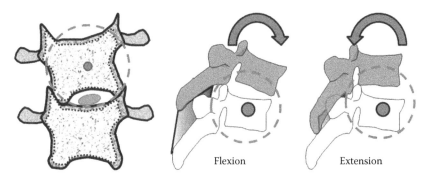

Flexion   Extension

FIGURE 11.31
COR shifts from the superior (or proximal) vertebral body to the inferior (distal) vertebral body in going from lateral bending to flexion–extension of the cervical spine.

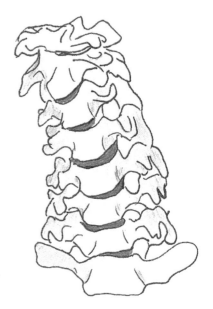

FIGURE 11.32
Effects of coupled motion. Motion in one plane
results in motion in other planes other than that of
primary motion.

for every 3° of lateral bending (ratio of 2:3) at C2 to 1° of coupled axial rotation for every 7.5° of lateral bending (ratio of 2:15) at C7 (Figure 11.32). In axial
rotation, a vertical distraction is observed and increases until additional structures are engaged to restrain motion.

Problems

11.1 You are asked to mechanically test a material to failure in compression. It is of uniform composition and is provided to you in the form
of a cube. Sketch and label an expected load versus deformation
curve.

11.2 An implant must be biocompatible as well as display good mechanical properties. List four reasons as to why an implant may fail despite
passing all biological and mechanical testing.

11.3 Unloading a painful hip joint can often be accomplished with a cane.
Using Figure P11.3, is there a difference if one uses the cane on one side
or the other? Pain on left. $F_{Abductor} = 300\,N$, BW $= 750\,N$

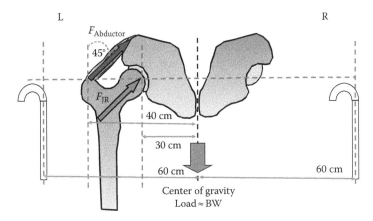

FIGURE P11.3

11.4 In normal activity, the living spine can sustain 2–3 times body weight. In fact, it has been reported that Olympic weight lifters place 25,000–30,000 N on their spine during competition. In cadaveric specimens, a human spine will begin to collapse under approximately 300 N of compressive load. Explain.

11.5 From the bone plots below (Figure P11.5), determine
(a) yield stress, (b) strain at failure, (c) elastic modulus for each bone type.

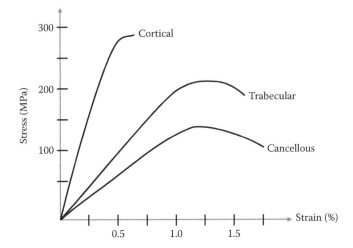

FIGURE P11.5

11.6 In his younger days, Smith was quite an athlete. Attending the institutions of higher learning expanded not only his mind but other features as well. Given that Smith entered college at 100 kg and graduated at 125 kg, determine the additional moment on the intervertebral disc located between lumbar vertebrae three and four. *Note:* [Length1 (L1) = 15 cm and Length2 (L2) = 25 cm]

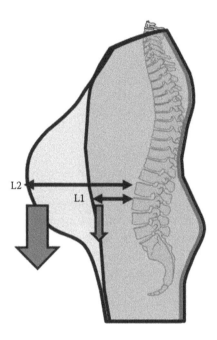

11.7 Draw the mechanical circuit diagram illustrating the general theory of isometric muscle action. If one applies the stimulation and resultant load T_0 is achieved for a time period C, as shown in Figure P11.7, show that

$$\frac{dT}{dt} = (T - T_0)J, \quad 0 \leq t \leq C$$

where
    T is the muscle tension
    t is time
    J is a constant

Assume that the muscle mass is negligible and that the muscle goes under isometric contraction.

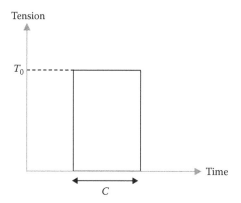

FIGURE P11.7

11.8 a. Compute the bracialis force F_B and joint reaction force F_{JR} on the elbow when one holds out the arms as shown below at 90°.

b. If you now hold a 1 kg mass in your hand what is the increase in the joint reaction force (JRF) with respect to part a?

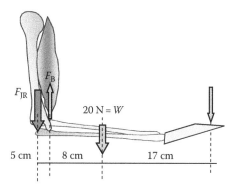

11.9 a. During stair climbing the patella lies at approximately 30° with respect to the vertical. That being the case, compute the JRF at the tibiofemoral joint given that the patellar tendon force $F_{PT} = 3\,BW$ and the ground reaction force = BW.

b. At some point, the angle between the tibiofemoral JRF will be at 20° with respect to the ground reaction force. How much shear is generated?

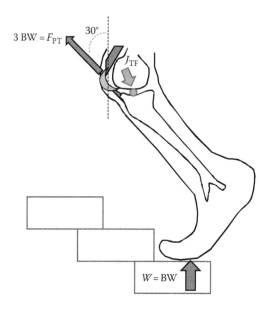

11.10 With the image given below (Figure P11.10)
   a. Draw the free body diagram representing the patellofemoral contact force, F_{PF}.
   b. If F_P is 1000 N= F_Q, compute F_{PF}.

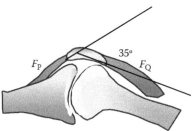

Assume the knee is in the neutral position

FIGURE P11.10

11.11 Hockey is a great way to stay in shape over the long winter months. Sooner or later one would get a slash across the tibia with an opposing player's hockey stick. A penalty, but it does not stop the pain in your shin. On one such slash you sustained a 0.01 mm tibial crack (see Figure P11.11). Being young, we can assume that the cortical thickness is on the average approximately 15 mm and that the tibia in this region can be represented by a tube. You have decided that perhaps running off the pain but participating in a marathon. Using the loading levels from the graph and data below, can you physically sustain the tibia for the duration of a marathon (>26 miles)?

$C = 2.6 \times 10^{-6}$ m $(MN/m^{3/2})^{-2.5}$ (cortical bone)

$m = 2.5$

$a_0 = 0.01 \times 10^{-3}$ m

$Kc = 2.2$ $(MN/m^{3/2})$

Stride length $= 2$ m

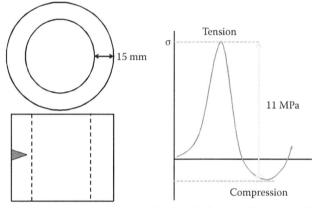

Stress on the human tibia over a stride

FIGURE P11.11

11.12 Consider a small element of bone material of thickness $= t$, under uniaxial stress $= \sigma$, and resulting in a strain $= \varepsilon$ in the same direction. If the undeformed surface area of the face where the stress is applied is $= A_0$ and the undeformed length of the material in the direction of the stress is $= L_0$, compute the work done (or energy stored) per unit volume. Surface Area $= A_0$.

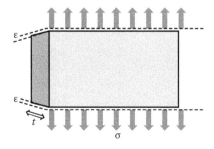

11.13 The adjacent table represents the deformation in millimeters of a cube of bone taken from a vertebral body in each of the three orthogonal loading directions from −5 N to −50 N. From the data provided

  a. Compute

      Mean and standard deviation for each sample

      Plot a bar graph of the mean ± standard deviation of the deflection

  b. Determine which sample number corresponds to the X, Y, or Z direction. Explain your rationale.

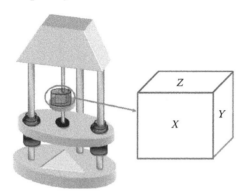

Sample1	Sample2	Sample3
0.033	0.09	0.048
0.031	0.089	0.05
0.032	0.088	0.05
0.032	0.088	0.05
0.033	0.087	0.05
0.031	0.087	0.05
0.031	0.087	0.051
0.031	0.086	0.049
0.031	0.086	0.05
0.032	0.085	0.05
0.032	0.085	0.05
0.031	0.085	0.05
0.032	0.087	0.05
0.032	0.085	0.049
0.031	0.085	0.049
0.03	0.086	0.05
0.031	0.085	0.05
0.031	0.085	0.049
0.031	0.085	0.049
0.031	0.085	0.049
0.032	0.084	0.049
0.031	0.084	0.05

References

1. Nordin, M. and Frankel, V.H. *Basic Biomechanics of the Musculoskeletal System,* Lea and Febiger, Malvern, PA, 1989.
2. Nigg, B.M. and Herzog, W. *Biomechanics of the Musculo-Skeletal System,* John Wiley, Chichester, UK, 1989.
3. Butler, D.L., Kay, M.D., and Stouffer, D.C. Comparison of material properties in fascicle-bone units from human patellar tendon and knee ligaments. *J. Biomech.* 29, 1131–1136, 1996.
4. Vamder, A.J., Sherman, J.H., and Luciano, D.S. *Human Physiology: The Mechanisms of Body Function,* 4th edition, McGraw-Hill, New York, 1985.
5. Desaki, J. and Uehara, Y. The overall morphology of neuromuscular junctions as revealed by scanning elelctron microscopy. *J. Neurocytology* 10, 101–110, 1981.
6. Eithier, C.R. and Simmons, C.A. *Introductory Biomechanics. From Cells to Organisms.* Cambridge University Press, Cambridge, 2007.
7. White, A. and Panjabi, M. *Clinical Biomechanics of the Spine.* Lippincott, PA, 1990.

Part IV

Capstone Design

12

Biomedical Engineering and Design

12.1 Why Biomedical Engineering Is Unique

12.1.1 Engineering for the Practice of Medicine

Medical problems are particularly tricky. They are as challenging as any other technical problem, plus, they involve people (or other living things). This means that you need to understand biology, or since there is so much that we really do not understand about biology, I prefer to say that you need to *respect biology*.

Biology is unlike the physical sciences upon which many other engineering disciplines are based. Biology is not always exact, and biological systems can adapt. As a result, it is much harder to establish a single set of rules for biological systems, and therefore much more difficult to find one good solution to an unmet biomedical need. Even when dealing with healthy people, there is a broad spectrum that must be considered; just think about how many different shapes and sizes of people there are in the world!

The other really important thing you need to *appreciate is how things work in the clinic*. Doctors, nurses, and other medical experts operate in a system that cares for the health of all those different people, and that system is quite different from the environment in which you learned many of your engineering practices. You will not truly appreciate this until you actually work in a clinical setting, but be prepared to find a way to address this issue, even if you are working in your classroom, lab, or office.

Why is the clinic so different? We are maintaining and repairing living systems! While the answer to this might be somewhat obvious, the ramifications are sometimes subtle. If you have a fender bender and take your car to the shop, the mechanics will either fix or put new parts on your car to make it as good as new. If you go to the hospital with a broken finger, the doctors cannot throw the old finger away and replace it with a new one (at least, not yet!). They can only fix it, and sometimes, if it is badly broken, they cannot make it as good as new or may have to remove it altogether.

Another big difference is the way clinicians have to do their job. Does your mechanic leave the motor running when he replaces your radiator?

Of course not; that would make the job much more complicated. Does your doctor "turn your heart off" when he performs surgery to repair your broken finger? Certainly not, and so this makes his work far more complex. And if the doctor is performing heart surgery, and he absolutely must "turn the motor off," he has to put a very complicated and rather risky auxiliary system in place. We know this is a bit trickier than jump-starting your car's engine.

Does your mechanic worry about your car getting sick, or getting an infection while it is being fixed? I do not think so. If your car is so badly damaged, your mechanic may recommend that it is cheaper to send it to the junkyard and just buy a new one. If you were badly hurt, would you want your doctor to tell your mother that it would just be cheaper to replace you? Of course not! Even if it costs a ton of money, your doctor would do everything he could to make you better. These examples represent issues that define some of the values and constraints that are unique to the practice of medicine. The complex biology construes a unique and challenging *physiologic system* that must be understood. Social values must also be considered.

12.1.2 Clinical Testing

If you continue along this line of thinking, you can begin to understand why biomedical engineering can be so complex. As in any engineering practice, you must perform tests on the product, but engineering tests for medical products have to be a lot more robust. Let us look at a simplified example: imagine that you work for a company that makes hinges for an airplane wing.

To start, somebody would need to test the components, such as the bolts. If the bolts are a standard item, it is likely that you would purchase them from a bolt manufacturer (vendor), and the vendor supplied them along with necessary *quality documentation* to demonstrate that they had passed all required tests. The hinge materials would also need to be tested. Again, this is probably done by the vendor, and is provided along with appropriate *material certifications*.

Now you, as the hinge manufacturer, would customize the use of those standard items to make a unique hinge. You would need to test that hinge before you sold it to Boeing. You would perform various *stress analyses*, both theoretical and practical. You would probably perform something like finite element analysis (FEA) for a *static test*, and you would probably have to customize a computer program to make a *dynamic model*.

Next, you would design a laboratory experiment to test the real hinge. You would perform cyclical loading and collect data to determine *fatigue*. You would also perform maximum loading to *failure* tests. You would follow established general engineering practices, such as American Society for Testing Materials (ASTM), and you would also have to follow specific industry regulations, such as those dictated by the Federal Aviation Administration (FAA) and associated bodies.

After you have demonstrated that your hinge meets all the performance requirements, you would test several more of them to ensure that you can obtain uniform results with all the same hinges that you make; basically making sure that your manufacturing process is consistent—this is referred to as *validation*. Once you are certain that you can consistently produce a hinge with the appropriate quality and performance characteristics, you can sell it to Boeing.

Now, Boeing would not instantly install the hinge on the wing and fly the plane. Rather, they would perform their own analyses and ultimately test the hinge on a wing in the laboratory. They would then run some test flights with experienced test pilots and crew but without passengers. They would probably test a few identical planes, and they would probably have defined certain extreme conditions and test to those conditions.

So, just how much more complicated could it be to test an artificial knee joint, for example? (Figure 12.1). Testing of a hinge for an airplane wing is certainly a rigorous and detailed process. It takes a lot of time and costs a lot of money, but it is necessary to ensure aircraft performance and passenger safety.

Our concerns are similar when designing medical products. After all, a knee joint is quite like a hinge, and, in the end, the joint's performance and patient's safeties are critical. Indeed, as a design engineer, you can identify many similarities between the two in the design process. As the engineer who designs an artificial knee joint, you would have to do much of the same type of testing, especially with regard to the stress analyses and laboratory testing. In fact, you would even follow ASTM testing standards for an artificial knee joint, which are similar to that of a hinge in some ways.

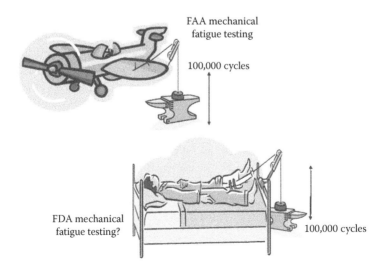

FIGURE 12.1
Because the system is living, a biomedical engineer's job is complex. Traditional engineering approaches must be tempered with biological, physiological, and social concerns.

Just as in airplane manufacturing, there are specific regulatory bodies that govern issues in the design of medical products. Certainly, you have heard of the Food and Drug Administration (FDA), which might be construed to be quite like the FAA in many ways. The differences become evident, however, when you look at all of the objectives in medical testing.

12.1.3 The System Is Living!

Although there are many commonalities between an airplane wing hinge and a knee joint, we need to appreciate that when designing prosthesis, we are designing for a living system. In addition to mechanical and functional performance, which we discussed in the context of the airplane wing hinge, we must also consider "biological performance" of an artificial knee joint. We need to assert that the prosthesis will perform once implanted, in vivo, that it will not cause the patient to become sick (or sicker), and that it will maintain its function over time.

In the broadest sense we could say that there are many more facets to "performance" in biomedical engineering, and there are variations among test subjects that limits our ability to generalize or standardize. You will often hear about "biocompatible" materials. However, you cannot assume that just because lab tests show that living tissue and that material "get along," it is safe and effective. That product must be evaluated in the context of the dynamic living system. In the case of a knee prosthetic, it must be evaluated in the context of the living muscle chain, which includes an understanding of the effects on the associated soft tissue, bone, and vasculature.

Not only do you need to understand the effects of combining a technology with a biological system, but you also need to account for the fact that there are significant variations among individuals within a population. People come in different shapes and sizes. Certainly, you would not give an 11-year-old girl the same drug dosage as you would to a 300 lb linebacker. Similarly, you would use a smaller knee implant on a small young woman than you would on that same linebacker. Some differences are subtler. For example, nickel titanium, NiTi, has proven to be a useful material for some medical devices. Still, a small percentage of the population has a "nickel allergy," and this must be accounted for when marketing a medical device. In the same way, ethnic and genetic differences play a big role in the safety and efficacy of a therapy or technology.

In the United States, the FDA regulates many of the requirements of medical technologies. In fact, when you review the FDA's Web site, you will see that the

FDA is responsible for [1]:

- Protecting the public health by assuring the safety, effectiveness, and security of human and veterinary drugs, vaccines, and other biological products, medical devices, our nation's food supply, cosmetics, dietary supplements, and products that give off radiation

- Regulating tobacco products
- Advancing the public health by helping to speed product innovations
- Helping the public get the accurate, science-based information they need to use medicines and foods to improve their health

You probably already know that in order to use a medical device in the United States, the device must be FDA "approved" (the correct term is "obtain FDA clearance"). You might have also heard about how much more time and money is needed in order to bring a medical technology to market. These responsibilities are very significant and quite unique to biomedical engineering. You need to develop a solid appreciation for this complex relationship between engineering and living systems in order to be successful.

You will learn more about the uniqueness, testing requirements, and clinical challenges related to your biomedical design project as you progress through your project. Now let us ensure that you have the tools and skills you need to work through your project effectively.

12.2 Getting Started with Your Design Project: Design Tools and Processes

12.2.1 A Design Challenge

Picture yourself a year from now. You and your old fraternity brother, Bob Valdictoran, are in your first jobs. You have not seen each other in a while, and arrange to meet for lunch. While eating your burgers, you ask Bob, "So, how's your job going?" and he proceeds to tell you what happened to him the other day.

Bob was highly recruited, and he accepted a top position as a junior design engineer for a Fortune 500 company, Hot Shot Innovations, that takes pride in being a leader in innovation. Last week, he was sitting in his cubicle, working on a CAD drawing at his computer, and his boss, Mr. Swift, walked over to his desk. Mr. Swift said, "You've been doing such a good job on those drawings that I think you are ready for your first assignment as project leader."

Then he proceeded to give Bob the assignment. "There's a small upscale restaurant on the riverbank that would like to serve a fresh 'catch of the day,' but they cannot always catch enough fish. Let us see if we can use your innovative engineering to help them." At that moment, Swift's blackberry buzzed on his belt clip; he grabbed it and ran off.

Bob tells you about how excited he was; this was a big opportunity for him. He tells you that his mind started racing. He had a thousand ideas; Bob always does. He knew his boss was busy, and so he decided to lay out his top three ideas, in general terms, to present to him. To make a good impression, he prepared a detailed PowerPoint presentation describing his three ideas.

Idea number 1: "build a better fishing pole." Then, since Bob is so good with CAD, he drew a few design options. Idea number 2: "design some lures that will catch more fish," and then he displayed his engineering knowledge describing how to make a fishing lure skip and bob underwater while being reeled. Idea number 3: "build the fisherman a boat, so that he can go across the river to get more fish," and he prepared design concept options for a rowboat and a motorboat.

To make the best impression, Bob worked late to polish up his presentation and scheduled a meeting with Mr. Swift for the next morning. For extra power he dressed in his Brooks Brothers suit, and though nervous, he was ready to knock his boss over with his ideas. After he completed his presentation, he looked over to Swift, who was checking his text messages. "So, Mr. Swift, what do you think?" he asked. Swift glanced up from his blackberry and said, "That section of the river is frozen 5 months in the year; I think you better take some more time to understand the problem," and then he grabbed his buzzing blackberry and ran out of the room.

12.2.2 The Unmet Need

Wow, that was a real blow. Bob worked so hard; he skipped the gym last evening and now needs to get his suit dry cleaned, and he still did not make the impression he thought he was going to make. What happened? Well, Swift gave him a clue; Bob did not *understand the unmet need*. While he recognized that there was a problem, he did not understand the problem; there is a difference. If Bob had taken time to understand the problem, he would have known that his top three ideas would not be useful to this fisherman.

Well, you can tell Bob not to feel so bad; this scenario is a simple example of what happens all the time in biomedical engineering. After you complete this chapter, you will be able to help him.

It takes more than creativity and engineering skills to effectively solve a problem. I cannot tell you how many times in my career I have seen engineers present what they believed to be a very well thought out, technical solution to a medical problem, only to be told by the doctor that it will never work. To be effective, you need to take many other aspects into consideration. Most importantly, as a biomedical engineer, you need to understand the practice of medicine.

The design process does not have to be complicated, but you need to make sure that you do not fall into traps that cause you to go too far down a pathway that ultimately will not work. A small turn in the wrong direction can waste precious time, money, and resources, and if not corrected quickly, can prevent you from achieving your design goal.

12.2.3 Stage Gate [2]

Let us go back and take a look at what happened to Bob. He did not apply an appropriate process to understand the unmet need, and wound up wasting a lot of time and energy. Fortunately, he asked for feedback right away, and he quickly obtained feedback to prevent him from traveling too far along that path. While the experience might have been a little humiliating for him, the chances to reach his design goal are still good. Imagine if he took a few months and spent thousands of dollars before he presented his ideas. That would have been a much bigger problem.

Good designers know how to learn about a problem sufficiently to understand the unmet need. They follow processes and use tools to ensure the best result. Sometimes their company provides certain structures, for example, reports, meeting formats, etc., which are designed to help plan and move through the design process. Other times they develop their own. Some of these tools and processes are formal, and others are project specific or ad hoc. Usually, engineers use a mix of formal and informal procedures to manage their project. This chapter will give you some suggested tools and processes that can help you with your design project.

12.2.4 The Design Cycle

Your design project will require you to first go through a *product development* process and then a *project execution* process using *project management* skills and tools. Design is usually an iterative process, where you will go through many cycles. The first cycle is often the most challenging where you need to develop a *proof of concept* in an efficient and rational manner, as shown in Figure 12.2. Let us discuss some of these tools and processes, and maybe see if we can help Bob win his boss over.

12.2.4.1 Voice of Customer

Clearly, Bob neglected to obtain *Voice of Customer* (VOC). This is an extremely common mistake. Bob assumed that he understood the problem and he

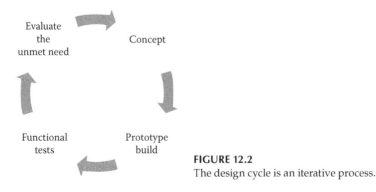

FIGURE 12.2
The design cycle is an iterative process.

engineered solutions to the problem that he imagined. The trouble with this approach was obvious in Bob's case. If he had gone to the restaurant owner and had spoken with him, he would have not made the error in judging the needs of his customer.

Most times, the errors committed are much more subtle. Still they can be equally as devastating. Sometimes, you might even understand the unmet need initially, but can drift offtrack or misalign your priorities if you neglect to maintain VOC throughout your project. It is imperative that you consult with your customer frequently to be sure that you are addressing his needs thoroughly as you develop your technical solutions.

If you take a marketing class, you will learn a lot more about VOC. At this time suffice it to say that, as a biomedical engineer, you need to be in regular and complete contact with a clinical expert during the entire course of your project.

Having said that VOC is necessary, keep in mind that you are the engineer. Your customer will tell you what he needs, and he may also tell you how to build it. Though, without engineering expertise, he may not propose the best technical solution. It is your job to interpret the needs of your customer and to develop the best technical solution to the problem that he has identified. Henry Ford once said, "If I asked my customers what they needed, they'd tell me they need faster horses." But, Henry Ford did check with his customers to be sure that he had a good idea, "Would you consider a motorized vehicle instead of faster horses?" I guess we all know the answer to that question.

12.2.4.2 Teamwork

Usually, design engineers work in a team. The best teams take full advantage of all the skills of the various team members. In biomedical engineering, these teams are often *multi-disciplinary* and require that the team members operate in an *interdisciplinary* way. Your design team will likely involve engineers from multiple disciplines such as a mechanical or electrical engineer, and it will include a clinician (e.g., a doctor or nurse), and maybe a biologist or researcher, as well as someone who focuses on marketing.

While you each have different skills, you will need to interact seamlessly among the team. This is especially challenging because the knowledge, skills, communication methods, priorities, and culture vary significantly between the disciplines. It is often the biomedical engineer's responsibility to help *facilitate communication* between the disciplines, and you will need to develop interdisciplinary skills to help bridge the individual team contributors. While this is a skill that you will hone lifelong, the method and exercises in Chapter 13 will help get you started in developing effective communication skills for this role.

12.3 Entrepreneurship*

Recently, Harrison Ford starred in a movie that brought the issue of "orphaned" drugs to the attention of the general public. It is a problem reaching beyond pharmaceuticals. I believe any engineers watching the movie were wishing the same treatment could be made for orphaned technology.

The issue of orphaned medical technology should concern anyone truly wishing to change health care. There are many technologies that never make it to market for a variety of reasons and it is not necessarily because they are bad ideas either. Instead, they might never get to serve the public because regardless of good intentions, health care, like any other service to society, is at heart, a business sector. So a good idea that does not meet the criteria to make it a viable business venture may just remain a "good idea" rather than a reality. Without the business aspect solidly in place, the chance of it getting to market is very small. But, with some changes, significant contributions can be made to improve this situation.

I believe this is a pertinent topic as we talk about health care reform. It is a practical and tactical concern. There are technologies that can benefit society that are being held back because it is too challenging or expensive to bring a product to market.

I know it is certainly a frustration that I have experienced as an engineer. Expansive amounts of data need to be gathered, situations mapped out, and tests run—I am not talking about getting the technology in order, I am talking about gauging market interest. Unfortunately, those hurdles have often proven too much to clear for many. Although there may be a brilliant mind developing the technology, a different skill set is needed to convince investors that the idea is a good business venture.

Just like the technology they are working with, engineers have had to evolve to stay relevant. Teaching as a professor, I have also had the experience of seeing that firsthand. I work with my students to not only help them develop as engineers, but also as entrepreneurs—because that is what is required today.

To be clear, it is not always a death sentence for a product to not be a major revenue generator. Some can get by due to intervention and assistance by philanthropic organizations or individuals. Others manage to gain federal funding to help foot the bill. In either case, I believe there are still more good ideas than there is funding available from any of those sources to make them all reality.

That is why my experience of late has been with venture capitalists. It is an interesting arena. They want a product that will be in demand, has a patent,

---

* This commentary was recently published in a trade journal, *DOT med Business News* [3], which speaks to thought leader MDs and C-level executives in hospitals and in health care. It is reprinted here, with permission, in its original form. I believe it provides you with some reasonable insight and rationale regarding the role and significance of entrepreneurship in biomedical engineering education.

addresses a large audience, will provide a service that is reimbursable by insurers, can go to market in a short time and has disposable aspects to the system or designed obsolescence to ensure a continuous flow of funds.

After reading that laundry list, you might think coming up with a great idea to benefit society almost seems like the easy part. Personally, it was the easy part. As an engineer, I have the mindset of identifying a problem and trying to construct a solution.

For example, in addition to my biomedical, clinical, and industrial interface, I do volunteer to work with people suffering from spinal cord injuries (SCI). What I have noticed is that many of these individuals want to be more independent than they are—for reasons that can be fixed. If they were to fall out of their chair, for example, they would be stranded until someone came to help them, so that limits their independence.

We developed a wheelchair with a seat that lifted down. It was a "winner" with my SCI friends. It was not a viable idea though, because we tried to take the price point low enough to make the product obtainable by a group of people who on average may have limited funds. A low price point and lack of design obsolescence meant a nonprofitable idea—unless you consider the improvement in the quality-of-life for the individuals getting the equipment.

Similar decisions would be reached by most companies, but we invest in people's health differently. In other industries, if you want and cannot afford you go without. But for health, if you need and cannot afford, that causes a severe problem.

From a perspective based on national statistics, it is not difficult to justify a $300 investment in a wheelchair that provides an individual with more independence. Using it could ease our national health care burden by reducing care-giving costs and morbidities associated with sedentary behavior. It is, however, difficult to justify building those wheelchairs when viewing the situation from a business standpoint.

My students, they are scientists; they graduate and they want to contribute to the greater good, but they also need to train as entrepreneurs. The kid that will be most successful as a scientist is also the one who can develop a business plan, understand the value assessment that takes place and be able to support the valuation process in their science. I think that is good in some ways, but it is a shame that the orphans remain unadopted; they would certainly save us money and improve public health.

12.4 Message to Biomedical Engineering Students

You have selected a noble profession. Like many of you who have chosen to be a biomedical engineer, I am certain you want to help people. You want to be part of a team that saves lives, improves quality of life, reduces medical costs, extends the availability of health care to all, and contributes to the overall improvement of our public health. While biomedical engineering holds great promise to do just that in the coming years, there is something else you can do right now that will have a huge impact on your goals. The U.S. national disease burden could be reduced by more than half if everyone followed these simple preventive tips:

1. Eat sensibly
2. Exercise regularly
3. Wash your hands often and thoroughly
4. Get plenty of sleep
5. Brush and floss
6. Do not smoke
7. Practice safe sex
8. Wear your seatbelt
9. Moderate your alcohol, caffeine, TV, and video game intake
10. Do not drink and drive or text and drive
11. Protect your body when in harsh elements, for example, sunglasses, sunscreen, etc.; dress appropriately for severe cold or heat
12. Manage your stress

I encourage you to explore the Center for Disease Control Web site, www.cdc.gov to learn more about our nation's health challenges and to develop your own ideas regarding how you can help to address them.

References

1. www.fda.gov/AboutFDA/Transparency/Basics/ucm194877.htm
2. Cooper, R.G., Scott, J.E., and Kleinschmidt, E.J. Optimizing the stage-gate process, *Research Technology Management*, 45(6), 1–8, 2002.
3. Hazelwood, V. Healthcare Chronicles: A new solution for health care reform: adopt the orphan technology, DOTmed Business News, May 2010.

13

A Capstone Design Curriculum for Biomedical Engineers

13.1 From Idea to Proof of Concept in 32 Weeks

13.1.1 Introduction

Even though 32 weeks might sound like a long time, it is really not much, especially to complete all you wish to accomplish in this course. It is pretty typical for an industry project, too, to not have enough time to do all that you would like to do. That is why it is important to be organized and focused. Earlier, we mentioned that you need good project management tools and skills to be most effective. This chapter will describe a model containing such tools, skills, and advice that has been used by many biomedical engineering senior design teams, several of whom have written patents, or even started their own company based on their project.

The single-most important piece of advice I can offer you is *Do not let things go to the last minute*. College students are amazingly adept at "cramming" before an exam or presentation. I can promise you that this will not work well on a design project. To have the richest, most rewarding, and most successful experience, you should review your project frequently during each week. On some days, you will find that you want to spend a lot of time in the lab, or developing a presentation, but on others, it may be just that you need to take 10 or 15 min to make an inquiry to a vendor, place an order, or just get some advice from an advisor or expert. Keep your project moving! Remember, you are dependent on your team and lots of external support, and you need to accommodate the schedules and priorities of others—not just your own—to succeed.

13.1.2 Capstone Resources

One of the things that makes your capstone project challenging is that you will need to learn how to find and utilize new resources. The better you utilize them, the better your project will be. In most of your courses to date, you have probably grown accustomed to asking your professor or instructor for help with anything related to your assignment. And, for the most part, he or she has been able to give you all the answers you needed.

In your capstone design course, as in professional practice, there is no single individual who "has all the answers." You are not ever expected to have all the answers yourself, but you will be expected to find them. You will need to develop skills that enable you to locate resources (e.g., experts, suppliers, clinic, or lab facilities) and evaluate them, ultimately determining their capability and credibility as it pertains to your project's needs.

So who are the people you will work with on your capstone design project, and what can they provide?

Your capstone design professor is the person primarily responsible for your course, whose responsibilities include:

- Course design and development
- Design and delivery of lessons or lectures in the classroom
- Identification and creation of qualified projects, including travel and collaboration with external facilities (e.g., clinics and corporations)
- Coordination of the evaluation of your deliverables and assignment of your grades
- Identification, development, management, and maintenance of all clinical advisor relationships
- Authorization of your travel to present your work and also accompany you when appropriate

He or she will also be the person who should be accessible to you (or make arrangements for another qualified person to be accessible to you) within a few hours and for emergencies, if needed. I like to try to answer student questions within 24 h or less. This is all part of the philosophy, "keep the project moving." You should not lose precious days waiting for help from your primary professor or designee.

Also, he or she will likely have both industry and academic experience. He or she will probably have some general experience related to your project and will also have some depth or expertise in certain areas, but it may not be in all the specific areas where you need help.

While you should see your professor with your classmates during class meetings, you also need to meet with your *academic advisor* in a small team at least once a week, and preferably more. Your professor may fill this role or, if your class is large, you may also have an exclusive academic advisor. This person is a faculty member, who has experience related to your project topic and can help you find and evaluate those resources that we mentioned you would need. Your academic advisor will work closely with you and should be involved in the details of your project just like the rest of your team, and he or she will supervise and assist your weekly progress and project management. Your academic advisor will work closely with your professor to guide you and evaluate you.

If at all possible, you should also have a clinical advisor. This person may not be readily accessible because many doctors have rigorous schedules that do not allow for frequent meetings or correspondence. If you are fortunate enough to find a doctor or other clinician with expertise in your project topic, be sure that you are exceptionally well prepared for your meetings with him or her. You may only be able to spend a few minutes with him or her, and you may only be able to meet once or twice during your project.

Most of your clinical advisors will want to be more available to you than that, and many can keep in reasonable contact by e-mail. But, as with any very busy professional, you should be certain to communicate succinctly and intelligently. You will make a lot more progress during your meetings if you are well prepared for it. The busier the professional, the lesser may be the frequency of your interaction and follow-up.

There are lots of other people on campus who can help you. Graduate students, teaching assistants, fellow students, or professors who are in another major may each contribute valuable support to you and your project. You may also seek technical advice from industry professionals. Your campus technology transfer office can also provide valuable advice and assistance. Do not be shy; ask around. Remember, you are not only seeking help but you are also building your skills in resource development and evaluation.

Keep your advisor informed of your plans and let him know, in advance, that you plan to contact someone; this will help to keep your project on track and also protect your *intellectual property*. Take special care not to discuss your ideas with outside parties until you "clear" it with your advisor. You may need to obtain a *confidentiality agreement* before you hold your discussion.

13.1.2.1 A Snapshot of Your Capstone Design Course

In this section, you will find three tables that provide a "snapshot" of a structure for your capstone course. You do not have to be brilliant or exceptional to manage a design project; it is more important that you are serious about maintaining your discipline. Of course, this also means that you spend an appropriate amount of time on your project. I recommend that you budget to spend about 8–10h per week on your project. This includes class time, advisory meetings, team meetings, and independent work. But remember, your project will fare much better if you spend 10h a week over a 4 week period rather than spending 3h per week for 3 weeks and then trying to cram 31h of work into the fourth week.

Before we go into detail, look at these tables to get an overview of your project milestones and deliverables. Table 13.1 describes what you should do, as a minimum, to keep your project moving swiftly and in the right direction. This basically distills down to *good housekeeping*. Get in the habit of doing these things almost automatically, and much of your project will fall right into place. Table 13.2 describes your first semester, which takes you through the product development process and guides you through the development of your concept, preparation of your proposal, and plans your design. Table 13.3 describes your second semester, and guides you through the process of building and testing your prototype, as well as through the publication and invention process if your instructor requires these elements.

Summary of Major Milestones and Deliverables in a Two-Semester Capstone Design Sequence [1–3]

TABLE 13.1

Routine Course Deliverables

Item No.	Milestone	Period Due
1	Maintenance of a current lab book	Checked weekly
2	Written "Action Plan"	Weekly
3	Team meetings with faculty advisor	Weekly
4	"Project Review Meeting" presentation	Monthly
5	"Clinical Advisory" meetings	2–3 times per semester

TABLE 13.2

Course Deliverables during First Semester

Item No.	Milestone	Week/Mode[a]	
1	Problem definition	3	W
2	Market assessment	4	W
3	Preliminary intellectual property review	5	W
4	"Mission Statement"	5	W&O
5	"Practice" proposal presentation	6	O
6	Mid-semester "Formal Proposal"	8	W&O
7	Confidential team assessments	10	W
8	Project review meeting presentation	11	O
9	Draft "Invention Disclosure"	13	W
10	Formal "Execution Plan"	14	W&O

[a] W = written; W × O = written and oral; O = Oral.

TABLE 13.3

Course Deliverables during Second Semester

Item No.	Milestone	Week/Mode	
1	Completed materials order	1	W
2	Scientific abstract	3	W
3	Detailed test protocol	5	W
4	Working "Proof of Concept" prototype	8	Demo
5	Recorded "Invention Disclosure"	8	W
6	Poster presentation	9	W&O
7	Evaluation of test results	12	W
8	Participation in "Senior Day Exhibition"	13	Demo
9	Final report	14	W&O

13.2 Routine Course Deliverables

13.2.1 Maintaining a Lab Book

The first most important thing you need to do is to keep a lab book. Believe it or not, this is a chief complaint among hiring managers today, "Most students don't demonstrate that they can keep a proper lab book." You have probably been asked to keep a lab book for your chemistry course, but you may not have felt the need to keep the same type of records for other courses or projects.

While companies and various instructors may recommend different formats for your lab book, the goals are the same. You need to be sure that you have detailed and accurate records of your work so that you may review the work, reproduce it, and even protect it from an intellectual property perspective. You should view your lab book as a formal diary. Write everything in it!

There are many good guidelines available for writing a good lab book. Here are two good references to get you started:

http://www.niehs.nih.gov/health/docs/guide-notebooks.pdf [4]
http://www.urmc.rochester.edu/technology-transfer/inventors/Laboratory_Notebook.cfm [5]

You can go to your library or search online for other examples.

Here is a listing of minimum recommended rules for your lab book, including a sample lab page as shown in Figure 13.1.

1. Your lab notebook should be bound so that you cannot tear pages out. You can buy a blank lab notebook that has bound pages that are prenumbered, but they can be expensive. I recommend that you get a composition notebook, which works well if you prepare it properly.

2. Be sure to put your name and contact information on the cover, so that if it is lost, it can be returned to you.

3. Write on both front and backside of the pages. Begin with a table of contents. You can leave a couple of pages blank so that you can fill in your table of contents, but be sure to label them at the top as "table of contents" pages.

4. Do not skip any pages.

5. Always date and number the top of the page.

6. Write legibly in ink. Do not erase anything.

7. If you create a drawing, graph, etc., on your computer, print it out and tape it neatly onto a page. Fold it if you must, so that it fits into the book. Sign your name and date across the tape and be sure that your signature touches both the notebook page and the tape.

8. At the end of the day, if you have empty space on the page, leave it blank and draw a diagonal line through it.

9. Sign the bottom of each page, including the date.
10. You must also collect a witness's signature. One recommended way to do this is to ask your project leader to sign and date one page at the end of each week, including the statement, "I have read and understood the preceding entries."

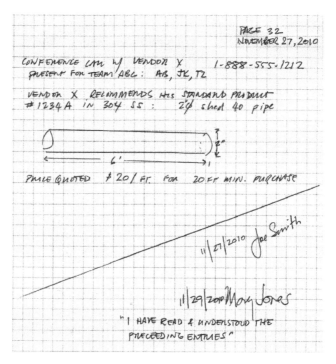

FIGURE 13.1
Sample lab book page.

Remember that your lab book is a professional diary. It is not only a place for you to record your data and observations. You should carry it with you at all times, and use it to record your ideas, notes from any correspondence or phone calls, and even use it as a place to write down contact information, phone numbers, addresses, Web sites, etc.

Here is a practical tip. I know this seems cumbersome to an electronic savvy student. But this process is necessary to properly preserve and protect your work. However, in addition to this process, I suggest that students set up a team electronic file to which all members have access. You should store all your electronic files, collectively, in that place. Be sure it is safe, and that your data is "backed-up" often. You should extract key documents for insertion into your lab book. For example, if you have multiple drawings for a manufactured part, extract the overall drawing and insert it. Bear in mind that if one aspect of that part is novel and potentially patentable, you should be sure to show that aspect in your lab book.

If you have any questions about what to include, you should ask your instructor or your school's tech transfer office.

13.2.2 Weekly Plan

The most significant distinction regarding this weekly plan is to note that it is a not a summary of what you already did, rather it is a *plan of action* for your upcoming week.

This plan will guide the entire team, as well as identify each individual's responsibilities for the week. It will also help your advisors to follow your work, and it gives them an opportunity to help you answer questions, find resources, or perhaps offer a more efficient method to accomplish a task.

Project managers and their teams commonly use a weekly plan in industry. Ultimately, the weekly plan should be a working tool for all project participants, and again, the format will vary, depending on the project requirements. You should develop your own weekly plan so that it best supports your team.

The weekly plan as shown in Table 13.4 should, at minimum, include the following information:
At the top of the page write

1. Team name (or Team ID)
2. "Weekly plan for the week ending _____"

Then set up a chart with columns for each:

3. Item/discussion point
4. Action required
5. Due date
6. Action owner

Try not to make the weekly plan too detailed or too brief. Write just enough so that all who are engaged in the project understand what each other plans to do in the next week. Also, do not list extremely routine items in the plan. For example, if it is understood by all members that everyone is required to go to class on Monday, do not put that information in the plan; it is superfluous and it detracts from important information.

TABLE 13.4

Sample Weekly Plan

Team ABC			
Weekly Plan for the Week Ending September 24, 2010			
Item/Discussion Point	Action Required	Due Date	Action Owner
Wheel size range	Research RESNA standards	September 22, 2010	JK
Maximum load and stress points	Define clinical needs scenario, determine appropriate SF analysis, finite element analysis (FEA)	September 22, 2010	AB, GH
Materials	Research and write short list of frame, wheel, and hub material suppliers	September 22, 2010	TZ
Optimum design	Team meeting to determine optimum design based upon outcomes from above	September 23, 2010	All

Note: RESNA = Rehabilitation engineering and assistive technology society of North America, www.resna.org;

13.2.3 Project Review Meetings

When you start your job, you will recognize that periodic *Project Review Meetings* are common and they take all forms. They can be extremely formal or very informal. You should learn to be prepared to participate and contribute during any type of project review meeting with minimal preparation. After all, your company has hired you to work more so than talk about it. Still, it is important to be able to communicate with other teams and other parts of the company, or even with vendors, and other external resources. So you must develop the skills to communicate effectively.

This can be difficult for engineers, especially when you need to communicate with others who do not have the same technical background that you do. It is very important to be able to communicate in *lay terms* or to provide an *executive summary* of your work. Still, in this process, you need to be careful to avoid oversimplification, or to skip important elements of your work. To do this effectively, it just takes practice. Each time you present your project and obtain questions, comments, or even "blank stares," you will gain good information regarding how to present your ideas better next time. You will also develop skills for future presentations, and it will become easier and easier for you to "hit your mark" on the first try.

To get the practice and skills, you need to communicate—your instructor will ask you to present your project in various ways. This will also serve as an opportunity for you to discuss your progress and to share you project development with your classmates, advisors and other faculty, and administration. Take these opportunities to both learn and teach.

In this curriculum, you will have an opportunity to present your project to the class in some form approximately every 4–6 weeks. Each time you present, be sure to incorporate what you learned from the last project review meeting. Pay attention not only to the critiques your team received, but also to those that other teams received. You can learn and borrow from each other's successes and missteps in class. Remember that this is a learning experience for you and your classmates. You are supposed to make mistakes and have successes along the way. Do not be intimidated by the process. Embrace it and take as much personally from it as you can.

Here are some suggested formats for your project review meetings, based upon real-life situations. You should try to practice any and all of these formats. You can even propose your own formats based upon your internship experiences.

Example: Project Review Meeting Formats/Exercises:

1. Traditional: 15 min presentation with 15 slides (PowerPoint) with additional 5 min for question and answer.
2. The "frantic cell phone call": Assume that you are driving to a client and your boss calls your cell to be sure that you are prepared for your meeting and to review what you are going to discuss. Your conversation is only verbal, without any visual aid.
3. "Impromptu": Imagine that you are walking out late from the office and CEO stops you in the hallway, and asks you how your project is going. You have only 5–10 min to get your point across well. You have nothing formally prepared, but you can draw on a piece of paper or a blackboard as you speak.
4. The "trade show": This is best done later in the program when you have built your prototype or model, or have built a portion of it. Use your model as part of your visual aid kit to demonstrate your project. Keep in mind that this exercise is not just a technical "show and tell." You must also communicate the significance of your work.

13.3 Design Proposal

Your *capstone design proposal* is one of the most critical deliverables. It sets the stage for all your work, going forward, in the rest of your project. The more thorough and accurate you can be in your proposal, the smoother your project will be. If your proposal is weak, you will need to invest extra time afterward in order to complete your project. As discussed earlier, you want to avoid cramming; it just does not work well in project management.

To help you prepare a strong proposal, your instructor and advisors will guide you through a process that enables you to research, identify, and make certain definitive statements regarding your proposal topic. In this process, you should be able to construct a basic *design concept*. This process may not be particularly formal. You will need to step into the driver's seat for your project! It is up to you to dig into the information your instructor and advisors provide, and then meet with them—and any other expert resources—often to ensure that you are preparing the best concept.

Normally, this assignment is due near the midterm of the first semester. At this stage, you may not have a final concept. You will certainly have an idea about what type of prototype you would like to build, and what concept you wish to demonstrate. But it may take more time and more research in order to determine the details. Still, you should have at least one basic concept reasonably well developed and if needed, include an alternate concept.

For example, let us say you know you want to build a rollator walker that has new safety features to prevent a person from accidentally falling when using them. You have determined, from your research and from conversations with your clinical advisor, that many users forget to engage the hand brake when needed. So, you have decided to install a fail-safe braking system on the walker. You are not quite sure whether you will put a brake that responds to excessive pressure on the wheels or whether you put a safety interlock on the seat. But, you have performed sufficient review of both options, you can describe that both are feasible, and you can describe how you might go forward to build either system.

13.3.1 Proof of Concept

It is important for you to recognize that your first prototype needs to *prove your concept*. You will not have enough time to make a perfect final product on your first try. This rarely happens anyway, neither in this class nor in industry. You will always find ways to improve your design after it is built.

As discussed in Chapter 12, design is an iterative process. When building your first prototype—*the alpha version*—you need to focus on proving that your main idea is valid and feasible. Do not try to meet all "final product" criteria. That will demand too much time and energy and will likely prevent you from completing the critical *proof of concept*. When such mistakes occur in industry, it "side rails" the project and the project is usually scrapped, even if it was a good idea.

Once you render an alpha prototype that, indeed, demonstrates your "proof of concept," you can garner support for yourself and your team or a successor to carry it forward. At that time, make the necessary improvements toward commercial readiness. For example, if you decide that you want to design a handheld electronic device that aids a physician in diagnosis and can be carried in his pocket, do not worry about making the device very small in the alpha version. Focus rather on demonstrating that the concept for the diagnostic tool may be viable. You can build a larger device with standard electronic components and demonstrate feasibility. This will provide your proof of concept and will cost less and take less time.

Once you do prove feasibility, then you can raise money and support to have custom electronic chips made so that you can meet your size criteria. This will likely be beyond the scope of your design course experience. But, I have had many students write patents and one team even started a company based upon their design projects. There are many ways for you to carry your project forward if you wish. If you are interested in doing so, you should talk to your advisors and your school's technology transfer officials to understand how they might help you with that.

13.3.2 Presentation Elements and Evaluation Criteria

Let us review the elements that comprise a good proposal in the following. If you can address these elements well, you will have a very sound basis for a strong proposal. Depending upon your specific project objectives, you may find that you must lend more weight to certain areas. Your classmates' projects may have the need to focus on other areas, so do not feel the need to be too uniform. These elements are guidelines that you should review with your team and advisors to determine how to establish priorities that best suit your topic and objectives.

Capstone Proposal Elements and Evaluation Criteria

A. Mission statement/statement of purpose:
   Unmet need
      Demographics
      Value summary
B. Background and precedent:
   Pathology
      Clinical practice today
      Alternates or competition
      Prior work
C. Concept description:
   Concept design
      Approach and use of engineering disciplines
      Budget
      Estimated deliverables with project schedule
      Expected outcomes
D. Strategy overview:
   Technical (Identify major problems to solve/tasks to execute)
      Regulatory
      Market

Legal/reimbursement/patent

Resource utilization

Challenges or obstacles

E. Performance observations:

Team dynamic

Individual contributions/performance

Resource utilization

Ability to define direction

Presentation of materials

13.3.3 Mission Statement/Statement of Purpose

In this section, you need to identify and explain the problem you plan to address and how you plan to solve it. This may seem simple, but in medicine, it can become very complicated for many reasons, including those described in Chapter 12. Recall the problem your friend Bob Valdictoran had with a simple design challenge; he did not appropriately identify the *unmet need*. He would have done a much better job if he did not assume that he knew the problem, but instead had asked the customer what the problem was.

Similarly, the best way for you be sure that you have identified the problem correctly is to be sure that your clinical advisor concurs with your understanding of the problem. You may require several discussions to get to this point; do not expect to do it all at once. You should engage in an initial discussion with your clinical advisor, then go back and do your research and review it with your advisors. Because this may require several cycles, be sure to meet with your advisors as frequently as possible during this process. Consider meeting with at least one of your advisors two to three times per week.

Once you are certain you have identified the unmet need, you should describe it, clinically, and also give an idea regarding the impact of the problem. Describe the demographics; who is affected by the problem, how many patients, doctors or hospitals are affected? Is the problem global or just in the United States, or just in a specific region?

You should also include a value summary. It is best when you have quantitative information. If you can find information describing market size, or incidence rates of disease, include it. Include information like the cost burden to public health, or cost per patient or per hospital.

It is important to note that there is no problem too small in medicine. If there is anything you can do, as an engineer, to help save or improve someone's life, that's valuable!

You will find that, because health care is a business sector, many companies or financiers will want to see *cost drivers* in your proposal. Some supporters may only be interested in potential *blockbuster* products, but there will be others who support a good project. Do not feel like you have to turn your project into a mega financial success. Still, you should become accustomed to developing a financial value statement to the best of your ability. If you were to carry your project forward toward commercialization, you will likely obtain help from experienced business people to write a business plan using this information, but your expertise is needed to help discern which information is applicable.

Mission Statement Exercise
In one paragraph (not exceeding four sentences) describe

1. The *unmet need* that you plan to address.
2. The *innovation* that you will contribute and the *results* you expect to produce.
3. How your project will save lives or improve quality of life? (Can you help to improve safety, efficacy? Reduce morbidity or mortality?)
   a. Who?
   b. How many?
   c. To what extent will the condition improve? (e.g., percentage reduction of disease prevalence, how many fewer days in hospital, etc.)
4. How your project will save costs or generate income?
   a. How much?
   b. In what areas and population segments (in the United States, globally, or in specific regions or population such as "adults under 70" or "women over 40")?

You should review this exercise often and rewrite your mission statement each time you prepare a presentation. As with everything, this will improve with practice and experience.

Keep in mind that, realistically, you will not be able to solve the problem completely in your 32 week course. So be sure that you identify what you will contribute, specifically toward your mission in the time frame allotted for your course. As discussed previously, your project can be carried forward in another form. Be careful not to overstate what you will contribute in the context of the current project.

13.3.4 Background and Precedent

It is likely that another researcher has addressed this problem before and that doctors or medical companies offer some kind of solution to the problem. You should perform research to understand the problem thoroughly from many perspectives. In your proposal, you should summarize any prior work such that you define your *step off point*, that is, just how far did other researchers, clinicians, and engineers come, prior to you, in developing a solution.

Be sure that you understand and can summarize the *pathology*, that is, the disease, injury, or other physiologic condition you plan to address. Also, be certain that you can describe clinical practice today. When there is a commonly accepted method of practice in medicine, it is referred to as *the gold standard*. However, there may be different current practices. Be sure that you understand all of them and assimilate them in your statement.

There may also be new potential devices, therapies, or drugs that are being developed. You should conduct market research to understand what alternate or competitive solutions may be on the horizon.

A good example is the topic of "Back Pain." This is a very big and complex problem and there are many different opinions regarding how to address the problem. You have probably seen or heard about people with back pain who have had surgery, or have had injections, or take pain-killing drugs, or see chiropractors. You might have even heard about special stretching machines, nerve control devices, or spa products that are marketed. Still, this is a very big problem and a better solution is needed.

If you use these guidelines for reviewing and defining the background and precedent in your proposal, you would learn that there are different types of back pain—some are better understood than others—and that there is probably a *niche* that you and some of the precedents address. Then you can perform an evaluation regarding the relative state of success to date in that niche, you can describe how effective the current practices are.

13.3.5 Concept Description

Now you can state what you plan to do that is innovative, and describe why it might yield a better result. You have probably learned by now that engineers communicate with drawings, graphs, charts, or tables. You can convey a significant amount of information with a properly prepared drawing or graphical depiction. You should prepare a drawing or sketch of your concept. It does not need a lot of detail. Start with a model that simply describes the key features that will prove your concept. It should be legible and depict the main ideas that you plan to demonstrate in your prototype.

Do not get bogged down in defining the detail of subordinate or dependent ideas at this time. After all, if you need to make an adjustment to your main idea, all the other, dependent aspects of your design will change too. Having said this, you should, to the best of your ability, understand how you will complete the detailed design once the main ideas are locked down.

If we go back to the example of installing a passive braking system on a rollator walker, you should prepare a drawing describing where the brake may be installed and how it might operate. It is too soon to know most of the detailed specifications. However, you should be able to define your *design parameters*, for example, "the system needs to fit onto a standard folding frame and therefore must be smaller than 2" × 5"." You will also identify other critical parameters, for example, the modified unit must not be heavier than 30 lbs so that it can be lifted and transported, etc.

Include equations, formulae, dimensions, typical or actual data. Identify any other critical issues or constraints. Show how you will verify your calculations once you obtain test data. Be as quantitative and specific as possible. You will find that, as you strive to provide specific quantitative information, you will identify the questions that outline your plan for your detailed design.

You should have a clear understanding of what work you must do in order to determine the details that you have not yet defined. What are the open questions you have yet to answer? If you have an option or alternate to consider, write a decision tree; be sure that you include a deadline for that decision so that you have enough time on the other side of that decision point to complete your project. What is your plan for obtaining those answers? You will certainly apply your engineering skills. Which skills and in what way?

You will also need to determine a budget. If you do not have detailed prices, yet, you should obtain some price estimates for parts and services. Your instructor will provide you with spending guidelines. It will be your responsibility to develop a solution, to the best of your ability, which falls within your budget. Remember, your model does not have to be commercially ready; it just needs to demonstrate the feasibility of your concept. You may need to get creative, but rest assured that you will deal with the same issue in your job, so it is good to learn how to work with a budget.

You should include your *Estimated Deliverables* with a *Project Schedule*. In Section 13.4, there is an exercise that helps you to write a project schedule. In that exercise, you will need to define the *major milestones* that you wish to achieve, and when you will reach them. You should summarize this information.

Assuming all goes according to your plan, what results will you deliver? What do you consider to be successful results? Define your expected outcomes. What will your prototype look like or what elements will you have improved? Describe how you expect it to perform.

13.3.6 Strategy Overview

In Section 13.3.5, we talked about your *plan*. To create an effective plan, it is helpful to look at your project from several perspectives, and then assemble those ideas into a *strategy*. There will be very specific strategic issues that will become critical to your project; many of these will be different from your classmates' or from other experiences or examples.

It will be up to you, with the help of your advisor, to evaluate your project and determine the best specific strategy for your project. This can be a daunting task, even for an experienced project manager, so it helps to have a process to guide you. For a biomedical engineering project, there are certain areas that will always contribute to your strategic plan. We already know that your timeline and budget are critical. There are other areas that you will need to evaluate: *technical, regulatory, market,* and *regulatory/reimbursement/intellectual property.* Depending upon the specific nature of the project, this listing may not be comprehensive, but when you use it as a tool for your strategic planning process, it will help you to identify most of your major ideas.

13.3.6.1 Technical

First, let us look at your project from a technical perspective. For an engineering project, this tends to be the major consideration and is also the most concrete, so it is often the easiest way to start building a strategy. Here, basically, you want to identify major problems to solve and define what must be done in order to solve them. While you may not know exactly how you are going to do this, you should know what must be done, and you should also have some ideas about how you might go about it.

This becomes a list of tasks that you must execute. Be sure to identify the *resources you will use* in order to complete the tasks you describe. Be as specific as possible. Make note of the tasks that need clarification or better development and state what is needed in order for you to fully define that task. Will it be straightforward, or is there a *challenge or an obstacle* that might prevent you from completing that task?

Referring to our rollator walker project as an example, you may have decided that you will modify an existing walker, but they are expensive and it is beyond your budget to purchase one. This is a challenge. Can you find a used one for less money, or can you get one donated from somewhere? Perhaps you cannot. Does this mean that you must scrap the project?

It would still be a valuable project if you could build a model of your novel brake mechanism using a typical walker caster in your model. You can even build your own frame from steel or even wood if your resources are limited. While you might not be able to prove overall efficacy of a walker, you can certainly still demonstrate the brake mechanism, and therefore you have taken the first step to prove your concept. If the idea is a good one, someone will find more money and more time to dedicate to further development in subsequent efforts. Remember, design is an iterative process.

13.3.6.2 Regulatory

While the execution of technical tasks might be foremost in your plan, there are often other factors that influence your plan. In biomedical engineering, you must always consider *regulatory* requirements, for example, FDA [United States], CE [Europe], and JMHW [Japan], as part of your strategy. In order to clear your project for the target market, it must meet specific regulatory requirements.

This can be a very complicated process, and companies who bring products to market utilize many experts to help guide them through the process.

At this stage, you are certainly not expected to prepare a detailed regulatory strategy. Still, you should understand some basic elements so that you make decisions regarding your design that will give you the best and most efficient opportunity to meet those qualifications.

There are many good resources for learning more about regulatory strategies. Your professors and advisors will guide you to the best ones for your project, or might arrange for someone with expertise in regulatory strategy to talk to your class. I believe it is useful for students to peruse the Web site www.fda.gov. Search for products and topics similar to yours and become familiar with the site and what it offers. This will grow more useful to you as you gain experience and will become an invaluable resource to you throughout your career.

13.3.6.3 Market

While regulatory requirements are a critical consideration for the marketability of your medical product, there are certainly other aspects to the market that you need to understand and incorporate into your strategy. These are some more traditional and general market considerations. You may have had a class in marketing that has given you tools and skills to define and evaluate your market and you can apply what you learned here.

If you have not had any experience in evaluating and defining your market, you can still be effective by answering some very basic questions. First, let us define the *demographics*.

- What is your target market? For example, white males over 50, or patients with type 2 diabetes, or doctors who practice pain management?
- What is the size of your market? How many patients, or what percentage of the population? What is the cost burden?

Be sure to qualify your definitions thoroughly. Many students make the mistake of mixing up regional and global data. Be sure you describe your target with the correct qualifiers, for example, 75,000 new cases occur annually in the United States, or 15% of the world population has this gene. Also, be sure to include a value statement and trends with references, for example, $100 billion per year is spent annually in the United States to treat obesity, and the prevalence of overweight and obese individuals has increased from 44.3% in the late 1970s to 52.6% in the early 1990s [6].

Check with your advisors and your library for help in finding references for the information you need. The information you seek may be stated directly in the references, or you may have to deduce the information from various pieces of data that you collect. If you are careful to pay attention to the basis and context of the data you collect, you can create a good picture of your market. Caution: if you misinterpret or mix-up the information, you can find yourself way off-base. You will depend as much upon your biomedical engineering expertise as your market analysis skills to make the right judgments and interpretations of the data. If you have a multidisciplinary team, be sure you work together. Do not simply "pass this off to the marketing guys"!

13.3.6.4 Legal/Reimbursement/Intellectual Property

Biomedical engineers must be keenly aware of legal, reimbursement, and intellectual property issues at the very beginning of the design process. Earlier, we discussed the regulatory issues from a marketing perspective. Regulatory guidelines are also key to ensuring that the design process is conducted in a legal manner. Most companies have regulatory specialists who are included in the design process for this reason. Those experts know the criteria for product development and testing in the lab and on live subjects, both animal and human. In addition to FDA guidelines, there are many national laws and institutional regulations regarding how clinical trials may be conducted, such as HIPAA (Health Insurance Portability and Accountability Act) and IRB (Institutional Review Board), for example, and these guidelines must be adhered to, strictly. They also offer a wealth of knowledge about *precedent* successes and failures, which will be invaluable to creating a design plan with the best chances for success.

This is a very broad and complex area. It takes a lot of experience to understand these processes thoroughly. You should arrange a meeting with your clinical advisor to identify what aspect of this field you should begin to learn about for your project. It is important for you to recognize that you must consider these aspects fully and you must have guidance in this area. If you ignore this in your design planning, you will have no chance of succeeding.

The topic of *Reimbursement* almost always goes "hand in glove" with this part of your design planning and evaluation. Very simply stated, you need to understand whether or not the patient's insurance will pay for the product or procedure that you are developing. Again, this is a very complex field. It ties in closely with both "legal" and "marketing" planning. For your basic purposes, you should find out whether there is an existing reimbursement code that "fits" your product. If there is, that is good for your near term market opportunity.

If your product is novel, there may not be an existing reimbursement code. In that case, you will have to include a plan for developing a reimbursement code. These codes are established by the AMA (American Medical Association) via a very formal and lengthy review process. It could take many years to get a new code. Oftentimes, you will find that the specialists who work in this area have a law background. At this stage of your project planning, you need to be aware that a reimbursement plan will need to be included in your later stages of product development, and that your time to market or, more accurately, collecting revenue in the market will be extended until this is satisfied.

It is quite possible that your idea is novel enough to be patented. You should perform as much research as possible to determine whether your idea has been "done before." Two good starting resources are www.uspto.gov and www.ncbi.nlm.nih.gov/pubmed (Pubmed). Your university or clinical advisor may have access to more specialized resources and you should inquire about what might be available to you.

While you might think that only a lawyer can write a patent, this is not really true. Some inventors have written their own patents and you too can, if you wish. The Web site www.uspto.gov provides instructions on how to do this. Certainly, it is best to utilize the expertise of a good patent attorney whenever possible. Bear in mind, however, that the attorney will rely upon you to identify and describe your novel idea as well as provide details regarding its applicability.

From a very basic view, you should be thinking about how your idea might be novel, and what type of patent might apply to your idea. There are four types of patents: *method*, *device*, *composition of matter*, and *utility*. So, if you have conceived of "a new way to do something, a new or improved device, a new material, or a new use for a technology," your idea may be patentable. I would highly recommend that you list as many novel aspects as you can, then consult with your university resources to obtain expert advice and guidance in writing a patent application.

In summary, you should begin by asking these three questions:

1. What regulatory guidelines apply to my project?
2. Is there an applicable reimbursement code already in place for what I am proposing?
3. What are the specific novel features to my project? (Prepare a list)

Then you should obtain guidance from your advisors who have expertise in these areas to guide you to resources and experts that will enable you to effectively build your plan and consider these elements.

Once you have compiled your strategy, you should prepare a summary and call it your *Strategy Overview*. Be sure to identify any challenges or obstacles that you anticipate, and identify how you might overcome them or alter your direction if you encounter them. Also, be prepared to take notes during the question period of your presentation; your listeners may be able to offer ideas or resources to help you address challenges that you might anticipate.

13.3.7 Performance Observations

It is important to your future success as a project manager that you develop certain skills in addition to your technical and clinical knowledge. Sometimes, these are referred to as *soft skills*, although they are sometimes the hardest and most critical aspect in a project. In fact, technical hiring managers frequently complain that such skills are in very short supply in the hiring pool, so if you develop these well, you will certainly gain a competitive edge in the workforce. Recall our friend, Bob Valdictoran, and think about how much more successful he would have been with his assignment had he developed these skills more thoroughly.

Your advisors will be evaluating your performance as if you were working on a project at your first job. This is a brief list of the soft skills you will need to apply and hone in your first job:

- Team dynamic
- Individual contributions/performance
- Resource utilization
- Ability to define direction
- Presentation of materials

It is very important that you demonstrate your effectiveness in these skills by the time you present your proposal. If you are experiencing challenges in team performance, be sure to address them early and do not be afraid to ask your advisor for help.

It has been my experience that many students in the capstone courses struggle with some or all of these items at times. In particular, I have often found that students need to understand the importance of a smoothly operating team. Your project will need each and every team member's full contribution—there is too much work and not enough time to ignore an individual or allow a team member to "slack off." The whole team will suffer if one or more individuals are not performing fully. You need to figure out how each individual member can contribute equally to the project. This sometimes requires a serious "heart-to-heart" meeting among team members. Your advisors are experienced in these matters and can help.

Be careful not to delegate assignments too narrowly. While a certain individual might lead on a certain task, all members are ultimately responsible for that task. Be sure to plan substantial "crossover" in your weekly plan assignments. One effective method for managing team dynamic and individual contributions is to "pair up" the two individuals with the most extremely different strengths. This way, each can learn from the other, or the very strong member can provide guidance and direction to the less experienced member; at the same time, the very strong member has "an extra pair of hands" to execute tasks more quickly.

Another common challenge for students is the ability to broaden your use of resources. Let us face it, until now, you have been trained, mostly, to go to one person—your teacher—for all the answers. In your job, there will not be just one person who has all the answers, or if there is, he or she will be way too busy to help you all the time. Nor will there be a "directory" of "who knows what" in your new employee handbook. You will have to find people and resources to help you. You will need to obtain the most current and expert support available, and that means that you cannot simply go to the same person or Web site for help each time.

It would be unusual if your advisor has sufficient expertise to "give you all the answers" on your capstone project. You should take the initiative to identify as many varied and expert resources as possible. Certainly, you can review those resources with your advisor, but you should not expect him or her to hand you a complete list of all you need to execute your project. The better you can develop some independence in your identification and use of resources at this time, the easier it will be for you on future projects. This will also help you to obtain the most effective results in your capstone project.

Once you have performed your research, using the varied resources, you should be able to describe your project needs and next steps. Again, do not expect your advisor to simply tell you what to do next. At each step, you should prepare a list of needs and potential next steps and then review them with your advisor who can help evaluate your list of ideas and also help you organize them. When you get into the workforce, you will be working to solve problems for which there are no existing solutions, and you will need to define the direction for your project. In fact, it is possible that you are already working on an unsolved problem in your capstone project, and you must define the best direction for developing a solution!

13.3.8 Proposal Format and Tips

Your advisor will provide you with certain presentation format requirements. One format I recommend is that your oral proposal presentation be 15 min long and allow 5 min for questions. For this format, you should prepare only 12–15 PowerPoint slides and no more. You should ensure that you capture an appropriate *summary* of all the recommended elements in this section.

Do not try to squeeze everything onto your slides. Anyone should be able to read your slides without a microscope and in a brief period of time. In fact, you should use as few words as possible on your slides. Charts, diagrams, photos, sketches, etc., are all very effective tools for communication. While you may wish to prepare these in very detailed form for a written proposal, you should simplify them for your oral presentation so that they are visually comfortable and convey the key points.

Your proposal presentations—both written and oral—should be well organized and should flow without "skipping around" or leaving gaps of critical information. Your proposal should be neatly prepared. Feel free to use props, or demo models, or other visual aids as well as traditional presentation slides. Most importantly, you should review and rehearse your proposal to ensure that your audience—classmates and professors—can clearly understand your concept and follow your development. You should practice your presentation with classmates, roommates, and advisors in advance. Each time you do, you will find a way to improve; the secret to a good presentation is preparation and practice.

13.4 Planning a Project Schedule

You have probably learned about project planning and scheduling in some of your courses. There are many good project-scheduling tools; one commonly used tool that you might have learned about is the Gantt Chart. This is a very effective method to plan project details when tasks and roles are well defined. I have found that for students who are in the early planning stages of their capstone design project, such a tool requires more detail than that which the student has available.

Students generally do not have enough previous experience to know just how long a task will take, or even how to organize the tasks that need to be done. If you attempt to use a detailed scheduling tool without the proper information, it becomes ineffective.

13.4.1 Critical Path Method

There is another tool that is commonly used by project managers. It is less detailed and more intuitive, but it helps by outlining a structure and time frame into manageable "compartments." This method is called the *Critical Path Method*. An exercise and example (Figure 13.2) for planning a capstone

• August 30	– Today	(1)
	– Research and concept development	(5)
• October 20	– Proposal due	(4)
	– Detailed design	(6)
• November 24–26	– Thanks giving break	(7)
• December 10	– Design freeze	(3)
	– All parts ordered	
• January 17	– Initiate prototype build	(8)
	– Prototype build	(12)
	– Prototype completed	(13)
• March 14–18	– Spring break	(10)
	– Testing	(14)
• April 27	– Final report due	(11)
• May 26	– Graduation day	(2)

FIGURE 13.2
Example page for the critical path scheduling method.

design project in a typical two-semester sequence is included. You should follow the exercise given in the following to develop your own project plan:

The Critical Path Method for Planning Your Capstone Design Course
Open your lab book to a blank page.

1. At the top of the page, write "Today" and include the date.
2. At the bottom of the page write "Graduation Day" with the date.

    You will spend your first semester preparing your design. You should have your detailed design completed by the end of the first semester.

3. In the middle of the page, write "Design Freeze" and write in the date of your final exam (or last day of class).

    We just discussed your design proposal. This should be completed somewhere near the midterm.

4. Halfway between "Today" and "Design Freeze," write "Proposal Due" and include the date of your midterm exam.

    You will recognize that you have two large compartments of about 8 weeks each before and after your "Proposal Due" milestone.

5. In the upper compartment, write "Research and Concept Development."
6. In the lower compartment, write "Detailed Design."

    In keeping with a typical academic calendar, you will have about 4 or 5 days off around Thanksgiving. Because of the short week, you will only have time to address essential items, and keep in mind that suppliers and other people will only be working part of the week as well. Check the date of your break.

7. Near the bottom of that lower compartment entitled "Detailed Design," write "Thanksgiving Break" and include those dates.

    Now look closely at that compartment. You will have only a week or two to complete your detailed design after you return from Thanksgiving break! This means that you should be ready to put the finishing touches on your project before you leave for that break. This milestone should help you define critical deliverables for your project.

    You will probably have a few weeks off for winter recess and return for your second semester sometime in early to mid-January. In your second semester, you will build and test your prototype. You will want to start building your model as soon as your return from winter recess.

8. Leave a small space below "Design Freeze" and then write, "Initiate Prototype Build" and include the date of your first day of class in the second semester.

    Think for a moment, what do you need in order to build your model? Of course, you need the parts! You should have ordered your parts and allowed about 3 weeks for delivery. Be careful to understand the time required for the ordering process at your institute. Just because a supplier has a part "on the shelf" does not mean you can always get it in a few days. Also, some parts take longer to deliver or manufacture. You should have a very specific understanding about how long it takes for you to receive your orders.

It is a good idea to build in some cushion too, in case there is a glitch in the ordering/shipping process, or if your part is incorrect. You will experience that "things don't always work on the first try." If you do not allot time in your project schedule to revise your design, you may not be able to complete it in the time allocated.

So, getting back to your schedule: in order to have your parts on hand when you need them, you will need to place your orders at the end of your first semester, before you leave for your break.

9. Just below "Design Freeze" write "All Parts Ordered" and use the same date as you used for "Design Freeze."

Let us define two major compartments in your second semester. As mentioned earlier, you will build and test your prototype in this period; you should allow about equal time for each. Since spring break usually falls near your spring midterm, we can use that to define compartments.

10. Midway between "Initiate Prototype Build" and "Graduation Day," write "Spring Break" and include those dates.

Even if you plan to work on your project over the break, be aware that your teammates or key faculty may be away, and your campus facilities, for example, shops and labs may not be accessible, so you should not assume that this is a fully productive workweek. To be realistic, you should also note that your completed project would be due a few weeks prior to graduation day.

11. Directly above "Graduation Day," write "Final Report Due" and include that date.

12. In the upper compartment of the second semester, write "Prototype Build."

13. Just above "Spring Break," write "Prototype Completed."

14. In the lower compartment of the second semester, write "Testing."

Now you have defined your milestones for the second half of your project, in which you plan to complete your prototype just prior to spring break and you will complete your testing before you issue your final report.

After completing the project scheduling exercise, you may have a reaction that is common among my students: "panic!" There is simply not enough time to do everything you want to do on this project. While this is true, you should not panic. Most projects must be completed with less time than is "needed" and often with less funding than is desired as well.

A critical skill in project management is to design and drive the project toward the most successful outcome, despite the tight time frame and sparse resources. This means that you must frequently assess the project plan and you may need to adjust or revise it several times. At all times, you must keep in mind that you should be working toward proving your concept in the best way possible, as I like to say, *keep a straight line to your proof of concept.*

As you go along with your project, you may need to simplify your tasks and apply some creative thinking about how you can still deliver a successful result. In some cases, you will need to decide that you cannot go as far as originally planned, and you will need to revise your endpoints so that you have some sensible goal. What you do not want to do is bluntly or abruptly run out of time or money in the middle of a task. The more closely you monitor your plan, the more effectively you will manage and utilize your precious resources.

13.5 Project Execution Plan

Once you have fully devised your concept, it is time for you to plan how to build it and test it. While you might be eager to pick up some parts, run to the lab and begin tinkering with your idea, such an approach could cause you to digress off course if you have not first prepared an *execution plan*. I require that my students submit an execution plan at the end of their first semester. If you do the same, this will allow you to order your parts before you leave for winter recess and you can focus upon building your project as soon as you return from your break.

What exactly do we need to include in an execution plan to be effective? Well, essentially, you must provide enough information so that you can order all the right parts, so that they will fit together as you build your prototype model. Also, you must understand how you will test your model once it is built. In essence, you need to create a plan that encompasses these aspects:

Execution Plan Components

- Detailed design
- Specification
- Bill of material
- Test protocols
- Detailed schedule

You have already learned how to prepare a detailed schedule. The following sections will describe how to prepare the other components listed. When you have completed each component, you can assemble them into one document that comprises your *Execution Plan* and use that document as a guide for your work in the next semester.

13.5.1 Detailed Design

This means that you must complete a detailed design of your model. You must be certain of all the dimensions and understand how it will be assembled. Remember that you will be spending money to purchase parts and raw materials. If you buy the wrong item, you will not only be wasting money, but you will also lose time if you have to place a new order for the correct item.

You should have prepared a detailed *assembly drawing* of your model as well as detailed drawing for each of the parts which you plan to build—or have built for you. You should also fully understand how your procured components would be integrated into your model. If, for example, you are including a motor in your model, you need to know the dimensions and performance criteria of that motor. Be sure to have reviewed motor catalogs thoroughly and learn about what is commercially offered. Usually, there are more options and criteria than you are aware of, and you need to fully evaluate those details before you commit to a design.

13.5.2 Specifications

To be certain of your detailed design, you must perform calculations that verify that the model will work properly. In order to do so, you must understand the conditions in which it will be used and be able to define those conditions quantitatively. You need to define a set of *operating conditions*. You also need to consider reasonable extreme conditions and be sure that you include sufficient *safety factors* to accommodate such conditions.

For example, when designing a home to be built by the ocean, you might apply standards for a "Category 5 (C5)" hurricane, even though most days in that home will be sunny and mild. The sunny and mild days are the operating conditions, while the C5 hurricane is a *design condition*. Note that it is not expected that your model will be subjected to constant design conditions, just like the occurrence of a C5 hurricane is rare, but it must be considered for short durations. Different industries and engineering practices have different standards, and your advisor will guide you to the appropriate ones.

It is also important that you understand how to define your dimensions and *tolerances*. We know that few things fit "perfectly," but we usually can permit some leeway. If you fully define your dimensions with the appropriate tolerances, you should be able to avoid dimensional fit errors.

When specifying items to be procured, such as the motor we mentioned earlier, be sure of the performance criteria. You need to be thorough across many disciplines. While you might, for example, be building a motorized wheelchair and believe that you have fully evaluated the service factor and other necessary engineering criteria for all available 10 hp motors, did you consider that the motor may need to meet certain standards for the FDA, or for the American Disabilities Act, or maybe it must be waterproof—and to meet any or all of these specifications, it would need a larger *footprint*.

It is not easy to define proper specifications. It is challenging to even the best engineers to fully understand how a load is actually applied in practice or how a complex material will actually perform. This is why you perform tests on your model to compare its performance to that which you have specified. In many cases, you will need to make some minor dimensional adjustments during assembly, even if you planned very carefully. Keep these practical challenges in mind during you planning, and try to give your team as much latitude as possible when detailing your design, for example, leave room to drill a different sized hole and purchase extra length in your raw materials.

13.5.3 Bill of Materials

Once you have completed your detailed design, and you know all the materials or parts that you need to purchase or have built, you should make a list. This list is known as a *bill of materials* (BOM). A BOM is a useful tool. In addition to being a list of parts and components, it can also be used to summarize your anticipated costs. It is essential for quality management, and in cases where quality audits are performed, it is a critical part of your record. In a large company, it is a document that can be shared by many departments.

Every organization has a slightly different way of presenting a BOM, but the general information on it is essentially the same. Originally and essentially, the BOM provides enough information so that an order can be placed for that item. Think of the forms that are in the center of your favorite clothing catalog. Let us take J. Crew's order form for example. If you were to order a shirt, you would include the following information about the item: How many do you want (quantity)? Which shirt (item description or number)? What color (material)? What size (dimensions)? Who is the supplier (in this case, J. Crew)? And, how much does each item cost (unit price)? Adding together, how much for all pieces of this type (extended price)?

In the same way, you would specify each of your components. If you were ordering some screws, for example, you would need to include a summary as outlined earlier:

Quantity = 6
Item/Description # = 12345
Size = 1/2" × 1" 1/2–20
Material of construction = 316 stainless steel (SS)
Vendor = Home Depot
Unit price = $0.60
Extended price = $ 3.60

A sample format for your BOM is provided in Table 13.5.

TABLE 13.5

Sample Bill of Materials

Quantity	Item/ Description	Size	Material of Construction	Vendor	Unit Price	Extended Price
6	# 12345	1/2" × 1" 1/2–20	316 SS	Home depot	$0.60	$3.60

13.5.4 Writing a Testing Protocol

Once you have built your model or prototype, you will need to test it to ensure that it functions properly and safely for the designated purpose. While you may have already performed some sample or anecdotal trials along the way, you must now perform a formal test and document those results. It is not unusual for a medical device or technology to undergo tens or hundreds of tests during the course of development—all of which requiring formal documentation.

You will not have enough time in the context of your capstone course to perform all the necessary tests on your model. Still, you should select one or a few tests that you can perform and that will best support your top priorities in your proof of concept. You and your advisor will determine what tests you should perform. Once you have done so, you can follow the outline provided in the following to write a *testing protocol*. This protocol should be written before you go to the lab to perform your tests. In fact, it should be written as part of your *Execution Plan*, and then used as a guide for your testing once you go to the lab with your model.

Writing a Test Protocol
In general all test protocols have a header section, which contains
Test title
Test number
Company/team name
Effective date
Approval date with signature section
Revision number with signature section

Page 1:
Executive Summary:
This is a short description of the test and the rationale behind the test. It also includes a results section and a sentence or two on how the results are related to your rationale. This section is generally less than 200 words.

Page 2:
Purpose:
The aim of the test. What will your test show and why do you need such a test?

Materials:
List the materials (manufacturer and/or model number, if applicable) that will be used.

Methods:
Provide in detail the test description including test parameters. This should be written in such a way that upon completion of the test, all you would do is to change grammar from "will do" to "was done." Pictures or drawings of experimental setups are very useful and valuable here.

This section will also describe the data reduction and statistical analyses that will be performed on the experimental data.

13.6 Final Report

If your project ran entirely as expected, your final report would look just like your *Execution Plan*. As you would expect, however, things change as you actually build and test your project. You also learn more about the project and begin to build your own expertise. Your final report should be an accurate record of your actual work. It should reflect any changes and it should also reflect your growth. You should fine-tune your mission, be more precise with your background and research, correct your drawings to match what you have done, or even show what should be done in the next cycle, and most importantly reflect upon lessons learned.

As mentioned earlier, *the design process* is iterative, and engineers will almost inevitably figure out a better way to do something after they have tried it once. Now that you have been through a design cycle yourself, you too can apply your experiences and expertise to make recommendations for improvement and future development. This is a very important part of your capstone experience. When you have completed the design conceptualization, building, and testing process as outlined in this chapter, and you can make thoughtful recommendations for the future, you have succeeded. Congratulations!

References

1. Hazelwood, V., Wisniewski, H., and Ritter, A.B. (2007). Entrepreneurship in biomedical senior design, from classroom to corporation. In *International Conference on Engineering Education, Annual Symposia*, Coimbra Portugal, September 2007.
2. Hazelwood, V. (2009). The value in linking entrepreneurship and undergraduate engineering education. In *The 2nd International Multi-Conference on Engineering and Technological Innovation: IMETI 2009*, Orlando FL, July 2009.

3. Hazelwood, V., Valdevit, A., and Ritter, A. (2010). A model for a biomedical engineering senior design capstone course, with assessment tools to satisfy ABET "Soft Skills". In *Capstone Design Conference*, Boulder, CO, June 2010.
4. Guide to keeping laboratory notebooks. National Institutes of Health. http://www.niehs.nih.gov/health/docs/guide-notebooks.pdf, Last accessed date July 1, 2011.
5. The bridge from science to success, technology transfer. University of Rochester, The Medical Center. http://www.urmc.rochester.edu/technology-transfer/inventors/Laboratory_Notebook.cfm, Last accessed date July 1, 2011.
6. Jakicic et al. (2001). Appropriate intervention strategies for weight loss and prevention of weight regain for adults. *Med. Sci. Sports. Exerc.* 33: 2145–2156.

Appendix A: *Selected Reference Values*

Blood chemistry

Ammonia (P)	15–120 µg/dL (12–65 µmol/L)
Amylase (S)	56–190 IU/L (25–125 U/L)
Aspartame aminotransferase (ALT) or glutamic-oxaloacetic transaminase (SGOT) (S)	10–40 U/mL (5–30 U/L)
Bilirubin (S)	Total: 0.1–1.0 mg/dL (5.1–17.0 µmol/L)
	Conjugated: <0.5 mg/dL (<5.0 µmol/L)
	Unconjugated: 0.2–1.0 mg/dL (18–20 µmol/L)
	Newborn: 1.0–12.0 mg/dL (<200 µmol/L)
Blood urea nitrogen (S)	7–26 mg/dL (2.5–9.3 mmol/L)
Cholesterol	<200 mg/dL (<6.5 mmol/L)
Creatine kinase (CK) (S)	Male: 55–170 U/L (12–70 U/mL)
	Female: 30–135 U/L (10–55 U/mL)
Creatinine (S)	0.5–1.2 mg/dL (44–97 µmol/L)

Gases

Bicarbonate	22–26 mEq/L (22–26 mmol/L)
Carbon dioxide content	Arterial: 19–24 mEq/L (19–24 mmol/L)
	Venous: 22–30 mEq/L (22–30 mmol/L)
Carbon dioxide partial pressure (P_{CO_2})	Arterial: 35–45 mm Hg (same)
	Venous: 45 mm Hg (same)
Oxygen (O_2) saturation	95%–98% (same)
Oxygen partial pressure (P_{O_2})	80–105 mm Hg
pH	7.35–7.45 (same)
Glucose (S)	Fasting: 70–120 mg/mL (3.9–6.1 mmol/L)

Immunoglobulins (S)

IgG	560–1800 mg/dL (5.6–18 g/L)
IgE	0.01–0.4 mg/dL (0.1–0.4 mg/L)
IgA	85–563 mg/dL (0.85–5.6 g/L)
IgM	55–375 mg/dL (0.5–3.8 g/L)
IgD	0.5–3.0 mg/dL (5–30 mg/L)
Ketone bodies (S or P)	Negative toxic level > 20 mg/dL (0.2 g/L)
Lactic acid (lactate) (WB)	Arterial: 3–7 mg/dL (0.3–0.7 mmol/L)
	Venous: 5–20 mg/dL (0.5–2.0 µmol/L)
Lactate dehydrogenase (LDH) (S)	115–225 IU/L (0.4–1.7 µmol/L)

(continued)

(continued)

Lipoproteins (S)

    Total 400–800 mg/dL

               4.0–8.0 g/L

    High-density lipoproteins (HDL) 25% of total

               Female: >55 mg/dL (>0.1 mmol/L)

               Male: >45 mg/dL (>0.1 mmol/L)

    Low-density lipoproteins (LDL) 75% of total

               60–180 mg/dL (<3.2 mmol/L)

    Very low-density lipoproteins (VLDL) 25–50% of total (same)

    Osmolality (S) 285–295 mOsm/kg H_2O (285–295 mmol/kg H_2O)

    Phosphate (phosphorus) (S) 2.5–4.5 mg/dL (0.8–1.5 mmol/L)

    Potassium (S) 3.5–5.1 mEq/L (3.5–5.1 mmol/L)

Protein

    Total (S) 6.0–8.0 g/dL (60–80 g/L)

    Albumin (S) 3.2–5.0 g/dL (32–50 g/L)

    Sodium (S) 136–145 mEq/L (136–145 mmol/L)

    Triglycerides (S) 10–190 mg/dL (0.1–1.9 g/L)

    Uric acid (S) Female: 2.0–7.3 mg/dL (0.09–0.36 nmol/L)

               Male: 2.1–8.5 mg/dL (0.15–0.48 nmol/L)

Hematology

    Hemoglobin (S) Female: 12–16 g/dL (7.4–9.9 mmol/L)

               Male: 14–18 g/dL (8.7–11.2 mmol/L)

    Hematocrit (WB) Female: 37%–47% (same)

               Male: 42%–54% (same)

    Partial thromboplastin time (activated) APTT: 30–40 s (same)

     (PTT or APTT) PTT: 60–70 s (same)

    Platelet count (WB) 150,000–400,000/mm³ (150–400 × 10⁹/L)

    Prothrombin time (PT) (WB) 11–12.5 s (same)

               1.5–2 × control (evaluating anticoagulant treatment)

    Red blood cell count (WB) Female: 4.2–5.4 million/mm³ (4.2–5.4 × 10¹²/L)

               Male: 4.7–6.1 million/mm³ (4.7–6.1 × 10¹²/L)

    Reticulocyte count (WB) 0.5%–2.0% (same)

    White blood cell count (WBC) (WB) 5,000–10,000/mm³ (5–10 × 10⁹/L)

     total (males)

WBC count, differential (WB)

    Neutrophils 55%–70% (same)

    Lymphocytes 20%–40% (same)

    Eosinophils 1%–4% (same)

    Monocytes 2%–8% (same)

    Basophils 0.5%–1.0%

(continued)

Urine	
Amylase (24 h)	3–35 IU/h
	<5000 Somogyi units/24 h (6.5–48 U/h)
Bilirubin (random)	Negative (same)
Blood (random)	Negative (same)
Osmolality (random or fasting)	Random: 50–1400 mOsm/kg H_2O (50–1400 mmol/kg H_2O)
	Fluid restriction: <850 mOsm/kg H_2O (<850 mmol/kg H_2O)
Phosphate (24 h)	80%–90% reabsorbed (same)
Potassium (24 h)	25–120 mEq/24 h (25–120 mmol/24 h)
Protein (random)	<8 mg/dL (<0.8 mg/L)
Sodium (24 h)	40–220 mEq/24 h (40–220 mmol/24 h)
Uric acid (24 h)	0.4–1.0 g/24 h (1.5–4.0 mmol/24 h)
Urinalysis (random)	
Color	Straw, yellow, amber
Odor	Aromatic
Specific gravity	1.005–1.030
Urobilinogen (24 h)	<4 mg/24 h
Volume (24 h)	1000–2000 mL/24 h (1.0–2.0 L/24 h)

Source: *Interns and Residents Handbook,* University Hospital, Newark, NJ, 2003.
Note: Common units are listed first, followed by SI units in parentheses.

Appendix B: The Continuity Equation and the Equations of Motion in Several Coordinate Systems

B.1 The Equation of Continuity in Several Coordinate Systems

Rectangular coordinates (x, y, z):

$$\frac{\partial \rho}{\partial t} + \frac{\partial}{\partial x}(\rho V_x) + \frac{\partial}{\partial y}(\rho V_y) + \frac{\partial}{\partial z}(\rho V_z) = 0$$

Cylindrical coordinates (r, θ, z):

$$\frac{\partial \rho}{\partial t} + \frac{1}{r}\frac{\partial}{\partial r}(\rho r V_r) + \frac{1}{r}\frac{\partial}{\partial \theta}(\rho V_\theta) + \frac{\partial}{\partial z}(\rho V_z) = 0$$

Spherical coordinates (r, θ, ϕ):

$$\frac{\partial \rho}{\partial t} + \frac{1}{r^2}\frac{\partial}{\partial r}(\rho r^2 V_r) + \frac{1}{r\sin\theta}\frac{\partial}{\partial \theta}(\rho V_\theta \sin\theta) + \frac{1}{r\sin\theta}\frac{\partial}{\partial \phi}(\rho V_\phi) = 0$$

B.2 The Equations of Motion

B.2.1 In Rectangular Coordinates (x, y, z):

1. In terms of τ:

$$x\text{-component: } \rho\left(\frac{\partial V_x}{\partial t} + V_x\frac{\partial V_x}{\partial x} + V_y\frac{\partial V_x}{\partial y} + V_z\frac{\partial V_x}{\partial z}\right)$$

$$= -\frac{\partial p}{\partial x} - \left(\frac{\partial \tau_{xx}}{\partial x} + \frac{\partial \tau_{yx}}{\partial y} + \frac{\partial \tau_{zx}}{\partial z}\right) + \rho g_x$$

$$y\text{-component: } \rho\left(\frac{\partial V_y}{\partial t} + V_x\frac{\partial V_y}{\partial x} + V_y\frac{\partial V_y}{\partial y} + V_z\frac{\partial V_y}{\partial z}\right)$$

$$= -\frac{\partial p}{\partial y} - \left(\frac{\partial \tau_{xy}}{\partial x} + \frac{\partial \tau_{yy}}{\partial y} + \frac{\partial \tau_{zy}}{\partial z}\right) + \rho g_y$$

and

$$z\text{-component: } \rho\left(\frac{\partial V_z}{\partial t} + V_x\frac{\partial V_z}{\partial x} + V_y z\frac{\partial V_z}{\partial y} + V_z\frac{\partial V_z}{\partial z}\right)$$

$$= -\frac{\partial p}{\partial z} - \left(\frac{\partial \tau_{xz}}{\partial x} + \frac{\partial \tau_{yz}}{\partial y} + \frac{\partial \tau_{zz}}{\partial z}\right) + \rho g_z$$

2. In terms of velocity gradients for a Newtonian fluid with constant ρ and μ:

$$x\text{-component: } \rho\left(\frac{\partial V_x}{\partial t} + V_x\frac{\partial V_x}{\partial x} + V_y\frac{\partial V_x}{\partial y} + V_z\frac{\partial V_x}{\partial z}\right)$$

$$= -\frac{\partial p}{\partial x} + \mu\left(\frac{\partial^2 V_x}{\partial x^2} + \frac{\partial^2 V_x}{\partial y^2} + \frac{\partial^2 V_x}{\partial z^2}\right) + \rho g_x$$

$$y\text{-component: } \rho\left(\frac{\partial V_y}{\partial t} + V_x\frac{\partial V_y}{\partial x} + V_y\frac{\partial V_y}{\partial y} + V_z\frac{\partial V_y}{\partial z}\right)$$

$$= -\frac{\partial p}{\partial y} + \mu\left(\frac{\partial^2 V_y}{\partial x^2} + \frac{\partial^2 V_y}{\partial y^2} + \frac{\partial^2 V_y}{\partial z^2}\right) + \rho g_y$$

and

$$z\text{-component: } \rho\left(\frac{\partial V_z}{\partial t} + V_x\frac{\partial V_z}{\partial x} + V_y\frac{\partial V_z}{\partial y} + V_z\frac{\partial V_z}{\partial z}\right)$$

$$= -\frac{\partial p}{\partial z} + \mu\left(\frac{\partial^2 V_z}{\partial x^2} + \frac{\partial^2 V_z}{\partial y^2} + \frac{\partial^2 V_z}{\partial z^2}\right) + \rho g_z$$

B.2.2 In Cylindrical Coordinates (*r*, θ, *z*)

1. In terms of τ:

r-component:* $\rho\left(\dfrac{\partial V_r}{\partial t}+V_r\dfrac{\partial V_r}{\partial r}+\dfrac{V_\theta}{r}\dfrac{\partial V_r}{\partial \theta}-\dfrac{V_\theta^2}{r}+V_z\dfrac{\partial V_r}{\partial z}\right)$

$$=-\dfrac{\partial p}{\partial r}-\left(\dfrac{1}{r}\dfrac{\partial\left(r\tau_{rr}\right)}{\partial r}+\dfrac{1}{r}\dfrac{\partial\tau_{r\theta}}{\partial\theta}-\dfrac{\tau_{\theta\theta}}{r}+\dfrac{\partial\tau_{rz}}{\partial z}\right)+\rho g_r$$

θ-component:† $\rho\left(\dfrac{\partial V_\theta}{\partial t}+V_r\dfrac{\partial V_\theta}{\partial r}+\dfrac{V_\theta}{r}\dfrac{\partial V_\theta}{\partial \theta}+\dfrac{V_r V_\theta}{r}+V_z\dfrac{\partial V_\theta}{\partial z}\right)$

$$=-\dfrac{1}{r}\dfrac{\partial p}{\partial\theta}-\left(\dfrac{1}{r^2}\dfrac{\partial\left(r^2\tau_{r\theta}\right)}{\partial r}+\dfrac{1}{r}\dfrac{\partial\tau_{\theta\theta}}{\partial\theta}+\dfrac{\partial\tau_{\theta z}}{\partial z}\right)+\rho g_\theta$$

and

z-component: $\rho\left(\dfrac{\partial V_z}{\partial t}+V_r\dfrac{\partial V_z}{\partial r}+\dfrac{V_\theta}{r}\dfrac{\partial V_z}{\partial \theta}+V_z\dfrac{\partial V_z}{\partial z}\right)$

$$=-\dfrac{\partial p}{\partial z}-\left(\dfrac{1}{r}\dfrac{\partial\left(r\tau_{rz}\right)}{\partial r}+\dfrac{1}{r}\dfrac{\partial\tau_{\theta z}}{\partial\theta}+\dfrac{\partial\tau_{zz}}{\partial z}\right)+\rho g_z$$

---

* The term $(\rho V_\theta^2/r)$ is the *centrifugal force*. It gives the effective force in the *r*-direction resulting from fluid motion in the θ-direction. This term arises automatically on transformation from rectangular to cylindrical coordinates. It does not have to be added on physical grounds.
† The term $(\rho V_r V_\theta/r)$ is the *Coriolis force*. It is an effective force in the θ-direction when there is flow in both the *r*- and θ-directions. This term also arises automatically in the transformation. The Coriolis force arises in the problem of flow near a rotating disk.

2. In terms of velocity gradients for a Newtonian fluid with constant ρ and μ:

r-component:*
$$\rho\left(\frac{\partial V_r}{\partial t}+V_r\frac{\partial V_r}{\partial r}+\frac{V_\theta}{r}\frac{\partial V_r}{\partial\theta}-\frac{V_\theta^2}{r}+V_z\frac{\partial V_r}{\partial z}\right)$$

$$=-\frac{\partial p}{\partial r}+\mu\left[\frac{\partial}{\partial r}\left(\frac{1}{r}\frac{\partial(rV_r)}{\partial r}\right)+\frac{1}{r^2}\frac{\partial^2 V_r}{\partial\theta^2}-\frac{2}{r^2}\frac{\partial V_\theta}{\partial\theta}+\frac{\partial^2 V_r}{\partial z^2}\right]+\rho g_r$$

θ-component:†
$$\rho\left(\frac{\partial V_\theta}{\partial t}+V_r\frac{\partial V_\theta}{\partial r}+\frac{V_\theta}{r}\frac{\partial V_\theta}{\partial\theta}+\frac{V_r V_\theta}{r}+V_z\frac{\partial V_\theta}{\partial z}\right)$$

$$=-\frac{1}{r}\frac{\partial p}{\partial\theta}+\mu\left[\frac{\partial}{\partial r}\left(\frac{1}{r}\frac{\partial}{\partial r}(rV_\theta)\right)+\frac{1}{r^2}\frac{\partial^2 V_\theta}{\partial\theta^2}+\frac{2}{r^2}\frac{\partial V_r}{\partial\theta}+\frac{\partial^2 V_\theta}{\partial z^2}\right]+\rho g_\theta$$

and

z-component:
$$\rho\left(\frac{\partial V_z}{\partial t}+V_r\frac{\partial V_z}{\partial r}+\frac{V_\theta}{r}\frac{\partial V_z}{\partial\theta}+V_z\frac{\partial V_z}{\partial z}\right)$$

$$=-\frac{\partial p}{\partial z}+\mu\left[\frac{1}{r}\frac{\partial}{\partial r}\left(r\frac{\partial V_z}{\partial r}\right)+\frac{1}{r^2}\frac{\partial^2 V_z}{\partial\theta^2}+\frac{\partial^2 V_z}{\partial z^2}\right]+\rho g_z$$

---

* The term $(\rho V_\theta^2/r)$ is the *centrifugal force*. It gives the effective force in the r-direction resulting from fluid motion in the θ-direction. This term arises automatically on transformation from rectangular to cylindrical coordinates. It does not have to be added on physical grounds.

† The term $(\rho V_r V_\theta/r)$ is the *Coriolis force*. It is an effective force in the θ-direction when there is flow in both the r- and θ-directions. This term also arises automatically in the transformation. The Coriolis force arises in the problem of flow near a rotating disk.

B.2.3 In Spherical Coordinates (*r*, θ, φ):

1. In terms of τ:

r-component: $\rho\left(\dfrac{\partial V_r}{\partial t}+V_r\dfrac{\partial V_r}{\partial r}+\dfrac{V_\theta}{r}\dfrac{\partial V_r}{\partial\theta}+\dfrac{V_\phi}{r\sin\theta}\dfrac{\partial V_r}{\partial\phi}-\dfrac{V_\theta^2+V_\phi^2}{r}\right)$

$$=-\dfrac{\partial p}{\partial r}-\left(\dfrac{1}{r^2}\dfrac{\partial}{\partial r}\left(r^2\tau_{rr}\right)+\dfrac{1}{r\sin\theta}\dfrac{\partial}{\partial\theta}\left(\tau_{r\theta}\sin\theta\right)\right.$$

$$\left.+\dfrac{1}{r\sin\theta}\dfrac{\partial\tau_{r\phi}}{\partial\phi}-\dfrac{\tau_{\theta\theta}+\tau_{\phi\phi}}{r}\right)+\rho g_r$$

θ-component: $\rho\left(\dfrac{\partial V_\theta}{\partial t}+V_r\dfrac{\partial V_\theta}{\partial r}+\dfrac{V_\theta}{r}\dfrac{\partial V_\theta}{\partial\theta}+\dfrac{V_\phi}{r\sin\theta}\dfrac{\partial V_\theta}{\partial\phi}+\dfrac{V_r V_\theta}{r}-\dfrac{V_\phi^2\cot\theta}{r}\right)$

$$=-\dfrac{1}{r}\dfrac{\partial p}{\partial\theta}-\left(\dfrac{1}{r^2}\dfrac{\partial}{\partial r}\left(r^2\tau_{r\theta}\right)+\dfrac{1}{r\sin\theta}\dfrac{\partial}{\partial\theta}\left(\tau_{\theta\theta}\sin\theta\right)\right.$$

$$\left.+\dfrac{1}{r\sin\theta}\dfrac{\partial\tau_{\theta\phi}}{\partial\phi}+\dfrac{\tau_{r\theta}}{r}-\dfrac{\cot\theta}{r}\tau_{\phi\phi}\right)+\rho g_\theta$$

and

φ-component: $\rho\left(\dfrac{\partial V_\phi}{\partial t}+V_r\dfrac{\partial V_\phi}{\partial r}+\dfrac{V_\theta}{r}\dfrac{\partial V_\phi}{\partial\theta}+\dfrac{V_\phi}{r\sin\theta}\dfrac{\partial V_\phi}{\partial\phi}+\dfrac{V_\phi V_r}{r}+\dfrac{V_\theta V_\phi}{r}\cot\theta\right)$

$$=-\dfrac{1}{r\sin\theta}\dfrac{\partial p}{\partial\phi}-\left(\dfrac{1}{r^2}\dfrac{\partial}{\partial r}\left(r^2\tau_{r\phi}\right)+\dfrac{1}{r}\dfrac{\partial\tau_{\theta\phi}}{\partial\theta}+\dfrac{1}{r\sin\theta}\dfrac{\partial\tau_{\phi\phi}}{\partial\phi}\right.$$

$$\left.+\dfrac{\tau_{r\phi}}{r}+\dfrac{2\cot\theta}{r}\tau_{\theta\phi}\right)+\rho g_\phi$$

2. In terms of velocity gradients for a Newtonian fluid with constant ρ and μ*:

r-component: $\rho\left(\dfrac{\partial V_r}{\partial t}+V_r\dfrac{\partial V_r}{\partial r}+\dfrac{V_\theta}{r}\dfrac{\partial V_r}{\partial\theta}+\dfrac{V_\phi}{r\sin\theta}\dfrac{\partial V_r}{\partial\phi}-\dfrac{V_\theta^2+V_\phi^2}{r}\right)$

$$=-\dfrac{\partial P}{\partial r}+\mu\left(\nabla^2 V_r-\dfrac{2}{r^2}V_r-\dfrac{2}{r^2}\dfrac{\partial V_\theta}{\partial\theta}\right.$$

$$\left.-\dfrac{2}{r^2}V_\theta\cot\theta-\dfrac{2}{r^2\sin\theta}\dfrac{\partial V_\phi}{\partial\phi}\right)+\rho g_r$$

θ-component: $\rho\left(\dfrac{\partial V_\theta}{\partial t}+V_r\dfrac{\partial V_\theta}{\partial r}+\dfrac{V_\theta}{r}\dfrac{\partial V_\theta}{\partial\theta}+\dfrac{V_\phi}{r\sin\theta}\dfrac{\partial V_\theta}{\partial\phi}+\dfrac{V_r V_\theta}{r}-\dfrac{V_\phi^2\cot\theta}{r}\right)$

$$=-\dfrac{1}{r}\dfrac{\partial P}{\partial\theta}+\mu\left(\nabla^2 V_\theta+\dfrac{2}{r^2}\dfrac{\partial V_r}{\partial\theta}-\dfrac{V_\theta}{r^2\sin^2\theta}-\dfrac{2\cos\theta}{r^2\sin^2\theta}\dfrac{\partial V_\phi}{\partial\phi}\right)+\rho g_\theta$$

and

ϕ-component: $\rho\left(\dfrac{\partial V_\phi}{\partial t}+V_r\dfrac{\partial V_\phi}{\partial r}+\dfrac{V_\theta}{r}\dfrac{\partial V_\phi}{\partial\theta}+\dfrac{V_\phi}{r\sin\theta}\dfrac{\partial V_\phi}{\partial\phi}+\dfrac{V_\phi V_r}{r}+\dfrac{V_\theta V_\phi}{r}\cot\theta\right)$

$$=-\dfrac{1}{r\sin\theta}\dfrac{\partial P}{\partial\phi}+\mu\left(\nabla^2 V_\phi-\dfrac{V_\phi}{r^2\sin^2\theta}+\dfrac{2}{r^2\sin^2\theta}\dfrac{\partial V_r}{\partial\phi}\right.$$

$$\left.+\dfrac{2\cos\theta}{r^2\sin^2\theta}\dfrac{\partial V_\theta}{\partial\phi}\right)+\rho g_\phi$$

---

* In these equations

$$\nabla^2=\dfrac{1}{r^2}\dfrac{\partial}{\partial r}\left(r^2\dfrac{\partial}{\partial r}\right)+\dfrac{1}{r^2\sin\theta}\dfrac{\partial}{\partial\theta}\left(\sin\theta\dfrac{\partial}{\partial\theta}\right)+\dfrac{1}{r^2\sin\theta}\left(\dfrac{\partial^2}{\partial\phi^2}\right)$$

Appendix C: Frequency Components of Pressure Wave

Using Fourier analysis, the original pressure wave can be separated into a fundamental (DC component) and components of increasing frequency (the harmonics). The following graph shows the fundamental plus the first six harmonics. Similarly, 20 components can be algebraically added together to produce the original waveform very closely.

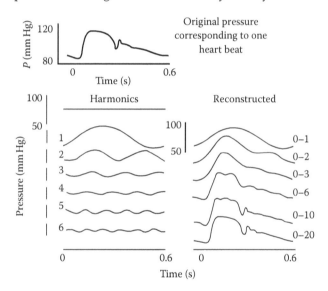

Index

.

Milton Keynes UK
Ingram Content Group UK Ltd.
UKHW030902141024
449569UK00025B/1267